16 0208052 2

TEL

AF804920

Two week loan

Please return on or before the last date stamped below.
Charges are made for late return.

10 FEB 1997

31 MAR 2000

22 NOV 1995

LF 114/0895 UNIVERSITY OF WALES, CARDIFF, PO BOX 430, CARDIFF CF1 3XT

WITHDRAWN

The Mechanism of Economic Development

The Mechanism of Economic Development

Growth in the Japanese and East Asian Economies

KEN-ICHI INADA
SUEO SEKIGUCHI
YASUTOYO SHODA

CLARENDON PRESS · OXFORD
1992

338.952 I

Oxford University Press, Walton Street, Oxford OX2 6DP
Oxford New York Toronto
Delhi Bombay Calcutta Madras Karachi
Petaling Jaya Singapore Hong Kong Tokyo
Nairobi Dar es Salaam Cape Town
Melbourne Auckland
and associated companies in
Berlin Ibadan

Oxford is a trade mark of Oxford University Press

Published in the United States
by Oxford University Press, New York

© Ken-ichi Inada, Sueo Sekiguchi, and Yasutoyo Shoda 1992

All rights reserved. No part of this publication may be reproduced, stored in a retrieval system, or transmitted, in any form or by any means, electronic, mechanical, photocopying, recording, or otherwise, without the prior permission of Oxford University Press

British Library Cataloguing in Publication Data
Data available

Library of Congress Cataloging in Publication Data
Inada, Ken-ichi, 1925–
[Keizai hatten no mekanizumu. English]
The mechanism of economic development / Ken-ichi Inada, Sueo Sekiguchi, Yasutoyo Shōda.
p. cm.
Translation of: Keizai hatten no mekanizumu.
Includes bibliographical references and index.
1. Japan—Economic conditions. 2. Economic development—Mathematical models. I. Sekiguchi, Sueo. II. Shōda, Yasutoyo. III. Title
HC462. I5713 1992 338.9—dc20 91–35067
ISBN 0–19–828628–7

Typeset by Alliance Phototypesetters
Printed in Great Britain by
Bookcraft (Bath) Ltd,
Midsomer Norton, Avon

CARDIFF
14 JAN 1993

Preface

Our economies have changed significantly, compared not only to one hundred years ago but also to fifty years or even ten years ago. How and why does such change or reform arise? This is an important subject for economics students, and almost all of them will be interested in this field of economics. But this subject—that is, research on the reform or development of economies—does not seem to be an appropriate field to which to apply the analytical methods employed in growth theory, which assume an intact economic structure. At least, we can make such an inference from the fact that few analyses employing such methods exist in this field. Our aim in this book is to challenge this preconception at least in the modest sense of finding some subfields in which the analytical methods employed in growth theory are applicable.

We begin by constructing a theoretical model and then examine whether it can describe or explain the process of development as it arises in reality using data for the Japanese economy. Of course, we have to impose some restrictive assumptions in the theoretical analysis because of the limitations of the mathematical tools. We also sometimes have to be satisfied with less than complete tests because of the limitations of the data. Nevertheless, we believe we have achieved results that enhance our understanding of the economic development process. We hope that many readers will share our opinion, though of course we would welcome critical comments on all parts of our book for their contribution to our future studies in this field of economics. Our book consists of three parts. Part I presents the theoretical model, Part II deals with the economic development of Japan pre–World War II, and Part III considers the development in the post-war period until 1960. Parts II and III attempt to analyse empirically the mechanism of economic development in the Japanese economy in line with the theoretical model of Part I.

Since our research involved empirical analyses, we needed to check the facts in the history of Japan. For this purpose Iwanami Shoten (1968) was very useful. We also received advice and assistance from many people, for which we express our gratitude. First we received unpublished statistical data from professor K. Akasaka, K. Emi, R. Minami, K. Ohkawa, A. Ono, Y. Shionoya, N. Takamatsu, and M. Umemura.

Without their generous assistance, our empirical analyses would have been impossible. We also acknowledge useful comments from M. Aoki, H. Aoyama, Y. Fujita, M. Hatanaka, Y. Hayami, Y. Kawashima, K. Kuga, S. Nishibe, M. Saitoh, N. Sakashita, Y. Shinkai, H. Uzawa, and T. Yasui.

During the stay of one of the authors K. Inada at Stanford University, he received very useful comments from Professor T. N. Srinivasan. He also wishes to thank the Ford Foundation for its financial support for his research and stay at Stanford University. The Original draft of Part I of this book was actually completed during this period, but needless to say, the research results expresses here are not necessarily the opinion of the Ford Foundation.

An earlier version of Part I was first published in the *International Economic Review* in 1971. We wish to express our sincere gratitude to the editors of the *Review* for permission to publish the material in the present form.

We also wish to express our thanks for the facilities provided by the Japan Centre for Economic Research, to which one of us is affiliated, and to the following people associated with the centre: Messrs. K. Abe, M. Hosokane, O. Matsuda, N. Nakanishi, and S. Ohkita. We likewise acknowledge the assistance rendered by Misses H. Kido, K. Sakurai, and K. Tokura in preparing the figures and tables and in proof-reading. Finally, we are grateful to the Santory Foundation for their financial support.

This book is a translation, somewhat amended, of the book published in Japanese by the same authors. We are indebted to Professor C. Y. Horioka, of the Institute of Social and Economic Research at Osaka University for his many valuable comments during the course of the translation work, especially of Part I. We are also greatly indebted to Mr Andrew Schuller, who initiated discussions regarding the publication of an English version of this book by the Oxford University Press, to the anonymous referees, whose comments were valuable in improving our arguments, and to the Press's desk-editor and copy-editor.

K.I.
S.S.
Y.S.

Suita, Osaka
April, 1988

Contents

PART I
The Theoretical Model

PART II

The Economic Development of Japan: Pre-World War II

PART III

Economic Development of Japan: Post-World War II

PART I

The Theoretical Model

1

Introduction and Summary

1.1. Introduction

The theory of economic development seems to have been located at the boundary between economic theory and economic history. This is the reason why the phenomena of economic development have been the focus of research of both theoretical economists and economic historians. So far, the approaches by historians seem to have been superior in quantity, but the comparison of quality is not an easy task; certainly, it is not the aim of this book.

If we define the theory of development in the narrow sense as rigorous analysis by model building, examples thereof are really scarce. The reason for it is summarized by the Nobel Prize economist Gunnar Myrdal in his celebrated book *Asian Drama* (1968) that model analysis, especially the methods and concepts developed in the developed Western economies, are not useful. The Keynesian concept of involuntary unemployment is the key to understanding the developed economies, but the situation this concept represents does not exist in the developing countries. That is, the structures of economies, societies, and cultures in the developing world—or, more basically, human attitudes, behavioural patterns, sense of values, and morals—are quite different from those in the Western world. The concepts, methods, and analytical tools that have proved successful for describing, analysing, and explaining Western economies are thus not effective for the economic analysis of the developing or less developed economies.

We understand this view. We admit the difference in human attitudes, behavioural patterns, and sense of values between the two worlds. But we are sceptical about the assertion that the difference is so conspicuous that the concepts, methods, and analytical tools developed in the Western world are completely ineffective for the analysis of developing economies. In fact, the attitudes, behavioural patterns, and sense of values in the developing economies may be viewed as being quite different from those with which Myrdal, who was born and brought up in the Western world, was familiar. But various aspects of the Asian societies, especially India, described by him have been observed in Japan before and even after its economy took off, and seem still to be present even in present-day Japan. That is, our impression is that the social aspects of the present-day Asian countries described by Myrdal as 'soft states' are not essentially different from those in Japan.

The development of the Japanese economy was initially due to government intervention. The process of development was then interrupted by World War II, but after the war a new period of vigorous growth enabled the Japanese economy to catch up with and eventually move to overtake the Western economies. This begs the question of whether other Asian countries, or indeed developing countries elsewhere in the world, could pursue the same path as that pursued by Japan.

The fact that the Japanese economy has caught up and is surpassing the countries of Western Europe does not necessarily mean that the concepts and analytical tools developed in the Western economies are also effective for describing, analysing, and explaining the development of the Japanese economy (although even Myrdal would agree that they are useful for analysing the present-day, fully developed Japanese economy). We believe, however, that the tools are effective for analysing the development of the Japanese economy, even for the early period of opening to the rest of the world that started in 1868.

It is true that the attitudes and ways of thinking of the Japanese people changed significantly after the country was opened up, and that Japanese society changed also. It adopted the social systems of the Western world—for example, the legal and education systems of Prussia, the military system of France and Prussia, the parliament and administrative system of Great Britain, and so on. Of course, the formal introduction of the social systems of developed countries does not immediately bring about a change in the attitudes and ways of thinking of the people. The reason why Japan could adjust to the change in its social systems is that the basic social conditions, attitudes, and sense of values needed for economic development had been nurtured even during the period when it had been closed to the rest of the world.

It may be true that economic incentives are not an important influence on behaviour in some Asian countries. The development of these countries will require the reform of their social systems, human attitudes, behavioural patterns, and sense of values. We agree that the various forces impeding the working of economic incentives are strong in conventional societies, but the essential character of every human being is the same. Thus, when Japan's social system was feudalistic, the movement of people, money, and resources was severely restricted. Around the end of the Edo era (1968), however, the merchant class had appropriated enough economic power to affect the political system of the society, and this increased the possibilities for freedom of movement in response to profitable opportunities.

The same, we believe, can be said of today's less developed countries too. That is, from Japan's experience, we believe that they may also be able to attain economic take-off. It is not a prerogative of westernized societies. Thus, although people, money, and resources became more mobile in post-war Japan, labour in particular is not as mobile as in Western countries. This is seen in the social custom of lifetime employment. The system of seniority wages also does not fit the economic logic of the Western countries. The labour employment policies of Japanese firms are also quite different from those of

Western countries. Discharge or lay-off during periods of recession is rare in the former compared to the latter, and thus the unemployment rate in Japan is much smaller than in Western countries. In Japan, workers—for instance, automobile assemblers—even engage in weeding the factory sites during recession periods, but they are not dismissed. Such labour policies result in a sacrifice of short-term profits, but increase mutual trust between managers and employees. This contributes to the development and success of firms and thus to that of the economy in the long run. This would suggest that countries in Asia and other regions—that is, countries not sensitive to short-term profits and with similar cultures to that of Japan—would be able to develop. Although the characteristic feature of the analytical tools of economics developed in Western countries seems to be the presumption of the rationality of the economic behaviour of agents, this does not preclude their applicability outside the Western world. Even the economic agents in pre-war Japan, the newly industrialized countries, or the less developed countries (LDCs) are rational in a sense. Thus, the analytical tools and implications derived from Western economies apply at least partly to these countries, with some modifications and some provisions. Of course, it is necessary at the same time to remove the conventional or vested interests preventing people, money, and resources from being utilized in and distributed to more efficient uses, for this lies at the basis of adequate capital accumulation. We of course agree with Myrdal that overall social reform is necessary for economic development, but we insist that it is this mobilization of resources that is the key factor for economic development. Thus, the mechanism of development cannot be understood or explained without providing an analysis of the capital accumulation process.

In short, while we partly agree with Myrdal's assertion, we are less pessimistic. The social institutions, customs, habits, sense of values, and attitudes of traditional countries with ancient civilizations, such as India or China, may be quite rigid, but we believe that they can still be changed. Even India and China are changing. Once countries begin to change, reform becomes semi-automatic; and the pace of change also accelerates. Japan's experience shows this. W. A. Lewis, a Noble Prize winner like Myrdal, is more optimistic than Myrdal. This is apparent in their attitudes to the problem of the shortage of skilled labour. According to Myrdal, this is a symptom of a traditional society: it is the family system, attitudes to work, sense of values, and other conventional customs that cause the shortage of skilled labour. In such societies, it is not enough to treat symptoms, one has to treat causes. Lewis (1955), on the other hand, wrote:

> There may at any time be a shortage of skilled workers of any grade—ranging from masons, electricians or welders to engineers, biologists or administrators. Skilled labour may be the bottleneck in expansion, just like capital or land. Skilled labour, however, is only what Marshall might have called a 'quasi-bottleneck', if he had not had so nice a sense of elegant language. For it is only a very temporary bottleneck, in the sense

that if the capital is available for development, the capitalists or their government will soon provide the facilities for training more skilled people. The real bottlenecks to expansion are therefore capital and natural resources, and we can proceed on the assumption that so long as these are available the necessary skills will be provided as well, though perhaps with some time lag.

Lewis seems to have recognized that factors relating to social systems other than the shortage of capital and natural resources would adapt almost automatically with economic development. Even though the content of his paper is stated verbally, the logic is one which can easily be expressed in exact mathematical terms and can be regarded as employing the analytical tools developed in Western Europe. Myrdal, in contrast, insisted that even if there are profitable job opportunities in the modern factories located in cities, poor farm workers in traditional rural areas are not eager to acquire the skills needed to work in modern factories, nor do they wish to leave their home area. They do not want to change their traditional life-styles for discipline, punctuality, diligence, and aspiration. For Myrdal, it was the conservative attitude of rural workers that explains why capital accumulation does not automatically entail economic development. For him, the bottleneck was not capital accumulation but rather human attitudes.

How do we account for this difference in the views of these two great economists? One possibility is differences in the countries in which they grew up. The Swedish-born Myrdal might well have viewed the people and the social system of less developed countries as very backward, whereas the Jamaican-born Lewis might not have done. The front-runner usually exaggerates the difference between himself and the next runner; the runner behind has a more realistic view.

But the difference in their opinions may be better understood by another interpretation. In a great country like India, which was once the world's most advanced and civilized country and has its own culture with long traditions, the institutions, customs, habits, sense of values, and attitudes are so tightly entangled with each other that it is very difficult to untie the knots. In fact these social characteristics may well not be suitable for economic development, even though at one time they might have had a *raison d'être* and enabled the society to reach a coveted position. In such cases, it might be necessary to admit Myrdal's views; that is, if his pessimistic views are derived from his stay and experiences in India, and Indian society is viewed as a typical model of a less developed country, his views might be correct. In fact, *Asian Drama* might have been better named *Indian Drama*, because his description of less developed countries pertains in very large part to India. But India is not the only LDC in the world, or even in Asia. Japan, in contrast, has been absorbing or assimilating culture from abroad. It is much smaller than India, so communication within the nation is relatively easy. This makes it easier to accomplish a shift in the sense of values. The same can be said of some other Asian countries, such as Korea, Taiwan, and Singapore, which have recently come to be known collectively as 'newly industrializing economies' (NIEs). In these countries, Lewis's approach will be more applicable.

Although we have to concede that there are cases where Myrdal's approach should be applied, our intention here is to analyse the mechanism of development of the economies where Lewis's views are valid. That is, our main purpose is to analyse theoretically how capital accumulation proceeds without limitation and thus take-off is possible, provided appropriate policies are adopted. However, our analysis is also applicable to economies where Myrdal's views are valid; because even if Myrdal is right that social reform is necessary prior to the accumulation of capital, development is still impossible without capital accumulation. That is, capital accumulation is always necessary at some stage of development, and without some efforts in this area, development will not occur.

When we began our analysis, we believed that it was countries such as Malaysia, Brazil, Mexico, Ceylon, and the above-mentioned NIEs to which our model would be applicable. We now think that it can be applied to a wider range of countries. This is one reason why our empirical validation of the model is focused on the economic development of Japan. We also thought that an analysis of Japan's success would be helpful to countries of non-European cultures wishing to follow the same success course.

Our model is meant to be useful for analysing and describing the process of economic development. This does not mean, however, that it analyses the process from first principles—how capitalistic enterprises are initially introduced into a primitive economy, how they begin to grow, how their number gradually increases, and how the economic sector they comprise eventually comes to occupy the dominant proportion of national output and employment. That is, our plan is not to reconstruct the whole process of economic development. Rather, we presuppose a starting point where there already exist some light industries, such as textile and food processing, which have already reached a high enough level to compete with similar sectors in the developed countries. Our aim is to study whether an economy that has somehow reached this level can 'take off' to become an economy with heavy and chemical industries, and how this take-off can be achieved.

1.2. Summary of the Theoretical Model

Less developed economies are often characterized by two conspicuous features: unemployment of unskilled labour, and one or a small number of major exportable products.[1] Of course, almost all of them produce some light industrial products too, but their weight in the national output is not high. Generally speaking, these economies export a large proportion of their primary products, and sometimes also the products of light industries, and import almost every kind of consumer good and capital good, mainly from the developed countries. We can cite some examples of this type of country: Japan one hundred years ago (exporting silk and tea), Egypt (cotton), Sri Lanka (tea), Thailand (rice), Brazil and Colombia (coffee), Malaysia (rubber and tin), Chile (copper), Saudi Arabia, Kuwait, and Venezuela (oil), and so on.

With a few exceptions, these countries were colonies or dependencies of European countries until around the end of World War II. The merits of low wage rates and/or their climatical, geographical, and geological suitability for primary products made it possible for these countries to supply such products cheaply to the developed, especially suzerain, countries. On the other hand, the latter countries, initially as well as thereafter, have continuously supplied capital to the less developed countries' more modern, capitalistic sectors—for example, plantations such as rubber, coffee, and tea farms, and various metal and oil mining firms. After World War II, many of the less developed countries, aspiring to develop economically to escape from long-lasting poverty gained political independence. So far, few economies have succeeded or are succeeding in achieving this aim.

Of course, it cannot be said that the production of a small number of primary products and light industrial products is the unique cause of the low average income of less developed economies. Australia and New Zealand fit this description too, but their average income is comparable to those of the developed countries. One reason might be the low density of population, but population density was even lower before the migration from the European countries, yet average income was lower. What, then, is the cause of low income levels? Many causes and factors are conceivable. This means that any simple answer cannot be perfect. Conversely, any single cause or small number of causes cannot be deterministic or unique. That is, the poverty of each less developed country may be explicable by its own innate causes and factors. Cultural, social, and political structures, for example, have a number of causes. They may be determined largely by history—for example, the religion, manners, customs, existence or non-existence and type of social class system (i.e. caste); or by the kinds and quantities of natural resources, climate, geological features, geographical and geopolitical circumstances; or by the international environment—for example, location at the border between the camps of East and West, or use as an important military base by a large country; and also by national traits. Each of these factors is somehow related to the country's economic situation. Intrinsically, no social aspect is determined independently of other social aspects; in some cases, even the natural environment will be affected by social factors. In fact, ecologists suggest that desertification of the Sahara is being accelerated by the way of life in the area, and especially by grazing practices.

To explain the mechanism of low income, we have to untangle a knot that has become interwined because of various causes and factors. This is very complicated. Even worse, some social aspects may be the result of poverty as well as a cause thereof. For example, while the extended family system may function as a kind of necessary security system in cases where per capita income is very low, it weakens the economic incentive to move to high-income jobs because the higher income will be shared by a large number of people, thus reducing the benefit for the earner himself. This, then, is a vicious circle, we cannot distinguish cause from result. The best expression to explain this situation may be the tautological dictum 'The cause of poverty is poverty and the result of poverty is poverty'. Given the importance of economic progress in the less developed economies, however, we should not give up in the face of such difficulties. We may say that world peace depends on it: many of today's international conflicts in fact occur in these countries.

Faced by this challenge, we are emboldened to use some very daring assumptions to try to explain the mechanism of economic development. We identify two main causes of poverty as the monocultural structure of the economy and the existence of unskilled labour, and then consider what mechanisms will allow the economy to escape from such a structure (or, conversely, will trap the economy in it), and study the policy measures for the escape. Note that we do not see a monocultural economy as precluding take-off. For example, Japan succeeded in spite of its poor endowment of natural resources. The Japanese are said to be diligent, but then should other peoples be less diligent? As they are also human beings, it cannot be due to their innate nature, but must be due rather to the restrictions in the society. If, therefore, the reform or even the revolution of such a restrictive social system can be achieved, take-off should not be impossible. Of course, such reforms will inevitably be disadvantageous to many vested interests—the wealthy class, and also sometimes to the leadership class of the society. It is thus not easy to achieve. We shall consider the case where take-off is structurally impossible for reasons relating to the capital accumulation process i.e. where the social background is one that inevitably obstructs capital accumulation. In such cases, foreign aid, unless it actually converts the social structure will not bring about self-sustaining development, no matter how much it increases capital accumulation. Thus, although our model concentrates on the capital accumulation process, it nevertheless recognizes that development is impossible unless the social background, the stage on which the development drama is played, is reformed.

Let us provide a general view of the analysis underlying our model. Suppose a country continues the economic structure of monoculture. If it aspires to elevate its average income significantly, it must produce more. It may be possible, paradoxically, to adopt a policy of producing less, improving the terms of trade, and as a result earning more from the sale, provided that the price elasticity of demand is less than one and that the country is a monopolistic supplier of the product. This is similar to the strategy of withholding overproduced goods in order to prevent a price decline. Usually, however, the country is not a monopolistic supplier of the product, so average income would not be significantly increased by this policy. On the other hand, if the output of the products increases, the price of the products will decline because, generally speaking, the price elasticity of demand for primary products is low, so an increase in supply will bring about a price decline unless the demand structure of the importing countries changes sharply, due, say, to rapid growth. Other factors, such as the invention of substitutes for the primary products, will usually have an adverse effect on demand. The invention of synthetic fabrics and rubber are typical examples.

If a supply increase brings about a price decline, an output increase will not necessarily result in an income increase. That is, the output increase may bring about an increase in the amount of profit but not in the profit rate, because output does not increase proportionately with capital input. As an inevitable result, the growth rate of this economic sector will decline. That is, even if the profit of the sector is wholly re-invested in the sector, the growth rate of the sector will decline because the growth rate depends on the incremental ratio of capital investment. Furthermore, in general, if the profit rate declines, a portion of the profit obtained in the sector will flow out of the sector, and sometimes flow out of the country, seeking more profitable opportunities in foreign countries. This will be especially serious in the case of plantations. The foreign funds invested in plantations are usually very sensitive to short-term profit opportunities. Thus, they will be very sensitive to the decline in the profit rate, and the scale of the plantation will not expand beyond a certain level. The growth of an important economic sector will thus stop, and consequently the national income will level off also. If the country has a small population, however, even if total income does reach a ceiling, it may still be high enough for *per capita* income to be comparable to the European industrialized countries; Australia and New Zealand are examples. In most LDCs, however, the large population will make the divided pie very small. This is the situation anticipated for an economy of monocultural structure.

But if there should exist some manufacturing industries, particularly heavy and chemical industries, producing some import substitutes and/or export goods and providing profitable opportunities for capital investment, then what happens? Ideally, the low profit rate of the plantation sector would stimulate investment in these industries. However, the real world does not operate like this. In reality, there are few such industries in the less developed countries. Why do they not exist? The most important reason is technological. The technologies employed in the heavy and chemical industry sector generally follow the law of increasing returns to scale, at least up to a certain point. In other words, the cost per unit of the infant industry's products is high compared to that of the mature industries in the developed foreign countries. Free competition of the former against the latter will be a futile struggle. Thus, the low profit rate in the plantation sector does not necessarily stimulate the emergence of facilities for producing import-substitutes or export goods as an alternative investment channel. For this to occur, the prospects for the latter industries must be bright. This rarely happens as the proposed scale of development is generally too small for economies of scale to be realized. Since the profit rate would be low, the sector is not attractive to investment; if investment in the sector is not high, the growth rate cannot be high. If the scale of the industry does not attain the level where economies of scale operate fully, investment is not attracted. That is, investment is not attracted because it is not attracted. This is 'the vicious circle' from which many LDCs cannot escape. Worse, if support from the government or foreign countries stops, the industries will collapse completely. That is, self-sustaining growth is impossible.

On the other hand, if the heavy and chemical industry sector can exceed a certain critical level—through tremendous endeavours, through a decline in the competitive power of the foreign industries (due for example, to wars,[2] internal disturbances and disasters, or to changes in the industrial structure of the foreign countries caused by economic development, or wage increases), or through government support and foreign aid—then there may be cases where the industries concerned can continue to grow through their own efforts. Our theoretical model will show where such self-sustaining growth is possible. In that sense, our theory can be called a 'minimum critical effort theory',[3] because it identifies the minimum critical efforts needed for self-sustaining growth to become possible and identifies the point at which this is attained.

Of course, our model also allows for the possibility that the heavy industry will eventually disappear no matter how much government support and/or foreign aid is provided initially if that support and/or aid is terminated. This is Myrdal's case where the structural reform of the society is a prerequisite for take-off.

The foregoing vision of development has been presented by others before us, and has been cherished by many as an image of the developing economies. The purpose of our book is to clarify this image in a rigid and precisely formulated framework. The method used is to express the investment patterns both of the light industry sector (including plantations), and of the heavy and chemical industry sector by means of a differential equation system and to show and analyse the characteristic features of the solution to this system. That is, we study the possibility of take-off, and the possibility of the existence of a minimal critical effort line. For this we need a certain level of abstraction, but the economic implications of any abstractions are carefully explained, in the text or in footnotes. We shall start from a simple and limited model abstracted from economic reality and then make the analysis more general and more realistic by eliminating the abstractions.

Notes

1. See Nurkse (1952) and Meier (1953).
2. For example, World War I provided a chance for the heavy and chemical industries in Japan to attain self-sustaining growth because the same types of industries in Europe were seriously damaged by the war. Japanese industry was destroyed in World War II, but quickly picked up again due to the stimulus provided by the Korean War (1950–3). However, the experiences of the two world wars should not be viewed as the main cause of the take-off or the revival of the heavy and chemical industries in Japan but rather as a cue therefor.
3. On the theory of minimum critical effort, see Leibenstein (1957) or Ranis and Fei (1961).

2

Notation and Assumptions

We start by explaining the notation of the model.

X	Output of domestic light industries, including plantations, measured in real terms. We call this sector the *X*-sector.
K_X	Accumulated capital stock of the *X*-sector, measured in real terms.
$\dot{K}_X$	Net investment in the *X*-sector.
N_X	Employment in the *X*-sector.
$F_X\,(K_X, N_X)$	Production function of the *X*-sector.
R_X	Gross profit rate of the *X*-sector, measured in terms of the goods of the subsistence sector as defined later.
$p(X)$	Terms of trade of the *X*-sector, i.e. the price level of *X*-sector goods measured in terms of subsistence sector goods.
$\overline{w}$	Subsistence wage rate, or institutional wage rate,[1] measured in terms of subsistence sector goods.
Z	Output of domestic heavy and chemical industries, measured in real terms. We call this sector the *Z*-sector.
K_Z	Accumulated capital stock of the *Z*-sector, measured in real terms.
$\dot{K}_Z$	Net investment in the *Z*-sector.
N_Z	Employment in the *Z*-sector.
$F_Z\,(K_Z, N_Z)$	Production function of the *Z*-sector.
R_Z	Gross profit rate of the *Z*-sector, measured in terms of subsistence sector goods.

Next, we explain the basic assumptions:

1. *The economy consists of three sectors. They are the subsistence sector (the S-sector), the light industry sector, including plantations (the X-sector), and the heavy and chemical industry sector (the Z-sector).*[2] Here the term *S*-sector is employed in the same sense in which it is used in Lewis (1955), but in our model it mainly plays the role of unlimited supplier of low-wage labour. Our concerns are mainly focused in the other two sectors, i.e. the *X*-sector and the *Z*-sector, which together comprise the capitalist sector in Lewis's paper.

2. *Disguised unemployment always exists in the S-sector, and thus the marginal productivity of labour in this sector is always lower than the subsistence wage level.* We do not assume the existence of surplus labour in the *S*-sector. That

is, we do not assume that the marginal productivity of labour in this sector is zero. This assumption of zero marginal productivity is a controversial one, and our analysis is free from this assumption.[3] At a later stage, however, we analyse the case where the X- and/or the Z-sectors, especially the latter, has grown sufficiently to absorb considerable labour from the S-sector, so that as a result a scarcity of labour arises in the S-sector—i.e. the marginal productivity of labour exceeds the subsistence wage level. Assumption 2 is thus violated. Until that point, however, we examine under what conditions and in what situations the economy can get out of the trap of a subsistence economy. That is, our analysis is carried out to determine under what situations and circumstances the dynamic path of the economy proceeds so as to violate assumption 2. This method of analysis is somewhat paradoxical because we assume that assumption 2 remains valid in order to examine the mechanism through which it is violated.[4]

3. *The price of X-sector goods relative to S-sector goods is a decreasing function of X-sector output. Further, the price elasticity of demand worldwide for X-sector goods is less than 1 for at least one level of X-sector output.* The relative price of X-sector goods is a function of X, say $p(X)$. The first half of this assumption is expressed as $p'(X) < 0$. The latter half is expressed by $\left|\frac{X}{p(X)}\frac{dp(X)}{dX}\right| = \left|\frac{X}{p(X)}\cdot p'(X)\right| < 1$ for a certain value of $X > 0$. Noting that $p'(X) < 0$ this inequality can be written as $p(X) + p'(X)X > 0$. If this condition is not satisfied, it may become optimal for the X-sector to cease production. The reason why we assume this condition is to exclude this trivial case. This assumption does not violate reality.[5]

4. *The price of Z-sector goods relative to S-sector goods is assumed to be fixed.[6] We express the price of Z-sector goods in whatever unit is needed to make the relative price 1.* At the initial stage of a country's economic development, we may assume that the scale of Z-sector output is not large enough to influence the world market price of Z-sector goods.[7] As already mentioned, our main concern is to examine how existing small-scale heavy and chemical industries can begin to achieve self-sustaining growth. Take, for example, the case of the Japanese ship-building industry. In the initial stage of Japan's economic growth, the Japanese ship-building industry was of negligible size. Japanese engineers studied technology from foreign engineers who were invited to Japan, or went abroad to study, and Japan went on to become the world's largest ship-building power, accounting at one time for almost 50 per cent of world ship-building. Recently, however, a newcomer, Korea, has come to rival Japan. Our concern is thus to analyse how an infant industry gains or can gain the momentum to grow. The dynamic analysis of mature industries is not our concern here, although we do briefly analyse the situation where the industry has grown to become a medium power in the world and assumption 4 is thus violated. However, our purpose is to see how a deterioration in the terms of

trade of *Z*-sector goods affects the take-off of the economy, and not to perfect the growth theory of developed economies. For example, the terms of trade for *Z*-sector goods may deteriorate even if the *Z*-sector is still short of medium size, depending on the worldwide market for Z-sector goods. This may have a negative impact on the prospects for take-off as our analysis will confirm.

If the *X*- and/or *Z*-sectors grow sufficiently to decrease the employment level in the *S*-sector, then the domestic food supply will decrease unless counter-measures are taken, such as the mechanization of agriculture and/or the increase of fertilizer input. To simplify the analysis we initially assume that any amount of food can be imported without a price increase. However, we later study the case where the growth of the *X*- and/or *Z*-sectors brings about an increase in the price of food.

5. *The sum of investment in the X- and Z-sectors is equal to the sum of profits in the X- and Z-sectors.*[8] *How total profits are divided between the two sectors will be explained later. Wage incomes in the S-, X-, and Z-sectors are spent exclusively on consumption. This assumption, and particularly the rule allocating profits* to investment, is the most important assumption in our model. This rule will be described and explained precisely in Chapter 3. We initially assume that the profits obtained in the country are re-invested in the country and are not remitted to a foreign country, even to a suzerain country. It should be noted, however, that in our model, the *X*-sector includes plantation farms although in both former and present colonies quite a large proportion of the profits obtained from plantation farms were or are remitted to the suzerain country; we therefore study the latter possibility too. Likewise, although we initially ignore the inflow of capital from abroad, we recognize that this inflow is non-negligible at both the government level and the private level in the case of the post-war developing countries. We therefore consider this possibility later especially in Chapter 4.

As regards the disposition of non-wage income in the *S*-sector, that is, land rent income, we assume that this income is spent exclusively on consumption and on investment in the *S*-sector.[9]

It should be noted that one important reason for the development of the Japanese economy is that excess income in the *S*-sector, whether originating in land rent income or elsewhere, was directed to investment in the *X*- or *Z*-sectors, especially the latter. Heavy taxes, especially the land tax system, and the forced saving caused by the intentional or unintentional inflationary policies, forcibly induced the excess income to flow to government-owned enterprises and/or to heavily protected and supported private enterprises. Furthermore, Japanese landowners were thrifty, contrary to Ricardo's spendthrift stereotype, and invested their rent income in the industrial sectors, either directly or indirectly through the banking system. But such landowners must be very rare in world history, so we focus on cases that are more widely applicable. In other words, we focus on explaining how development proceeds, or is impeded, when such tax and/or forced saving policies are not adopted (whether intentionally or for political reasons) and landowners are not thrifty. But again, we later study the case where a portion of non-wage income in the S-sector is directed to investment in the X- and Z-sectors.

6. *The production function of the X-sector satisfies the following conditions*:

$$\frac{\partial F_X}{\partial K_X} > 0, \quad \frac{\partial^2 F_X}{\partial K_X^2} < 0,$$
$$\frac{\partial^2 F_X}{\partial K_X \partial N_X} > 0,$$
$$\frac{\partial F_X}{\partial N_X} > 0, \quad \frac{\partial^2 F_X}{\partial N_X^2} < 0,$$
$$F_X(\alpha K_X, \alpha N_X) = \alpha F_X(K_X, N_X) . \quad (\alpha > 0)$$

That is, the marginal productivity of each productive factor is positive and diminishing and the function exhibits constant returns to scale. This assumption can be relaxed slightly. Even if we assume increasing returns to scale for the production function of the X-sector, the analysis will hold without any modification, provided the increasing returns to scale are offset by a worsening in the terms of trade, or that terms of trade are constant but returns to scale are decreasing. We discuss this again later.

7. *The production function of the Z-sector satisfies the following conditions:*

$$\frac{\partial F_Z}{\partial K_Z} > 0, \quad \frac{\partial^2 F_Z}{\partial K_Z^2} < 0,$$
$$\frac{\partial^2 F_Z}{\partial K_Z \partial N_Z} > 0,$$
$$\frac{\partial F_Z}{\partial N_Z} > 0, \quad \frac{\partial^2 F_Z}{\partial N_Z^2} < 0,$$
$$F_Z(\alpha K_Z \, \alpha N_Z) > \alpha F_Z(K_Z, N_Z) . \quad (\alpha > 1)$$

That is, the marginal productivity of each factor is positive and diminishing, and production exhibits increasing returns to scale. Here it should be noted that the diminishing marginal productivity of each productive factor is not necessarily inconsistent with increasing returns to scale. For example, consider the following Cobb–Douglas-type production function $AK^{\alpha}L^{\beta}$.

Here, K, denotes capital input and L the labour input. If $\alpha + \beta > 1$, there exist economies of scale. The marginal productivity of capital is in this case $\alpha AK^{\alpha - 1}L^{\beta}$.

Thus, if $\alpha - 1 < 0$ or $\alpha < 1$, this function is decreasing in K. That is, the marginal productivity of capital is diminishing. Similarly, if $\beta < 1$, the marginal productivity of labour is decreasing also. Thus, if $\alpha + \beta > 1$, $\alpha < 1$, $\beta < 1$, the above Cobb–Douglas-type production function satisfies all the conditions of assumption 7.

8. *The output level of the X-sector is determined so as to maximize profits, given the amount of accumulated capital at each moment, the subsistence wage level, and the demand function for X-sector goods.* Since we assumed that the amount of accumulated capital is given, maximizing profits is the same as maximizing the profit rate. This assumes that the X-sector consists of some representative firms, each of which knows from experience how an increase in its output will bring about similar increases in the outputs of the other firms, and also what the total effect will be on the price level. If the X-sector consists of only one or a few co-operative firms, the assumption will hold, but this unreasonably limits the range within which the analysis is valid. Our analysis is also exactly valid for the case where production in the X-sector is carried out by the so-called *tâtonnement* process.[10] If the X-sector consists of numerous small firms, such that each firm accepts the prevailing market price as given and behaves so as to maximize profits subject to this market condition, and if as a whole the output level is quite large compared to world output, then the market price may fall sharply. The new output level will not maximize the profit rate at the lower price, so each firm will have to adjust its output. This approach is called the disequilibrium approach and would require another differential equation to show the movement of the world price over time. But, dynamic theory expressed by a system of three differential equations has never been successful, because, we cannot write down a three-dimensional solution path on a two-dimensional sheet of paper. Intuitively, the general pattern will not be very different from that in our analysis, but any exact statement is impossible.

Our analysis, however, does hold when $p'(X)$ is sufficiently close to zero—that is, when the total output of the numerous small firms in the X-sector is low compared to world output. In this case, the terms of trade remain constant, so it is not necessary to introduce another differential equation. The empirical research on Japan presented later will show this.

9. *The output level of the Z-sector is determined so as to maximize the profit rate, given the amount of accumulated capital at each moment and the subsistence wage level.*[11] The terms of trade of the Z-sector are taken as fixed. The world market is competitive. Since the amount of capital is fixed at any given moment, there exists an optimum output level even though returns to scale are increasing.

Our model assumes the maximization of profit rates in both the X- and Z-sectors. This means that labour and resources move to more profitable opportunities. It is assumed that where labour in the S-sector is in surplus, or the marginal productivity of labour is lower than the subsistence wage, labour will move without strong friction from the S-sector to the X- or Z-sectors as long as the employment opportunities exist.

In this sense, we assume the Lewis-type economy described in Chapter 1, or an economy in which development has already advanced to the stage where the above mentioned movement of labour is not socially hindered. We do, however, assume some friction with respect to the movement of funds between the X- and Z-sectors, but as the degree of friction in our model can vary, this assumption does not restrict its generality.

The assumption pertaining to labour mobility is more restrictive for the model to be general and applicable to the real situation in less developed economies. To accommodate the case where some friction exists with respect to labour mobility, we therefore consider the case where the movement of labour incurs a cost.

Notes

1. For an explanation of the subsistence or institutional wage rate, see Lewis (1954) or Ranis and Fei (1961). As will be seen in the following we assume three production sectors: the light industry sector including plantations; the heavy and chemical industry sector; and the subsistence sector. It is often said that unless there is a marked difference between the industrial wage and the subsistence wage, labour migration will not occur. In other words without compensation for the cost, demerits, and disadvantages of city life, and factory work, migration will not occur. The following analysis will hold even if we assume that such a difference exists, but for the sake of simplicity, we assume that such a difference does not exist. Or we can take account of this by defining the production function in such a way that this extra cost is subtracted from output.
2. As will be discussed in ch. 5, if the agricultural sector adopts modern, capitalistic production processes and management methods, it should be viewed as a branch of the X-sector, or even the Z-sector. That is, the employment level will be determined so as to maximize the profit rate, and hence the wage rate will be equated to the marginal productivity of labour, and investment in the sector will be decided through a comparison of the profit rate with that in the other sector. The following analysis would hold even if the situation were changed in this way.
3. For example, Ranis and Fei (1961) assert that at the stage of pre-development, the marginal productivity of labour in the S-sector is zero. Contrary to this, Schultz (1964) tried to show through empirical research that even in a country with a very low level of income such as India this assumption does not hold. There is a refutation to Schultz's analysis, but we do not enter deeply into this dispute.
4. The Ranis–Fei model does not assume an explicit capital accumulation process. Their model assumes that capital accumulation proceeds, but there is no mechanism to explain how the funds for it are managed or raised. That is, capital accumulation is an exogenously given variable in their model. But the most important mechanism in dynamic models is that of capital accumulation. That is, it is vitally important to show what proportions of the various types of income are directed to new capital formation in the various sectors, and also how the accumulation of capital will in turn affect the various types of incomes as this will in turn affect new investment. and so on in an endless chain of cause and effect. It is this endless

process that determines the dynamic path. That is, unless capital accumulation is an endogenous variable, the model cannot be called dynamic, and hence the explanation and analysis of the development process cannot be complete. Ranis and Fei give no rule on how the various types of income are allocated to consumption and investment (saving), yet it is this which determines whether the economy has the potential for self-sustaining growth or whether it is destined to remain in a state of stagnation. Without the explicit introduction of the capital accumulation mechanism, the analysis can be a comparative static one but not a dynamic one.

5. We implicitly assume a slightly stronger assumption: the gross profit rate for a certain value of K_X is assumed to be greater than the depreciation rate on capital (see n. 5 and 6 in ch. 3). Note that in the short-term optimum production schedule, the gross profit rate can be less than that rate because it is better to have a negative net profit but a positive gross profit than a zero gross profit. Here we are discussing short-term optimality.
6. There may be a tendency for the terms of trade between primary goods and industrial goods to move unfavourably to the former, but this tendency has no direct relation to assumptions 3 and 4. Our assumptions are essentially made concerning demand functions. The above-mentioned tendency will be expressed by a continual downwards shift in the demand functions over time.
7. The Z-sector products need not be import substitutes. They can be exportable goods. In reality, some industrial products, such as bicycles, become important export goods, along with plantation goods and some *X*-sector goods, at an early stage of development. But iron and steel, manufacturing machinery, automobiles, and ocean-going ships are still imported at this stage.
8. We assume that profits obtained in the *X*- and *Z*-sectors are not spent on consumption. But if a constant proportion of accumulated capital is diverted from profits and spent on consumption, the following analysis will hold without modification, provided we view the capital depreciation rate as being elevated by this proportion. It should be noted that successful take-off may turn into failure due to this consumption from profit.

   As for the validity of the assumption that profits from the *X*- and *Z*-sectors (Lewis's capitalistic sector) are invested in these sectors while wage incomes and land rent incomes are spent on consumption and on investment in the *S*-sector, see Lewis (1955). Under the assumption of a subsistence wage level, the assumption that wage income is spent exclusively on consumption seems necessary for preserving consistency because if the wage earner could save, the wage level would, by definition, exceed the subsistence level. According to Lewis, almost all land rent income is spent lavishly on consumption, but landowners will imitate capitalist behaviour and direct some portion to saving and hence to investment as time goes on. However, this situation will arise only when landowners have had many chances to interact with successful capitalists after development has proceeded far enough to exceed the level of self-sustaining growth (see ch. 5).
9. This assumption was made to be consistent with the assumptions concerning investment. We assumed that profits obtained in the *X*- and *Z*-sectors are wholly reinvested in these sectors, and do not flow out to foreign countries or to the *S*-sector. Conversely, we assumed that there is no *S*-sector investment in the *X*- and *Z*-sectors. Investment in the *X*- and *Z*-sectors from income originating in the *S*-sector would constitute a contradiction, hence our assumption. But even if we do not make this

assumption explicit, the above-mentioned assumptions automatically exclude the possibility of income originating in the *S*-sector being invested in the *X*- and *Z*-sectors. Thus, this assumption becomes redundant. But some redundancies may have the merit of clarifying the way in which income is disposed in our model. See also n. 8 above.

Related to this point, we assume that the trade balance is always balanced before we take account of foreign aid and the introduction of an import tariff. Of course, even if some portion of the profits of the *X*- and *Z*-sectors flows out to finance investment or consumption in the *S*-sector or to foreign countries, provided that there is an equivalent inflow the analysis will still hold. But such a generalization is really trivial.

The following points, however, are not trivial. There may exist a cross-flow of consumption between sectors. Funds directed to consumption are the entire wage incomes in the three sectors and a portion of the rental income in the *S*-sector. But consumption goods can be any of the following: *X*-sector goods, *Z*-sector goods, *S*-sector goods, and imported goods. If the trade balance is maintained, the cross-flow of goods presents no problem. The assumption that profits in the *X*- and *Z*-sectors are wholly re-invested in these sectors does not mean that *X*- and *Z*-sector goods are investment goods. In fact, we rather assume in the following chapter that *X*-sector goods cannot be investment goods. It is possible that *X*- and *Z*-sector goods are exported and that goods from foreign countries are imported as investment goods. When some portion of the rental income in the *S*-sector is spent on investment in the *S*-sector, this does not mean that *S*-sector goods are investment goods. In that case, *S*-sector goods might be exported and foreign investment goods in return might be imported. Of course, these are not unique cases. We can conceive of many other cases where there is a cross-flow of goods among sectors.

10. The *tâtonnement* process is the one in which actual production is not carried out until demand and supply coincide. The process proceed, as follows: for a tentatively determined price of the good, producers and buyers state their plans. If the demand and supply plans coincide, then the buying and selling contract is made at that price, and the plans are actually carried out. If the selling plan exceeds the buying plan, the tentative price is reduced, and both sides propose new plans at the new price. If the former is short of the latter, the tentative price is raised and new plans are proposed. This process is continued until both plans coincide. Then the tentative price for such plans becomes the contract price, and the plans are actually carried out.

    Our assumption is that the above process is carried out at a moment in time.

11. As already mentioned, it is not necessary in the following analysis for the subsistence wage level to be the same in the *X*- and *Z*-sectors. Gerschenkron (1952) pointed out that when the *X*-sector consists mainly of plantations, it may be more realistic to assume that the *Z*-sector wage is higher than that in the *X*-sector due to the shortage of factory labour. But our assumption does not correspond perfectly with Gerschenkron's opinion. He posits that a characteristic feature of less developed countries is the extreme shortage of labour who are both willing to cut the chain of convention tying them to the land and to possess the skills and attitudes suitable for factory work. If we want to express this exactly, we should assume that the wage difference between the *X*- and *Z*-sectors is not constant but that it depends on the amount of employment in the *Z*-sector. We touch on this point later; n. in ch. 4.

3

The Model

3.1. The Amount of Accumulated Capital and the Profit Rate

Before describing the model, we shall establish the properties of the profit rates in the X- and Z-sectors respectively, as derived from assumptions 8 and 9.

3.1.1. The *X*-sector

From assumption 8, the values of X and N_X are determined as the solution to the following maximization problem:

$$\begin{aligned} &\text{Maximize} && p(X)X - \bar{w}N_X, \\ &\text{subject to} && X = F_X(K_X, N_X). \end{aligned}$$

Here, X and N_X are variables, and K_X is a parameter assumed to be given. Using Lagrange's method of multipliers, the values of X and N_X are determined so as to satisfy the following relations:

$$X = F_X(K_X, N_X) = \frac{\partial F_X}{\partial K_X} K_X + \frac{\partial F_X}{\partial N_X} N_X, \tag{3.1}$$

and

$$\frac{\partial F_X}{\partial N_X} = \frac{\bar{w}}{p'(X)X + p(X)}. \tag{3.2}$$

[1]

Equation (3.1) derives from the assumption that the X-sector exhibits constant returns to scale.

Next, the profit rate is expressed as

$$R_X = \frac{1}{K_X}[p(X)X - \bar{w}N_X]. \tag{3.3}$$

When the value of parameter K_X changes, the solution of (3.1) and (3.2), and hence the value of R_X changes. That is, X, N_X, and R_X are functions of K_X.

It is impossible to express explicitly the functional forms of $X(K_X)$ and $N_X(K_X)$ because the relations (3.1), (3.2), and (3.3) involve general functional forms such as $F_X(K_N, N_X)$ and $p(X)$, but for our purposes it is not necessary to have a complete knowledge of the forms of these functions. In fact, the only property we need to know is the sign of the rate of change of the profit rate corresponding to a change in K_X—that is, the sign of dR_X/dK_X.

To examine this, differentiate (3.1) with respect to K_X.

$$\frac{dX}{dK_X} = \frac{\partial F_X}{\partial K_X} + \frac{\partial F_X}{\partial N_X}\frac{dN_X}{dK_X}. \tag{3.4}$$

Then, differentiate (3.3) with respect to K_X.

$$\frac{dR_X}{dK_X} = \frac{1}{K_X^2}\left[\left\{p'(X)\frac{dX}{dK_X}X + p(X)\frac{dX}{dK_X} - \overline{w}\frac{dN_X}{dK_X}\right\}K_X - (p(X)X - \overline{w}N_X)\right]$$

$$= \frac{1}{K_X^2}\left[\left\{(p'(X)X + p(X))\frac{dX}{dK_X} - \overline{w}\frac{dN_X}{dK_X}\right\}K_X - (p(X)X - \overline{w}N_X)\right]. \tag{3.5}$$

Substituting (3.2) and (3.4) into (3.5), we obtain

$$\begin{aligned}\frac{dR_X}{dK_X} &= \frac{1}{K_X^2}\left[\left\{(p'(X)X + p(X))\left(\frac{\partial F_X}{\partial K_X} + \frac{\partial F_X}{\partial N_X}\frac{dN_X}{dK_X}\right)\right.\right.\\ &\quad \left. - (p'(X)X + p(X))\frac{\partial F_X}{\partial N_X}\frac{dN_X}{dK_X}\right\}K_X\\ &\quad \left. - \left\{p(X)X - (p'(X)X + p(X))\frac{\partial F_X}{\partial N_X}N_X\right\}\right]\\ &= \frac{1}{K_X^2}\left[(p'(X)X + p(X))\left(\frac{\partial F_X}{\partial K_X}K_X + \frac{\partial F_X}{\partial N_X}N_X\right) - p(X)X\right]\\ &= \frac{1}{K_X^2}p'(X)X^2.\end{aligned} \tag{3.6}$$

This last relation is derived from (3.1). Since $p'(X) < 0$,

$$\frac{dR_X}{dK_X} < 0. \tag{3.7}$$

If the amount of accumulated capital in the X-sector increases, the profit rate of the sector declines. This property plays an important role in our analysis.[2]

3.1.2. The *Z*-sector

From assumption 9, if the amount of accumulated capital is given, the solution to the following maximization problem will give us the output level, Z, and labour demand, N_Z. Since labour will be supplied without limit at the subsistence wage, labour demand at $\overline{w}$ becomes the employment level. That is, the supply function of labour is perpendicular to the wage axis.

$$\text{Maximize} \quad Z - \overline{w}N_Z,$$
$$\text{subject to} \quad Z = F_Z(K_Z, N_Z).$$

Using Lagrange's method of multipliers again, we obtain

$$Z = F_Z(K_Z, N_Z), \tag{3.8}$$

and

$$\frac{\partial F_Z}{\partial N_Z} = \overline{w}. \tag{3.9}$$

Here, Z and N_Z are variables, and K_Z is a parameter. The profit rate is

$$R_Z = \frac{1}{K_Z}(Z - \overline{w}N_Z). \tag{3.10}$$

Here, we need to know the sign of $\frac{dR_Z}{dK_Z}$. Differentiating (3.10) with respect to K_Z, we obtain

$$\begin{aligned}\frac{dR_Z}{dK_Z} &= \frac{1}{K_Z^2}\left[\left(\frac{\partial F_Z}{\partial K_Z} + \frac{\partial F_Z}{\partial N_Z}\frac{dN_Z}{dK_Z} - \overline{w}\frac{dN_Z}{dK_Z}\right)K_Z - (Z - \overline{w}N_Z)\right] \\ &= \frac{1}{K_Z^2}\left[\frac{\partial F_Z}{\partial K_Z}K_Z + \frac{\partial F_Z}{\partial N_Z}N_Z - Z\right].\end{aligned} \tag{3.11}$$

The last relation is derived from (3.9). From the assumption of increasing returns to scale, we obtain

$$\frac{\partial F_Z}{\partial K_Z}K_Z + \frac{\partial F_Z}{\partial N_Z}N_Z - Z \geq 0. \tag{3.12}$$

The mathematical procedure used to obtain this inequality will be explained in Appendix A. Here, we note merely that the equality sign is necessary. Then, from (3.11) and (3.12), we obtain

$$\frac{dR_Z}{dK_Z} \geq 0. \tag{3.13}$$

That is, if the amount of accumulated capital increases, the profit rate rises. Note that in spite of the necessity of the equality sign in (3.13), an increase in K_Z will bring about an increase in R_Z—not merely a non-decline, but an increase. The reason for this will also be explained in Appendix A.[3]

Now, the properties obtained in 3.1.1 and 3.1.2—i.e. that an increase in the amount of accumulated capital will bring about a decline in the X-sector profit rate and an increase in the Z-sector profit rate—are central features of our model, in that it is this property that makes economic development possible in some cases and impossible in others. In other words, the character of the dynamic paths in our model derives from these two properties, and the assumptions made earlier are nothing but the economically meaningful stage-setting needed to derive them. Of course, if we were to assume (3.7) and (3.13) at the outset, the analysis would become more general but the economic situation satisfying these properties would not be made clear. Specification is thus not necessarily a drawback because it can clarify the economic situation that we want to study.

3.2. The Mathematical Formulation of the Model

Our model is expressed by the following relations:

$$X = F_X(K_X, N_X), \tag{i}$$

$$Z = F_Z(K_Z, N_Z), \tag{ii}$$

$$\frac{\partial F_X}{\partial N_X} = \frac{\overline{w}}{p'(X)X + p(X)} \tag{iii}$$

$$\frac{\partial F_Z}{\partial N_Z} = \overline{w}, \tag{iv}$$

$$R_X = \frac{1}{K_X}\{p(X)X - \overline{w}N_X\}, \tag{v}$$

$$R_Z = \frac{1}{K_Z}\{Z - \overline{w}N_Z\}, \tag{vi}$$

$$\dot{K}_X + \mu K_X = \xi_X(R_X - R_Z)\cdot R_X \cdot K_X + \zeta_X(R_X - R_Z)\cdot R_Z \cdot K_Z, \tag{vii}$$

$$\dot{K}_Z + \mu K_Z = \xi_Z(R_X - R_Z)\cdot R_X \cdot K_X + \zeta_Z(R_X - R_Z)\cdot R_Z \cdot K_Z. \tag{viii}$$

Here, μ denotes the depreciation rate on capital goods.

Relations (i)–(vi) are a short-term equilibrium system—a so-called miniature Walrasian system—in which the values of K_X and K_Z are viewed as parameters. From relations (i)–(vi), variables X, Z, N_X, N_Z, R_X, and R_Z are solved as functions of the parameters K_X and K_Z. If these functions are substituted into (vii) and (viii), we obtain two differential equations in which the variables are K_X and K_Z. These two differential equations express the constraints that the dynamic paths of the capital accumulation process should satisfy. The right-hand sides of (vii) and (viii) show the rule for how profits obtained at each moment of time in each sector are directed to each sector. Here, we explain the meaning of the functions $\xi_X(R_X - R_Z)$, $\xi_Z(R_X - R_Z)$, $\zeta_X(R_X - R_Z)$, and $\zeta_Z(R_X - R_Z)$. Note that these are functions of $(R_X - R_Z)$ and not the products of ξ_X, ξ_Z, ζ_X, ζ_Z, and $R_X - R_Z$. $\xi_X(R_X - R_Z)$ is the proportion of profits obtained in the X-sector and reinvested in the same sector; $\xi_Z(R_X - R_Z)$ is the proportion of profits obtained in the X-sector and invested in the Z-sector. Since we assume that profits obtained in the X-sector and the Z-sector are exclusively reinvested in the two sectors and are not directed to consumption or remitted to, or invested in, foreign countries, and that no foreign funds are invested in the X- and Z-sectors,

$$\xi_X(R_X - R_Z) + \xi_Z(R_X - R_Z) = 1.$$

The proportion of profits obtained in the X-sector and reinvested in the same sector is assumed to depend on the magnitude of the difference between the profit rates of the two sectors. If $R_X \geq R_Z$, then the profits obtained in the X-sector are assumed to be entirely reinvested in the same sector. That is, $\xi_X(R_X - R_Z) = 1$. If $R_X < R_Z$, some portion of the profits in the X-sector are assumed to be invested in the Z-sector. It may be realistic to assume that the bigger the difference in profit rates, the larger this portion is. That is, $\xi_Z(R_X - R_Z) > 0$ and $\xi'_Z(R_X - R_Z) < 0$ for $R_X - R_Z < 0$.[4] On the other hand, it may not be realistic to assume that if the profit rate in the Z-sector is larger than that in the X-sector, all the profits from the X-sector will be invested in the Z-sector. The reason why we assume the above investment distribution functions is that in this model, we do not explicitly separate the stockholders and managers of the firms, but suppose that this separation exists. Given this separation, a considerable proportion of profits will be retained in the firm as reserves, and these reserves will be reinvested in the firm and not directed to other firms unless the profit difference becomes quite large.

If funds move more freely between the sectors, the shape of the investment distribution curve shown in Fig. 3.1 will shift from the solid line curve to the dotted line curve. The extreme case is $\xi_X(R_X - R_Z) = 1$ for $R_X - R_Z \geq 0$, and $\xi_X(R_X - R_Z) = 0$ for $R_X - R_Z < 0$, but this extreme case is excluded from our analysis because we assume the continuity of the $\xi_X(R_X - R_Z)$ function. We do not think this seriously hurts the reality of our model because the extreme case can be approximated as closely as we want by a continuous function. Thus, our analysis may be viewed as holding for the extreme case also.

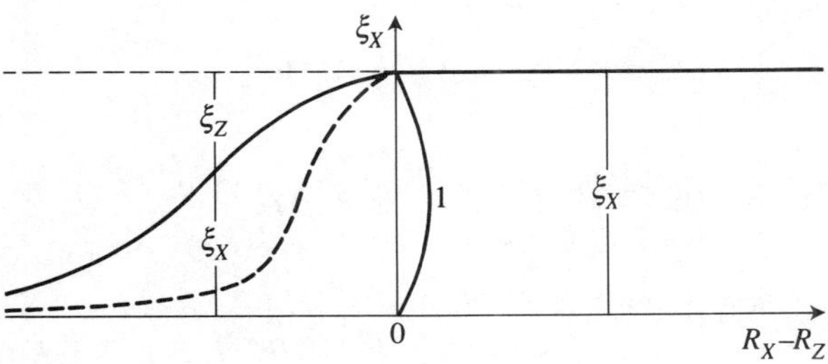

FIG. 3.1

The functions ζ_X, ζ_Z are defined in a similar way to ξ_X and ξ_Z. That is, they express the proportion of the profits obtained in the Z-sector that is invested in the X-sector and the Z-sector, respectively.

$$\zeta_X(R_X - R_Z) + \zeta_Z(R_X - R_Z) = 1.$$

Their shape is shown in Fig. 3.2.

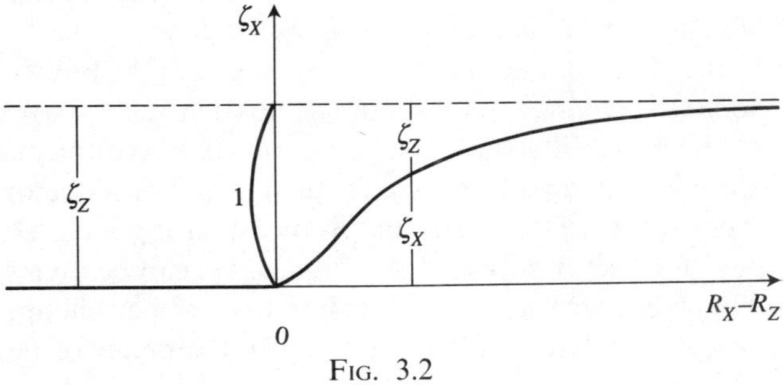

FIG. 3.2

As already mentioned, μ on the left-hand side of (vii) and (viii) denotes the rate of depreciation on capital goods. The rates in the two sectors are not necessarily identical, but for simplicity we assume that they are. If they are different, the profit rates discussed here should be taken as net profit rates. The following analysis then holds without further modification because it is net profit rates, rather than gross profit rates, that play the essential role in the analysis. The rule for allocating investment funds explained above assumes so-called static expectations, in the sense that the proportion of funds leaking from one sector to the other depends only on the difference in profit rates at the moment when investment decisions are made. A more general mechanism for expectations formation would of course be desirable, but it would make the analysis very much more difficult. We explain the reason for this difficulty and try to relax the limitation somewhat in Chapter 4.

Now, as mentioned above, R_X and R_Z are functions of K_X and K_Z, and we have shown (in 3.1.1 and 3.1.2, respectively) that R_X is decreasing in K_X and that R_Z is increasing in K_Z. Substituting these functions into (vii) and (viii), we obtain two functions of K_X and K_Z, respectively. Denote these as $\varphi_X(K_X, K_Z)$ and $\varphi_Z(K_X, K_Z)$, respectively. That is,

$$\dot{K}_X = \varphi_X(K_X, K_Z) - \mu K_X, \qquad \text{(vii)}'$$

$$\dot{K}_Z = \varphi_Z(K_X, K_Z) - \mu K_Z. \qquad \text{(viii)}'$$

Here,

$$\begin{aligned}\varphi_X(K_X, K_Z) \equiv\ & \xi_X(R_X(K_X) - R_Z(K_Z)) \cdot R_X(K_X) \cdot K_X, \\ & + \zeta_X(R_X(K_X) - R_Z(K_Z)) \cdot R_Z(K_Z) \cdot K_Z, \end{aligned} \qquad (3.14)$$

$$\begin{aligned}\varphi_Z(K_X, K_Z) \equiv\ & \xi_Z(R_X(K_X) - R_Z(K_Z)) \cdot R_X(K_X) \cdot K_X \\ & + \zeta_Z(R_X(K_X) - R_Z(K_Z)) \cdot R_Z(K_Z) \cdot K_Z. \end{aligned} \qquad (3.15)$$

Our aim now is to examine the dynamic characteristics of the differential equation system (vii)′ and (viii)′. Every dynamic path is expressed as the movement over time of point (K_X, K_Z) on the coordinate plane whose axes are the K_X-axis and the K_Z-axis. Equations (vii)′ and (viii)′ show the rule or constraint that controls this movement. Our aim is to examine what this rule means.

It should be noted that our investment mechanism implicitly assumes that X-sector goods are precluded from being investment goods, but imported goods are not. In our model, therefore, S-sector goods, Z-sector goods, and imported S- and Z-sector goods can be investment goods. But the inclusion of S-sector goods as investment goods is not necessarily realistic. We should also explain that the possibility that X-sector goods are investment goods is excluded because it is assumed that the terms of trade of X-sector goods—the price ratio between X-sector goods and S- or Z-sector goods—are a function of the output of X-sector goods. If X-sector goods can be investment goods, the amount of investment measured in real terms will then change as the terms of trade change. In relations (3.14) and (3.15), the prices of the investment goods are implicitly assumed to be fixed, which therefore rules out the possibility that X-sector goods are investment goods. If the possibility is admitted, the distribution of funds to each sector as shown by (3.14) and (3.15) is insufficient to express the dynamic path because the amount of funds is measured in terms of S- or Z-sector goods. If X-sector goods can be investment goods, it is necessary to introduce a mechanism to determine how much investment funds are spent on X-sector goods in each sector. That is, instead of the two investment mechanisms (3.14) and (3.15), four would be required. The differential equation system would then become so much more complicated that analysis of the dynamic paths would become impossible. This is why we have assumed the extremely simple case, but this simplification does not seriously affect the realism of the model.

3.3. The Direction of the Dynamic Path

If we can determine the values of the right-hand sides of (vii)′ and (viii)′ at each point (K_X, K_Z) in the $K_X - K_Z$ plane, we can determine the magnitude of variations in K_X and K_Z, that is, the values of $\dot{K}_X$ and $\dot{K}_Z$. We can then know the direction of movement of the dynamic path at that point. However, it is impossible to calculate these values explicitly. Is it then impossible to analyse the dynamic paths in our model? Not necessarily. If we can determine at least the *signs* of the right-hand sides of (vii)′ and (viii)′ at each point in the $K_X - K_Z$ plane, we can at least estimate the *directions* of the dynamic paths.

This global bird's-eye view of the dynamic paths can be obtained in the following manner. First, we divide the first quadrant of the $K_X - K_Z$ plane into several regions in such a way that the direction of the dynamic paths in each region can be unambiguously determined. (The detailed explanation of this division is discussed in Appendix B. Here we explain only the result.) The $\dot{K}_X = 0$ curve in Fig. 3.3 is the locus of points for which the right-hand side of (vii)′ is zero. That is, the $\dot{K}_X = 0$ curve is the point set defined by

$$\varphi_X(K_X, K_Z) - \mu K_X = 0.$$

Since $\dot{K}_X = 0$ for points on this curve, K_X neither increases nor decreases, and only K_Z changes; thus, the direction of infinitesimal movements of the dynamic path at these points is vertical. If $\dot{K}_Z > 0$ ($\dot{K}_Z < 0$), the dynamic path moves upwards (downwards) vertically.

The $\dot{K}_Z = 0$ curve in Fig. 3.3 is the set of points for which $\dot{K}_Z = 0$. Thus, the direction of the dynamic path at points on this curve is horizontal. If $\dot{K}_X > 0$ at a given point, the movement is rightwards horizontal. If $\dot{K}_X < 0$, the movement is leftwards horizontal. Where the $\dot{K}_X = 0$ and $\dot{K}_Z = 0$ curves intersect, label the intersection point as (K_X^*, K_Z^*). The first quadrant is divided into four regions.[5] The signs of $\dot{K}_X$ and $\dot{K}_Z$ are shown in Fig. 3.3; for example, in region I, $\dot{K}_X > 0$ and $\dot{K}_Z < 0$, meaning that K_X increases and K_Z decreases. That is, the

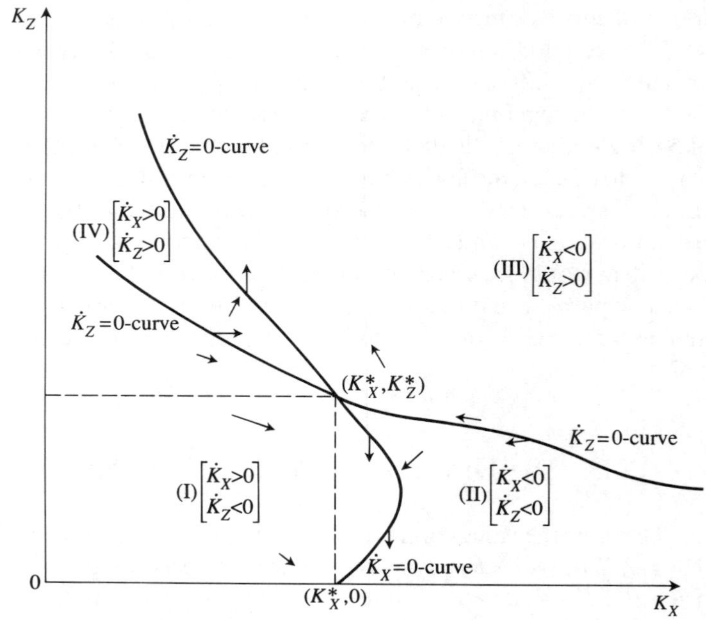

FIG. 3.3

direction of movement of the dynamic path at this point is southeasterly. As mentioned, the precise proof will be given in Appendix B. Here we merely outline why this is so.

Suppose the values of K_X and K_Z are both small. The profit rate in the X-sector will then be high, and that in the Z-sector low. In this case, even if the profits obtained in each sector are reinvested in the same sector, the X-sector will grow and the Z-sector will contract. Any transfer of profits from the Z-sector to the X-sector will only reinforce this tendency.

Next, suppose that K_X is large and K_Z is small. The profit rates of both the X-sector and the Z-sector will be small and not enough to offset capital depreciation. Thus, $\dot{K}_X < 0$ and $\dot{K}_Z < 0$. This is the case of region II. Suppose both K_X and K_Z are large. Then the profit rate in the Z-sector will be high and that in the X-sector low, and some portion of the profits obtained in the X-sector will leak out of that sector. That is, $\dot{K}_Z > 0$ and $\dot{K}_X < 0$. This is the case of region III.

The last case is the one where K_X is small and K_Z is large. Then the profit rates in both sectors will be high, and thus $\dot{K}_X > 0$ and $\dot{K}_Z > 0$.[6] This is the case of region IV.

The $\dot{K}_Z = 0$ curve is downward-sloping, but the shape of the $\dot{K}_X = 0$ curve is somewhat complicated:[7] it is downward-sloping for large values of K_Z, but bends back for smaller values of K_Z. The reason for this is shown in Appendix B.

3.4. Features of Dynamic Paths

Having established the direction of movement at each point in the first quadrant of the $K_X - K_Z$ plane, we can now get the outline of the dynamic path by connecting the directions starting from a certain point. This is done in Fig. 3.4.

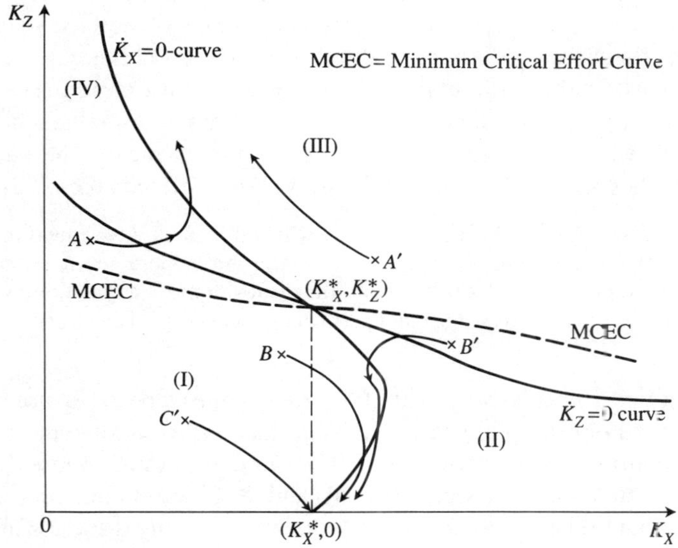

FIG. 3.4

Suppose the dynamic path starts from a point in region I. From this point, K_X increases and K_Z decreases. There are then two possibilities. One possibility is that the path does not intersect the $\dot{K}_Z = 0$ curve. In this case, two sub-cases are possible. Where the path starts from C, $\dot{K}_X > 0$ and $\dot{K}_Z < 0$ continue to hold, and thus it approaches infinitesimally close to point $(K^*_X, 0)$ in the figure. Another possibility is that the path starts from B, and after $\dot{K}_X > 0$ and $\dot{K}_Z < 0$ continue to hold for a while, it intersects the $\dot{K}_X = 0$ curve and thereafter enters region II; but because in region II both $\dot{K}_X < 0$ and $\dot{K}_Z < 0$, the final destination of this path is also $(K^*_X, 0)$. In these cases, the Z-sector will continue to contract and will sooner or later disappear. The long-term equilibrium of this economy is the existence of the S- and X-sectors but no Z-sector—that is, the economy continues to be dominated by subsistence cultivation, or subsistence and light industry. In such cases the economy fails to take off.

Now, consider the other possibility. This is the case where the path intersects the $\dot{K}_Z = 0$ curve, as shown by the path starting from point A. At the intersection point, $\dot{K}_Z = 0$—i.e. the direction of the path is horizontal—and the path enters region IV. Then, $\dot{K}_Z > 0$ and $\dot{K}_X > 0$. Both the X- and Z-sectors grow in this region. In this case the path sooner or later intersects the $\dot{K}_X = 0$

curve and enters region III, and thence $\dot{K}_X < 0$ and $\dot{K}_Z > 0$. That is, the Z-sector grows without limit, and the X-sector finally contracts. This is the case where the take-off of the economy is successful. Since we assumed an unlimited supply of labour at the subsistence level wage and constant terms of trade for Z-sector goods, the Z-sector can grow without limit. (Of course, in the real world, if the scale of the Z-sector becomes large, these assumptions will be violated and the growth of the Z-sector will thus be checked.)

In terms of the concepts in our model, we define take-off as the self-sustainable growth of the X- and/or Z-sector. But as we saw above, the X-sector cannot grow self-sustainably without limit. It may temporarily exceed the level of K_X^*, but eventually it will approach this level. On the other hand, the Z-sector can grow self-sustainably. Thus, take-off in our model means the sustainable growth of the Z-sector. By contrast, Rostow (1960) defined take-off as follows:

> The take-off is defined as the interval during which the rate of investment increases in such a way that real output *per capita* rises and this initial increase carries with it radical changes in production techniques and the disposition of income flows which perpetuate the new scale of investment and perpetuate thereby the rising trend in *per capita* output.

Our definition coincides with that of Rostow in one respect, but not necessarily perfectly. For example, a period during which the X-sector expands and the Z-sector contracts is not one of take-off by our definition. Suppose that, starting from A, B, or C, the X-sector expands and the Z-sector contracts in the first stage. We cannot judge at this stage whether the economy is successfully on the 'take-off' path or not. But at the stage when the Z-sector begins to grow, we can definitely say that the economy has taken off. Since Rostow's definition includes not only the change of investment or capital accumulation but also the change of production technology and the saving ratio, his definition cannot be translated into our concepts as it stands. But suppose we delete from his definition the part about investment and aggregate the X- and Z-sectors. Then, if the growth rate of the whole economy exceeds a certain level, the economy can be said to have taken off. In our model, however, it is possible that the X-sector grows very fast during a certain period while the Z-sector contracts slowly. The growth rate of the aggregate economy may then be sufficiently high that by Rostow's definition the economy has taken off. But if this situation does not last long, the Z-sector may sooner or later disappear and the *X-sector* may not be able to exceed a certain limit, so that the economy falls headlong to $(K_X^*, 0)$. The path starting at B is such a case: the economy has taken off for a period, but will sooner or later fall headlong. The reason why we define take-off as we did in our model is to exclude this possibility. That is, our definition is more restrictive. We say that the economy has taken off only when the Z-sector has reached the stage where it can grow by itself, without the assistance and protection of the government and foreign countries. Thus whereas Rostow judges that the Japanese economy took off around

1878–1900, our own empirical research shows that the Japanese economy took off around the latter half of the 1920s—that is, after the end of World War I.

Since successful cases of take-off are of particular interest, we shall describe the process in ordinary language. The profit rates at the starting point *A* are higher than the capital depreciation rate μ for both the *X*- and the *Z*-sectors, but that of the *X*-sector is higher than that of the *Z*-sector, and hence some portion of the profits in the *Z*-sector is not reinvested in that sector but leaks out to the *X*-sector. Thus the *X*-sector grows, while the *Z*-sector contracts. The *X*-sector growth rate is high because the profit rate is high; thus, the profits obtained are wholly reinvested in that sector, and some of the profits obtained in the *Z*-sector are also invested in the *X*-sector. The *Z*-sector, on the other hand, contracts, although the rate of contraction is initially rather low. This situation will continue for a while, but as the *X*-sector grows, the profit rate in that sector will decline sharply. On the other hand, the *Z*-sector profit rate also declines, due to a reduction in output, but the rate of decline is relatively low. The difference between the profit rates in the two sectors is thus reduced. The portion of the profits of the *Z*-sector leaking out to the *X*-sector is thus also reduced, and sooner or later the *Z*-sector stops contracting. The point at which this occurs is that at which the dynamic path intersects the $\dot{K}_Z = 0$ curve. Thereafter, the *Z*-sector begins to grow. Since the profit rate in the *X*-sector is still greater than that in the *Z*-sector, the leakage of profits obtained in the *Z*-sector continues for a while. The *X*-sector thus continues to expand, but as it does so the profit rate declines. In the *Z*-sector, on the other hand, the growth of the sector causes the profit rate to rise. This tendency continues until finally the profit rates in the two sectors become equal. After this point, the situation changes completely. The direction of the flow of profits is reversed: that is, the *Z*-sector profit rate becomes greater than that in the *X*-sector. The growth of the *Z*-sector accelerates, but that of the *X*-sector decelerates. The difference between the profit rates thus becomes even greater. Finally, the *X*-sector begins to contract because the proportion of profits obtained in the *X*-sector that is not invested in that sector becomes so large that the rate of investment falls short of the depreciation of accumulated capital. This happens just after the dynamic path intersects the $\dot{K}_X = 0$ curve. Thereafter, the *Z*-sector continues to grow without limit, and the *X*-sector continues to contract.

The foregoing description explains the typical case of successful take-off in our model. In the real world, however, the short-term reduction of the *Z*-sector cannot be observed, even around the starting point. Why is this? That *A* is the starting point means the existence of a *Z*-sector of some magnitude. How was this magnitude attained? In an economy where only the *S*- and *X*-sectors exist, and leaving everything to the profit incentive of free private enterprise, it is impossible that the initial point would be *A*. The fact that *A* is the starting point implies that there must have been some strong assistance and protection from the government or foreign aid, and the expectation of future prosperity by private firms. The path starting from *A* should thus be viewed as the path followed when such strong assistance and protection is abolished from point A on. In reality, however, it is not easy to judge at a specified point on the dynamic path whether take-off from the point is possible or not. For instance, at all points A, B, and C in Fig. 3.4, $\dot{K}_X > 0$, $\dot{K}_Z < 0$ and $R_X > R_Z$. Take-off is possible only for A, but not for B and C. This means that conditions $\dot{K}_X > 0$, $\dot{K}_Z < 0$, and $R_X > R_Z$ cannot be the criteria to judge whether the path is on the take-off course or not. As will be explained in the next section, the necessary and sufficient criterion for take-off is that the starting point locates above the minimum critical effort curve (MCE curve) in Fig. 3.4. But it is practically impossible for the policy agent to judge whether the specified point locates above the curve or not; for its shape and location depend on so many complex conditions such as the functional forms of production functions and investment functions. That is, it is practically impossible for the policy agent to judge whether the necessary and sufficient criterion is satisfied or not. Thus, the policy agent cannot rely on this criterion to decide to end assistance and/or protection.

We have, however, a simple sufficient criterion for it. It is that the starting point locates above the $\dot{K}_Z = 0$ curve in Fig. 3.4, or $\dot{K}_Z \geq 0$ is satisfied. This criterion does not require the information on the above complex conditions. It is natural that the policy agents eager for the success of development would continue assistance and/or protection to make $\dot{K}_Z \geq 0$ artificially, and would not abolish them until the condition $\dot{K}_Z \geq 0$ is satisfied with no assistance and/or protection. This attitude of the policy agents may be one of the reasons why the contract of the Z-sector is not observed for the take-off courses observed in the real developed economies. (Here it should be noted that when we say 'take-off', we mean the dynamic path on which the Z-sector never contracts without assistance and/or protection.) Let us not forget that foreign aid is sometimes granted less to facilitate take-off in the country assisted than to provide profitable opportunities for private firms in the donor countries. Such aid may well be less than effective in helping the developing countries to 'take-off'. In other cases, aid contributes more to the accumulation of private assets by powerful politicians in the receiver countries than to achieving take-off.

3.5. The Success or Failure of Take-off

We have seen above that attempts to achieve take-off may meet either with success or failure, even if the starting point is located in the same region I. What determines the outcome? Precisely speaking, however, there is also a third possibility. Consider the dotted line curve in Fig. 3.4. A dynamic path starting from a point on this curve eventually approaches point (K_X^*, K_Z^*). This is the third possibility. But this possibility is like proceeding on a razor's edge because if the dynamic curve deviates from this curve even slightly, it will not approach (K_X^*, K_Z^*). This possibility is unattainable in the real world because even a slight disturbance—and there are many possible disturbances in the real economy—will make the path deviate from the curve. In this sense, the interpretation of this curve as a dynamic path is almost nonsensical. But the curve has significance in another sense, however, in that it has the nature of a watershed. If the dynamic path starts from a point such as A in the figure, located above the curve, take-off is successful. If it starts from a point below the curve, such as B or C take-off is not successful. From this it can be seen that the starting point must be elevated above this curve through government measures and/or foreign aid; once the path enters the region above this curve, the economy can take off without any such 'artificial' measures—that is, self-sustaining growth is possible. We therefore call this curve the *minimum critical effort curve* because it shows the minimum height of the bar that must be cleared for the economy to achieve self-sustained growth. As shown above, (K_X^*, K_Z^*) is a long-term equilibrium point, but this equilibrium is not stable. That is, if the path deviates from this point, it can never return to it.[8]

In Fig. 3.4, we show dynamic paths starting from A' and B', but these paths are of no interest in the real world. The interesting cases are those starting from A, B, and C. It should be strongly emphasized that the less developed countries are characterized by the situations shown in B and C in the figure. This suggests that strong assistance and protection is needed for the Z-sector in such countries to elevate the starting point above the minimum critical effort curve and hence to succeed in achieving self-sustained growth.

There exists a case, however, where 'take-off' is structurally impossible. Fig. 3.5 shows this case: the $\dot{K}_X = 0$ and $\dot{K}_Z = 0$ curves do not intersect. There is no minimum critical effort curve—or, rather, the height of the bar is infinite. All dynamic paths sooner or later approach $(K_X^*, 0)$. That is, policies aimed at elevating the initial point are not effective for take-off. In this case, it will be necessary to introduce policies that can alter the production functions of the X- and Z-sectors—policies that can reform the structure of the economy, and even of society or human attitudes, in such a way as to cause the $\dot{K}_X = 0$ and $\dot{K}_Z = 0$ curves to intersect—as well as policies to elevate the starting point above the minimum critical effort curve. Otherwise, no amount of foreign aid and investment will have any effect. It would be like trying to scoop up water with a sieve.

The situation described in Fig. 3.5 is that of many of the less developed economies. Bringing about economic development in such countries is thus a truly formidable operation, because policies designed to bring about structural reform are not easy to enforce and they are sometimes not effective. Moreover, they require the modification of the culture, attitudes, morals, customs, and sense of values. This is the point made by Myrdal. Of course, we do not mean to suggest that economies with the structure described by Fig. 3.4 would not benefit from reform of the social structure. Even in these countries, policies aimed at reform will shift the $\dot{K}_X = 0$ and $\dot{K}_Z = 0$ curves, and thus the minimum critical effort curve, downwards, thereby making take-off easier. Thus, the reform of the social structure can be said to be necessary for all less developed economies to achieve take-off.

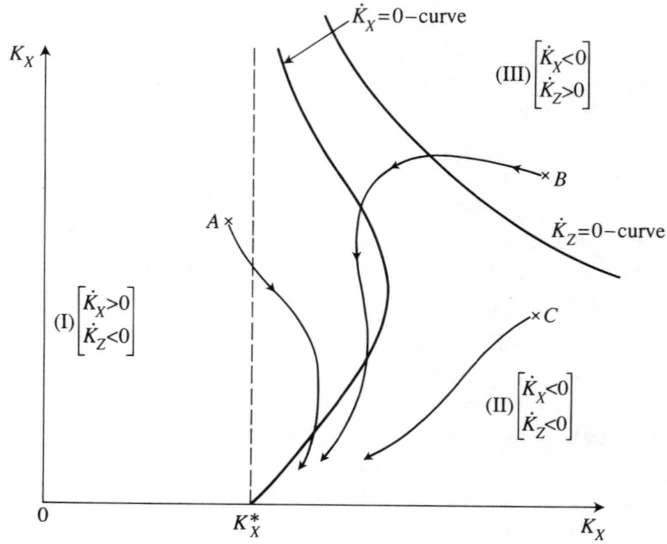

FIG. 3.5

We shall now explain why the $\dot{K}_X = 0$ and $\dot{K}_Z = 0$ curves do not intersect in some cases. In Fig. 3.4, these curves intersect at (K_X^*, K_Z^*). Then $R_X(K_X^*) = R_Z(K_Z^*) = \mu$. If the last relation does not hold, that is, if $R_Z(K_Z) < \mu$ for any value of K_Z, there is no K_Z^* such that $R_Z(K_Z^*) = \mu$. If the Z-sector profit rate is less than μ, or the net Z-sector profit rate is less than zero for any value of K_Z, such a case will arise. Here we mention merely one factor which may cause this. The wage level in less developed economies is low. In our model we assumed a subsistence wage level, but even this low wage level is high relative to the labour ability or skills supplied to the Z-sector. It is often the case that the nominal wage in a less developed country is about one-tenth of that in a developed country, but the labour productivity of the former is sometimes one-twentieth of the latter. The quality of labour required in the Z-sector is quite different from that required in the S-sector or the plantation-type X-sector. In such cases, the labour supplied is not suited for working in the Z-sector. Myrdal indicated this possibility.

As examples of policies required for structural reform in the case of Fig. 3.5 we can cite the expansion of social overhead capital, including investment in railroads, roads, harbours, hospitals, educational facilities, electricity, the water system, irrigation, and so on, and expenditures to elevate labour skills, and the general education level. These investments or expenditures will also be effective in shifting the $\dot{K}_X = 0$ and $\dot{K}_Z = 0$ curves downwards, even in the case of Fig. 3.4.

The conspicuous difference between Fig. 3.4 and Fig. 3.5 is that in Fig. 3.4 there exists a phase in which both the X- and Z-sectors grow, but not in Fig. 3.5. In other words, region IV—in which $\dot{K}_X > 0$ and $\dot{K}_Z > 0$—exists in the former case but not in the latter.

If $\dot{K}_X > 0$ and $\dot{K}_Z > 0$ are realized in the course of a dynamic path, it is certain that the path will sooner or later take off, or that it has already taken off. On the other hand, if $\dot{K}_X > 0$ and $\dot{K}_Z > 0$ are not realized, it is not certain whether the path will take off, or whether the path will enter the phase of $\dot{K}_X > 0$ and $\dot{K}_Z > 0$ in the future.

3.6. Import Tariffs

Protectionist policies toward the *Z*-sector are conceivable as a means of promoting development—subsidies to private corporations by the government, low-interest finance, a reduction in corporate taxes, the establishment of government-owned enterprises, and so on. Such policies aim at elevating the dynamic path at each point relative to where it would be if private enterprises behaved only in response to the profit motive, and thus at lowering the minimum critical effort curve so as to make it easier for the path to clear it.

Another reason why the *Z*-sector may be unable to clear the hurdle is that it cannot withstand international competition. In such a case, one way of protecting the *Z*-sector is to impose a tariff on imported *Z*-sector goods so that the domestic *Z*-sector is able to compete. However, this will cause the prices of both domestic and imported *Z*-sector goods to rise. Since some portion of wage income has been spent on the consumption of *Z*-sector goods before the imposition of the tariff and will also be spent thereon after the imposition, the subsistence wage level will also rise. This will have the effect of pushing down the profit rates in both the *X*- and *Z*-sectors. Since the price of *Z*-sector goods rises, the net effect of the import tariff on the profit rate in the *Z*-sector will be positive. This will be shown precisely in Appendix C. Now, policy measures favourable to the *Z*-sector and unfavourable to the *X*-sector will in general contribute to development. The reason for this will also be shown in Appendix C. The assumption there is that a constant proportion of consumption is spent on *Z*-sector goods, meaning that substitution does not occur in response to changes in the relative prices of *X*- and *Z*-sector goods. We also assume that *Z*-sector goods are not exported: export prices are lower than domestic prices by the amount of the import tariff, so if some proportion of *Z*-sector output is exported, the profit rate will decline. The amount of the decline will depend on the proportion of *Z*-sector goods that is exported. Since a rise in the wage level due to the imposition of an import tariff may have unfavourable effects on the *Z*-sector, which together with the additional unfavourable effect through exports may offset the favourable effects if the proportion of *Z*-sector goods exported is very high. We touch on this possibility in Appendix C.

We have so far assumed that no international imbalance exists. But the imposition of an import tariff will upset the international balance by the amount of tariff revenue. The way in which the revenue is disposed will affect the course of the dynamic path. We shall discuss this effect later.

Likewise, we have so far taken into account only short-term effects—that is, the values of K_X and K_Z are assumed to be fixed. But the investment goods in both the X-sector and the Z-sector are assumed in our model to be S- or Z-sector goods. The imposition of an import tariff will thus raise the market price of Z-sector goods, and investment, measured in physical units, will become smaller. Thus, the dynamic path will be influenced by the proportion of investment goods occupied by Z-sector goods. (In reality, the proportion will be close to 1, but we shall not discuss this complicated problem here. Appendix C shows that the Z-sector profit rate will rise after an import tariff is imposed even if profit is measured in terms of Z-sector goods. That is, even in the usual case where investment goods are entirely Z-sector goods, the amount of investment, measured in physical terms, will increase in the Z-sector and will decrease in the X-sector. Our exposition therefore relates only to this case, although it should be noted that there are some reservations, as set out in Appendix C.)

Now compare the case where an import tariff is imposed and the case where it is not. The imposition of a tariff will shift the $\dot{K}_X = 0$ and $\dot{K}_Z = 0$ curves south-west. The reason will be explained in Appendix C. Fig. 3.6 shows the situation. The solid line curves are those for the no-tariff case, while the dotted line curves are those for the tariff case. The minimum critical effort curve shifts downwards, so the effort needed to push the starting point above the curve will be less than in the original location. But if the economy does not succeed in taking off, its long-term stability will worsen because the long-term equilibrium level of the X-sector becomes smaller than in the original case. The reason is that the imposition of the tariff raises the wage level and thus the profit rate of the X-sector becomes lower. Of course, if the take-off fails, the import tariff would (or should) be abolished because it has no use after the Z-sector has

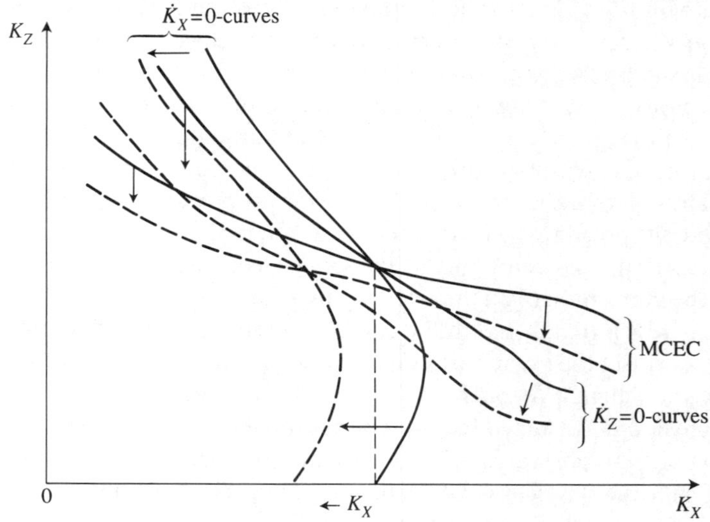

FIG. 3.6

disappeared. The long-term equilibrium level of the *X*-sector would then return to its original location.

The above discussion has ignored the *S*-sector. The damage from the imposition of the tariff is felt in the profit income of the *X*- and the *S*-sectors and in the rental income of the *S*-sector because, given the assumption of constancy in the market basket, the subsistence level wage will automatically rise.

The impact of the imposition of a tariff can be summarized as follows. *Z*-sector profits increase. *X*-sector profits decline. *S*-sector profits and rental income decline. Wage income remains constant. Since the wage level is already at the subsistence level, it cannot be reduced further. If the wage level is cut beyond this level, the supply of labour will decline in the long run and labour quality will decline in the short run because of the ill effect on the health of workers.

The government will receive the tariff revenue. The shift of income between productive factors can sometimes be a useful measure for promoting development, but in the real world, vested interest groups will oppose policies causing a redistribution of income. We may say that vested interests are the origin of all evils impeding development. In this sense we share the view of Myrdal and many other economists.

We now examine the effects of the disbursement of the government tariff revenue. If the revenue is invested in the *Z*-sector, the growth of this sector will be accelerated. If it is directed towards developing social overhead capital, the production functions of the *X*-, *Z*-, and *S*-sectors will shift upwards. This has the same effect as technological progress, which we discuss in the next section. If the revenue is used to finance lump-sum subsidies to the firms in the *X*-sector, two main cases are conceivable. One is that the production function shifts; using the subsidy to improve labour quality is one example of this case. Another is that the only result is an increase in profits. (A mixture of the two cases is also possible.) The case of a shift in the production function is the same as that of technological progress. A subsidy to the *X*-sector will accelerate the growth of the *Z*-sector. This is because a subsidy to the *X*-sector without a shift in the production function will shift the dynamic path to the right; since the minimum critical effort curve is downward-sloping, the shift might change the location of the path from below the curve to above it.

A subsidy to the *S*-sector will have no effect on the growth of the *Z*-sector. The reason is that non-wage income is assumed to be spent on consumption or on investment in the *S*-sector, so it has no effect on the growth of the *X*- and *Z*-sectors. But if the non-wage income earned in the *S*-sector is spent on investment in the *X*- and/or *Z*-sectors, as happened in Japan, even a subsidy to the *S*-sector will have positive effects. We discuss this possibility in chapters 4 and 5. Moreover, a subsidy to the *S*-sector may have other effects, such as to reform the society's sense of values, attitudes, and so on.

Regarding the statement that the protective tariff will facilitate take-off, we have to note that some reservations are described in Appendix C. This is why we stated that the protective tariff will *usually* facilitate take-off and not that it will *always* do so. The minimum critical effort curve will invariably shift downwards in the vicinity of the intersection of the $\dot{K}_X = 0$ and $\dot{K}_Z = 0$ curves, but it may shift upwards in a location far from the intersection. Such a shift (partly upwards and partly downwards) will be rather exceptional, however, which is why we stated that the protective tariff will *usually* facilitate take-off.

Many economists, including ourselves, are called promoters of free trade or of the principle of comparative advantage. In many cases, however, the efficiency of free trade refers to the short-term efficiency of resource allocation. In the case of the long-term development of the less developed countries, the short-term efficiency argument becomes tenuous. Our model shows that a protective tariff will accelerate take-off; it can therefore be seen as offering theoretical support for protectionism for latecomers. Of course, it may be seen as demonstrating the desirability of protecting infant industries. Yet we conducted our analysis not on an impressionistic basis or in the limited framework of a comparative static analysis, but rather in the framework of precise dynamic model. The imposition of a tariff on imported Z-sector goods will, as mentioned, have the same effect as reducing non-wage income in the X- and S-sectors and subsidizing the Z-sector. Conversely, the imposition of a tax on non-wage income in the X- and S-sectors and subsidizing the Z-sector using these funds will have the same effect as the imposition of a protective tariff.

3.7. Technological Progress

Economic development is achieved over a very long period. As it would be unrealistic to assume no technological progress over such a period, we must therefore examine the likely effects of technological progress.

Suppose the Z-sector profit rate increases due to technological progress in the Z-sector. The minimum critical effort curve will then shift downwards as shown in Fig. 3.7, and the possibility of take-off will increase because the downward shift of the minimum critical effort curve will reduce the effort needed to bring the starting point above it.

Next, what happens if the X-sector profit rate increases due to technological progress in that sector? In this case the intersection of the $\dot{K}_X = 0$ and $\dot{K}_Z = 0$ curves will shift horizontally rightwards. The dotted line curves in Fig. 3.6 can thus be viewed as the situation before technological progress occurs, and the solid lines as the situation after it occurs. The minimum critical effort curve shifts upwards. Thus, take-off becomes more difficult after technological progress in the X-sector. It may appear that the minimum critical effort curve might not necessarily shift upwards, but it will be shown in Appendix C that this is impossible.

So far we have studied the effect of technological progress of the once-and-for-all type. If, however, technological progress takes place on a continuing

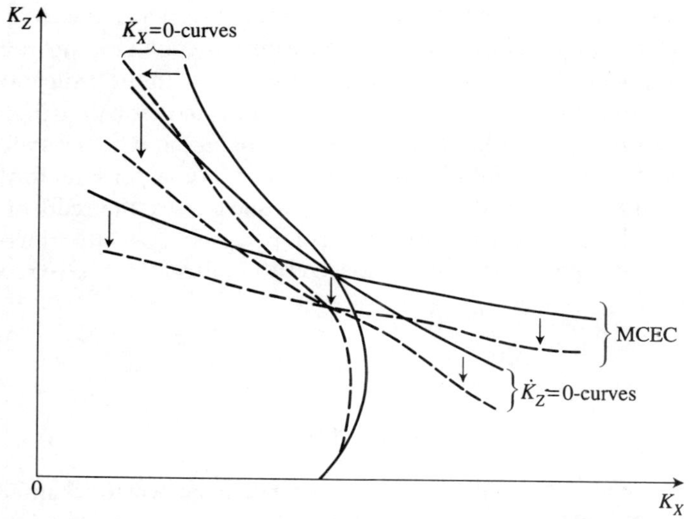

FIG. 3.7

basis over a long period, the shift of the minimum critical effort curve will also take place on a continuing basis. The pattern of its shift will then depend on whether the speed of technological progress is faster in the *X*- or in the *Z*-sectors, where speed of progress is measured by the rate of increase of the profit rate. In general, however, it will be faster in the *Z*-sector, and technological progress will then facilitate take-off.

Here, however, we want to note that where other countries are experiencing rapid technological progress, new products in the *Z*-sector of those countries may become substitutes for *X*-sector goods. Synthetic rubber and fibres (e.g. nylon) are such examples. In such cases, the *X*-sector profit rate will decrease. It may be possible that even the long-term equilibrium, shown as point $(K_X^*, 0)$ in Fig. 3.4, will shift horizontally leftwards. This actually happened in economies whose main products were primary ones and caused serious problems for them. Also, if technological progress in the *Z*-sectors of foreign countries is faster than that in the country in question, if considerable economies of scale are present, and if the scale of the *Z*-sector in the country in question is smaller than in the former, then the terms of trade of domestic *Z*-sector goods will deteriorate and take-off will become more difficult. Competition with foreign countries will become more violent.

Summarizing, the effects of technological progress depend largely on the relative speed of progress compared to that in foreign countries. If technological progress in the latecomer country is achieved by the introduction of technology from abroad, progress will be relatively slow and things will proceed as described above. In post-war Japan, however, the technologies accumulated in the developed industrial countries of the US and Europe were introduced with surprising speed, while further technological progress in the latter countries had to depend on the invention of new technologies. Thus, the technological gap between Japan and the US and the European countries was reduced remarkably quickly. This closing of the gap, together with the protectionist policies of the government, was a major factor in the rapid growth of the post-war Japanese economy. Another was the existence of labour whose wage level was low compared to its quality. Low wages sometimes compensates for the technological gap in worldwide competition. In such cases, the reduction of the technological gap serves to improve the competitiveness of latecomers. This is verified by the high growth rates achieved by countries that have recently industrialized.

Notes

1. For the second-order condition of the maximization problem to be satisfied, an additional condition is necessary. We do not enter deeply into this problem, but it is noted that the condition $p''(X) < 0$ is sufficient for this.
2. When deriving $R'_X(K_X) < 0$, we assumed that $\overline{w}$ is fixed. This means that wage income is exclusively spent on S- and Z-sector goods. In reality, however, some portion of wage income is surely spent on X-sector goods, meaning that the subsistence-level wage rate changes in response to changes in the price of X-sector goods. In that case, however, we can derive $R'_X(K_X) < 0$ by a similar but somewhat more complicated procedure than that in the text, provided the gross profit rate of the X-sector is positive, that is, $R_X(K_X) > 0$.
3. Here we assumed that $\overline{w}$ is constant, but if the output level of the X-sector becomes larger, the terms of trade of X-sector goods will decline, and thus $\overline{w}$ will also decline provided some portion of wage income is spent on X-sector goods. That is, $R_Z(K_Z)$ also depends on the level of K_X, and we can easily see that
$$\frac{\partial R_Z}{\partial K_X} > 0.$$
The following analyses and the statements in Part I will still hold with only slight modifications. Thus, in order to simplify the analysis, we hereafter assume that $\overline{w}$ is constant, and hence that R_Z is a function only of K_Z.
4. It is noted that if $R_X - R_Z < 0$, the greater the value of $|R_X - R_Z|$, the greater the value of $\xi_Z(R_X - R_Z)$. Thus, $\xi'_Z(R_X - R_Z) < 0$.
5. Here, we assume that there are values of K_X and K_Z for which $R_X(K_X) = \mu$ and $R_Z(K_Z) = \mu$, respectively. In the following we study the case where such values of K_X and K_Z, especially the latter, do not exist. This case seems to be very important from a realistic viewpoint because the case is one where take-off is structurally impossible.
6. See n. 5 above. We assume the existence of a K_X such that $R(K_X) = \mu$. Denote it as K_X^*. Then, $R(K_X) > \mu$ for values of K_X smaller than K_X^*.
7. Strictly speaking, the $\dot{K}_Z = 0$ curve also has some irregularities when $K_X \geq K_X^*$. See Appendix B.
8. Point (K_X^*, K_Z^*) is called the saddle point in the place diagram of the dynamic paths.

4

Modification of the Model

4.1. Introduction

We have hitherto assumed economies of scale in the Z-sector. Under this assumption, if the Z-sector succeeds in taking off it will expand without limit. This would be true if subsistence wage labour were supplied to the Z-sector without limit, if the terms of trade of Z-sector goods were fixed, if the marginal productivity of Z-sector were to increase with increases in the scale of the sector, and if the demand for Z-sector goods also expanded without limit. But in the real world, any one of these 'ifs' may not be satisfied. In this chapter, we try to make our model more realistic through the modification of the restrictive assumptions mentioned above.

4.2. Modification of the Assumption of Economies of Scale

In this section we assume increasing returns to scale in the case of relatively small and medium-scale operation and decreasing returns to scale in the case of relatively large-scale operation. This assumption seems realistic.

The Z-sector also requires land as a productive factor. In our model, land inputs have not hitherto been explicitly introduced in the production function, meaning that land inputs have been taken as fixed. If, however, the capital and labour inputs become proportionately very large but the land inputs remain fixed, then output will increase but less than proportionately to the increase in capital and labour.

If land rents are zero, economies of scale will still exist because land inputs can be increased without increasing costs and hence in this case without limit. Otherwise, however, the cost of land rents will become too large to be ignored, and the economies of scale regarding capital and labour will be offset by this cost. Now what happens if the assumption of economies of scale in the Z-sector is modified? Fig. 4.1 shows this situation. The profit rate R_Z of the Z-sector is a function of K_Z, and $R_Z(K_Z)$ will increase for relatively small and medium values of K_Z and will decrease for relatively large values of K_Z. In Fig. 4.2, the first quadrant of the $K_X - K_Z$ plane is divided into six regions. The magnitudes of R_X and R_Z are shown in each region. Since some portion of the profits in each sector will be invested in the more profitable sector, the $\dot{K}_X = 0$ and $\dot{K}_Z = 0$ curves take the shapes shown in Fig. 4.3. The reason why will be easily understood

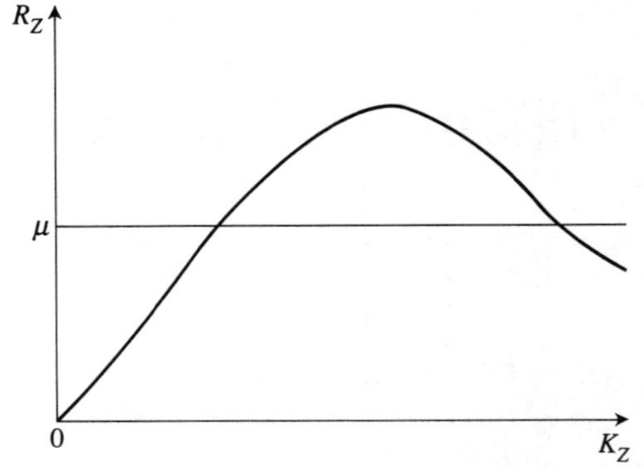

FIG. 4.1

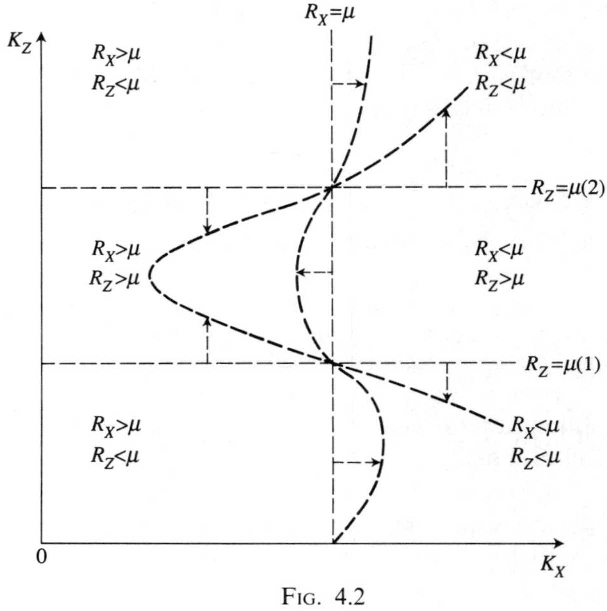

FIG. 4.2

using the same explanation given for Fig. B.2 in Appendix B. Here we give the intuitive explanation. Suppose the profits in each sector are wholly reinvested in the same sector. Then the $\dot{K}_X = 0$ curve becomes the vertical straight line shown as $R_X = \mu$ in Fig. 4.2, while the $\dot{K}_Z = 0$ curve is divided into the two horizontal straight lines shown as $R_Z = \mu$ (1) and $R_Z = \mu$ (2). The reason why the $\dot{K}_X = 0$ and $\dot{K}_Z = 0$ curves in Fig. 4.3 are bent is that some portion of profits in each sector is directed to investment in a more profitable sector rather than to the same sector. How they are bent is shown in Fig. 4.2 by the bent dotted line curves. Arrow signs show the direction of the shift of the $\dot{K}_X = 0$ and $\dot{K}_Z = 0$ curves.

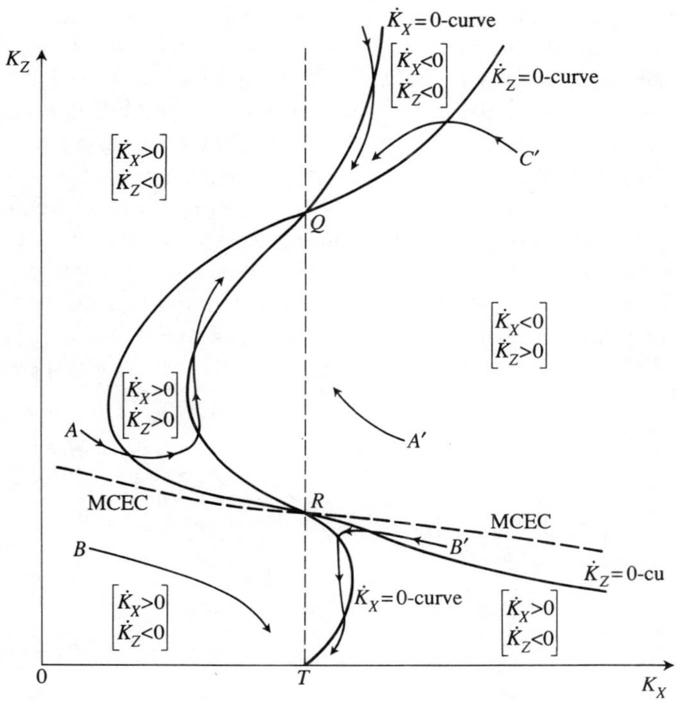

FIG. 4.3

How will the dynamic paths be affected by the modification of the assumption? Paths starting from a point on the minimum critical effort curve proceed along it and will eventually approach point R. But this path is easily seen as one proceeding along a razor's edge. If the path starts from a point slightly off the curve, it will eventually approach not R but T or Q. In this case, the development of the Z-sector without limit, as in the case of Fig. 3.4, becomes impossible. Fig. 4.3 shows several examples of paths starting from points in different regions, each with its own clearly characteristic pattern. Unlike the case shown in Fig. 3.4, where some paths diverge to infinity, in the present case every path eventually approaches some point and never diverges to infinity. Since the directions of the paths at points in the six regions are shown in the figure, we can draw the paths by tracing them in an ordinary phase diagram. Here we can easily see that the successful path for take-off is that starting from point A. In the first phase the X-sector expands and the Z-sector contracts. In the next stage both sectors expand. In the third stage the X-sector contracts and the Z-sector expands. The final stage is that both sectors expand again. But it is easily seen that both sectors eventually approach certain limits.

We do not assume continual technological progress in the above discussion. If continual technological progress were to occur in the Z-sector, point Q in the figure would shift ceaselessly upwards. In the familiar neoclassical growth models, per capita output level also approaches a limit when there is no technological progress. It should be noted that our model can be viewed as a two-sector neoclassical growth model for large values of K_Z. The dynamic behaviour of the growth paths is quite similar to that in neoclassical growth models. After the success of take-off, the path follows that of a neo-classical growth model.[1] This may be because it is natural to assume that the economic situation after take-off is expressed by a neoclassical growth model. Or conversely, neoclassical models generally aim only at explaining growth after successful take-off, that is, the process of growth in developed economies.[2, 3]

Under the assumption of this section, there also exists a case where the $\dot{K}_X = 0$ and $\dot{K}_Z = 0$ curves do not intersect with each other. This case is shown in Fig. 4.4. Then every path eventually approaches point T where the Z-sector disappears.

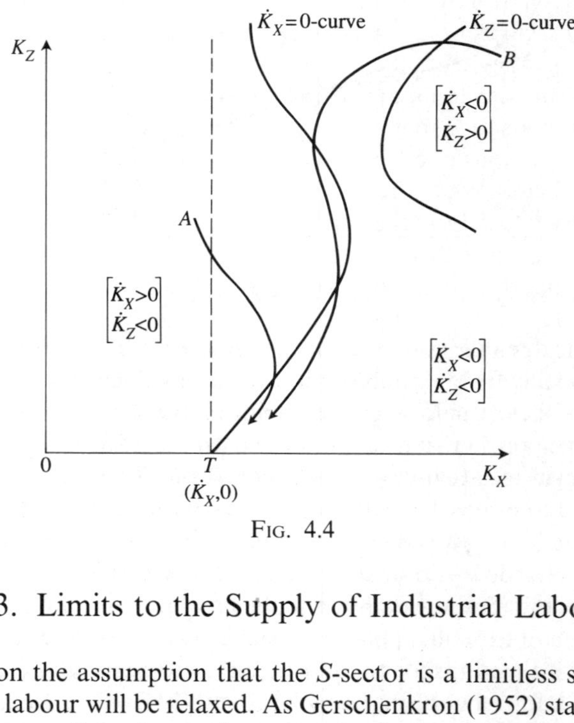

FIG. 4.4

4.3. Limits to the Supply of Industrial Labour

In this section the assumption that the S-sector is a limitless supply source of low-wage labour will be relaxed. As Gerschenkron (1952) stated, one conspicuous feature of the developing countries is the shortage of labour with the required skills and attitudes for modern factory work. If the economy consists mainly of a plantation-type X-sector and a conventional S-sector, the shortage in the supply of such labour will be serious. If so, even a small expansion in the scale of the Z-sector will bring about a steep increase in the wages of workers who already have the skills and attitudes required in heavy and chemical industries or who have received a basic education and are therefore able to acquire the necessary skills and attitudes in a relatively short period.[4]

In such cases, the profit rate will decline due to the steep increase of the wage level, regardless of whether the Z-sector exhibits technological economies of scale. The effects of this wage increase will be seen in Fig. 4.3—R will shift upwards and Q downwards, but the general patterns of the dynamic paths will not change in essence provided that the $\dot{K}_X = 0$ and $\dot{K}_Z = 0$ curves intersect.

However, if the shortage of factory labour is really serious, the two curves will not intersect—that is, the region where $\dot{K}_X > 0$ and $\dot{K}_Z > 0$ will disappear. The phase diagram becomes that of Fig. 4.4, confirming Myrdal's (1968) assertion that take-off is impossible without the reform of the overall structure of the society. The shortage of factory labour suggests that the improvement of labour quality is essential for development, and that this in turn will require investment in education. But as Myrdal stressed, old conventions and customs are very strong, and people hesitate and sometimes even resist leaving familiar social environments. In such cases investment for the improvement of labour quality will not be very effective, and development will be hampered. To achieve development, then, enormous investments will be required in order to reform conventional society; physical investment in the Z-sector or allocation of funds for a 'big push' will not be sufficient on their own. That is, here we find a vicious circle in which the economy is trapped and unable to escape. Another vicious circle is also conceivable. The improvements in labour quality needed for the expansion of the Z-sector can in fact only be effectively achieved if there are plentiful job opportunities in the Z-sector. That is, it is impossible to supply a large amount of labour possessing the ability and skills required for jobs in the Z-sector before the Z-sector expands. The availability of a large amount of such labour prior to expansion i.e. before skilled jobs are available would lead to disillusionment, social unrest, and/or the cessation of the upgrading. We do not intend to make value-judgements here—whether it is better to work in modern factories with restrictions and discipline and to receive higher wages or to work on traditional farms without discipline and to receive lower wages. But it may be stated as fact that the dissatisfactions of unemployed, educated and skilled labour will have damaging consequences for development. That is, Z-sector expansion will be hampered by the quality of labour available, and the labour quality cannot be improved until the Z-sector expands. This is certainly a vicious circle.

A convenient method for building in this improvement of labour quality into our model is to view the cost required for the improvement of labour quality as investment. That is, the increase in labour skills is viewed as accumulated educational investment, and the wage level of this skilled labour is viewed as the sum of the quasi-rent of this investment and the subsistence wage. If the portion of the wage income of skilled labour corresponding to the quasi-rent is spent on the improvement of his own skill and/or his children's skill or on investment in the X- or Z-sector, the situation fits our model perfectly. Japan's experiences before and immediately after World War II seem to fit this situation very well: adults commonly lived at subsistence level but sent their children, especially sons, into colleges and universities. This phenomenon is also seen in Korea, Taiwan, Singapore, and Hong Kong. The main cultural climate of these countries is Confucianism, one of the most important doctrines of which is education.

The speedy recovery of West Germany and Japan from the physical damage suffered during World War II is persuasively explained from this point of view when one remembers that although there were great human losses, there was still considerable accumulated capital embodied in the skill and knowledge of labour.

In addition to the shortage of skilled labour, there exists the problem of the shortage of management skills, or to use Schumpeter's term, 'entrepreneurs'. Economic development cannot be achieved merely by repeating the routine activities of everyday life. The mere existence of capital goods and of skilled labour will not result in development. New businesses would not be started if there did not exist a group of people capable of marshalling the above-mentioned productive factors with great resolve and executive ability. Of course, the appearance of such a group of entrepreneurs will become possible only with the emergence of a set of values that appreciates highly the role they play. At the same time, entrepreneurial ability can also be viewed as a form of human capital like skilled labour, whose development can only be fostered through investment. Of course, managers—whose main activities are not innovative, like those of entrepreneurs, but rather the routine management activities of enterprises—are also important for maintaining productivity at a certain level. Thus, if we include investment-fostering entrepreneurs and managers in K_Z too, this aspect can also be considered accounted for in our model. We can then explain how the supply of human capital such as entrepreneurs and managers can be increased, but we cannot explain how the demand for such human capital will increase. Here again we find another vicious circle: expansion of the Z-sector requires an increase in the supply of entrepreneurs and managers, but supply will increase only as demand increases, i.e. with the expansion of the Z-sector. From the above we should interpret K_Z as something combining entrepreneurship, managerial ability, the ability of skilled labour, and the intrinsic capacity of physical capital.

4.4. General Labour Shortage

In this section we study the effects of wage increases caused by the shortage of labour in general and not those caused by a labour shortage in the Z-sector. The expansion of the X- and/or Z-sectors may cause a shortage in the supply of subsistence wage labour from the S-sector. The marginal productivity of labour in the S-sector will then exceed the level of the subsistence wage. This is the situation of commercialization (Ranis and Fei 1961). An increase in the general wage level will occur even if the quality of labour is almost the same in the *S*-, *X*-, and *Z*-sectors.[5]

To understand what happens in this case, let us now consider Fig. 4.5. The general patterns of the dynamic paths are not largely different from those of Fig. 4.3. One thing to be noted, though, is that whereas in Fig. 4.3 *Q*, *T*, and *R* are located on a vertical straight line, that does not necessarily hold in the present case: *Q* is generally located farthest left, *T* is located farthest right, and *R* lies between them. We shall explain why the location of these points shifts.

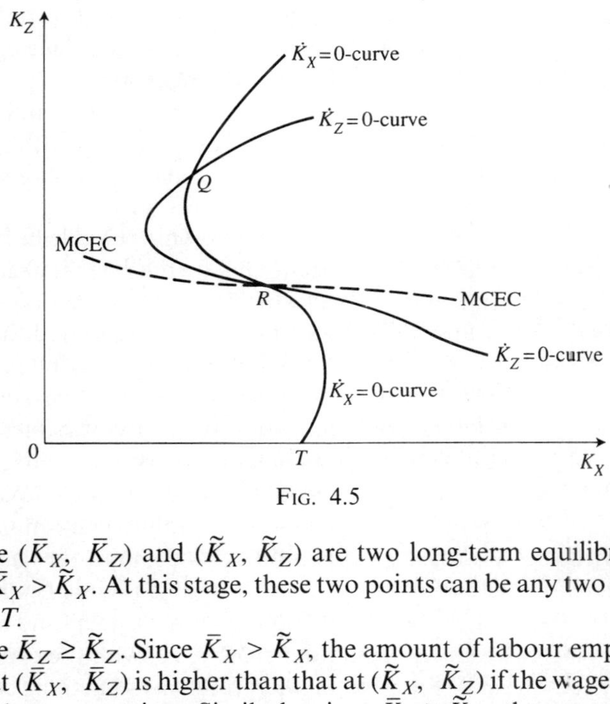

FIG. 4.5

Suppose $(\bar{K}_X, \bar{K}_Z)$ and $(\tilde{K}_X, \tilde{K}_Z)$ are two long-term equilibrium points. Suppose $\bar{K}_X > \tilde{K}_X$. At this stage, these two points can be any two of the points *Q*, *R*, and *T*.

Suppose $\bar{K}_Z \geq \tilde{K}_Z$. Since $\bar{K}_X > \tilde{K}_X$, the amount of labour employed in the *X*-sector at $(\bar{K}_X, \bar{K}_Z)$ is higher than that at $(\tilde{K}_X, \tilde{K}_Z)$ if the wage rates are the same for these two points. Similarly, since $\bar{K}_Z \geq \tilde{K}_Z$, the amount of labour employed in the *Z*-sector at $(\bar{K}_X, \bar{K}_Z)$ is not lower than that at $(\tilde{K}_X, \tilde{K}_Z)$. Thus the total employment of labour at $(\bar{K}_X, \bar{K}_Z)$ is higher than that at $(\tilde{K}_Z, \tilde{K}_Z)$ if the wage rates are the same. This means that the labour demand function at $(\bar{K}_X, \bar{K}_Z)$ shifts rightwards relative to that at $(\tilde{K}_X, \tilde{K}_Z)$.[6] The total labour demand functions in both the *X*- and *Z*-sectors at $(\bar{K}_X, \bar{K}_Z)$ and $(\tilde{K}_X, \tilde{K}_Z)$, respectively, are shown in Fig. 4.6. On the other hand, the labour supply from the *S*-sector is constant until the marginal productivity of labour exceeds the subsistence wage level. From this point on, the supply wage level becomes equal to the marginal productivity of labour in the *S*-sector. Thus, the supply function takes the form shown in Fig. 4.6. This pattern is the same as that shown by Lewis (1955). From the above, the wage level at $(\bar{K}_X, \bar{K}_Z)$ is higher than or equal to that at $(\tilde{K}_X, \tilde{K}_Z)$. Now $\bar{K}_X > \tilde{K}_X$ and the wage rate at

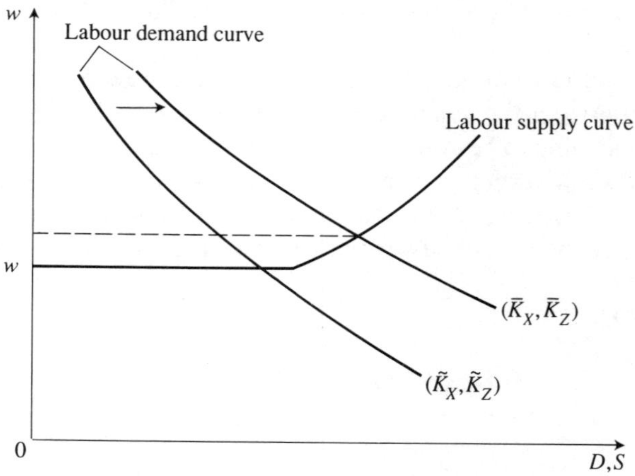

FIG. 4.6

$\bar{K}_X$ is not lower than that at $\tilde{K}_X$. As easily seen, R_X is a decreasing function of K_X and the wage rate. From this, $R_X(\bar{K}_X) < R_X(\tilde{K}_X)$. But both $(\bar{K}_X, \bar{K}_Z)$ and $(\tilde{K}_X, \tilde{K}_Z)$ are long-term equilibrium points. That is, both $R_X(\bar{K}_X)$ and $R_X(\tilde{K}_X)$ are equal to μ. This contradicts $R_X(\bar{K}_X) < R_X(\tilde{K}_X)$. Thus, if $\bar{K}_X > \tilde{K}_X$, then we must have $\bar{K}_Z < \tilde{K}_Z$. From this, we see that the three long-term equilibrium points are located from north-west to south-east in the order Q, R, T. The wage level is highest at Q and lowest at T. But it should be noted that $\bar{K}_X$ can be equal to $\tilde{K}_X$. In this case, both the demand curve at $(\bar{K}_X, \bar{K}_Z)$ and that at $(\tilde{K}_X, \tilde{K}_Z)$ intersect the supply curve at points located on the horizontal part of the curve, and thus the wage level is equal to the subsistence wage level. In this case both $(\bar{K}_X, \bar{K}_Z)$ and $(\tilde{K}_X, \tilde{K}_Z)$ are located on a vertical straight line.

If the absorption of labour from the *S*-sector reduces food production, and if food imports are not possible without an increase in the price of food, the subsistence wage level will rise. In this case, even if the marginal productivity of labour in the *S*-sector is still lower than the subsistence wage level—that is, the situation is still at the stage of pre-commercialization—the effect will be the same as that caused by the wage rise due to the expansion of the *X*- and *Z*-sectors. If these effects are serious, take-off will be hampered. The assertion by some economists that industrialization is impossible without agricultural reform might be seen as suggesting this situation. In the next chapter we consider how the agricultural structure is affected by increases in food prices.

4.5. Capital Flows to and from Abroad

We have so far implicitly assumed an absence of capital flows to or from foreign countries. We now try to relax this assumption, although the following analysis cannot be said to be precise: that would require exact and explicit formulation of the dependence between the outflow of profits from the country (or the inflow of investment funds into the country) and profit rates of capital in the various parts of the world, the profit rates in the domestic sectors, and even the political stability of domestic and foreign countries. This would not be an easy task, so for simplicity we assume that the profitable opportunities in foreign countries do not change.

Low profit rates in the two domestic industrial sectors will be a double obstacle to take-off. If profit rates are low, even if domestic profits are wholly invested in the country, capital accumulation will proceed only at a slow pace. Even worse, some portion of the profits will flow out of the country in search of more profitable opportunities. Lange (1961) identified this outflow of capital as the essential reason why developing countries cannot proceed along the development paths of the classical capitalist economies. In other words, the exploitative policy of the suzerain countries—in short, imperialism—is asserted to be the main reason for the stagnation of the colonial countries. This assertion is in line with the traditional Marxist–Leninist doctrine. Now, Lange quoted an example. In the Malaysian case, average annual outflow of interest per capita between 1949 and 1953 was $18.80. The per capita income during this period is not available, but that of 1961 was $218.00, so per capita income for the earlier period can be estimated at around $190, meaning that about 10% of per capita income flowed out of the country. If this amount of capital had been reinvested in the country, and assuming a capital coefficient of 2.5 to 3.0, the growth rate of per capita income would have been 3%–4% higher than that actually realized. If we note that the growth rates of the developing countries are around 1%–2%, we can estimate the magnitude of the damage caused by the outflow of capital. Lewis (1955) stated that the saving ratio needed for development is about 12%–13%. If 10% of per capita income is reinvested in the country, the condition for development is satisfied as regards capital accumulation.

On the other hand, if the profit rates in the *X*- and/or *Z*-sectors are high compared to the worldwide level, there may be inflows of capital seeking profitable opportunities. For example, in the case of Malaysia before World War II, there were considerable inflows of capital, especially to plantations. This is one reason why per capita income in Malaysia was higher than elsewhere in South-East Asia—that is, because the development of the *X*-sector was supported by the inflow of capital from abroad. But the development of the *X*-sector was already saturated. As shown in Chapter 3, the profit rate in the *X*-sector, including plantations, decreases with scale. Thus, sooner or later—and probably sooner—saturation will occur.[7]

The government of a country that aspires to a higher level of economic development should support its *Z*-sector by any means available until it can compete with that of developed countries. Once profit rates in the *Z*-sector are high, the inflow of capital will increase.

As above, when there are inflows and outflows of capital, inflow will be accelerated if the profit rate is high, and outflow will be accelerated if the profit rate is low. The general patterns of the dynamic paths are not modified very much.

We now turn to the policy implications of the possibility of capital movements between countries. As mentioned above, the direction in which the minimum critical effort curve will shift due to the liberalization of international capital movements cannot be determined definitively if we cannot determine definitively how international capital movements depend on the profit rates in the domestic sectors and on profitable opportunities in foreign countries. We therefore cannot judge whether the liberalization of international capital movements is favourable to the development of the less developed countries or not. But we can say that when the *Z*-sector is still at the stage of an infant industry, the profit rate will be low. Thus, even if there are capital inflows from abroad, they will be directed mainly to the *X*-sector, and when the *X*-sector becomes no longer profitable the profits obtained therein will not be invested in other domestic sectors but will be sent back to the suzerain country. In this case, the liberalization of international capital movements will be harmful for take-off: international capital will help expand the *X*-sector but not the *Z*-sector. Funds move in search of more profitable opportunities, not for the purpose of building the *X*-sector or the *Z*-sector in a given country. The expansion of plantations and mining in some less developed countries was only a by-product of this profit-seeking behaviour.

As mentioned in Section 3.6, the minimum critical effort curve is downward-sloping. This means that the minimum effort required for take-off can be smaller if the *X*-sector is larger. This suggests that if profits obtained in the domestic sectors are prohibited from flowing out of the country and are wholly reinvested in the domestic sectors, capital inflows will have a favourable effect on development even if they are not directed to the *Z*-sector. But the situation is not so simple. If the government adopts this measure, the inflow of capital will stop automatically. A policy of welcoming inflow but not outflow will be self-defeating. A Japanese proverb says 'Going is easy but coming back is dreadful.'

Next, we examine the effects of capital inflows from abroad and foreign aid on the dynamic path. If once-and-for-all aid takes the form of a donation of capital goods to the *Z*-sector, the dynamic path will jump vertically at that moment of time. If this donation is so great that the dynamic path crosses the minimum critical effort curve from below, this donation will have helped the economy to take-off so that thereafter the Z-sector will be able to grow without any aid. This case is shown in Fig. 4.7, where the jump from *D* to *E* represents just such a priming function. But where there is no minimal critical effort curve as in Fig. 4.4, donating capital goods would achieve nothing.

Where foreign aid is not once-and-for-all but is made to the *X*- and/or *Z*-sectors continuously over a certain period of time, the direction of the dynamic path will be shifted in a favourable direction at each point. This type of aid

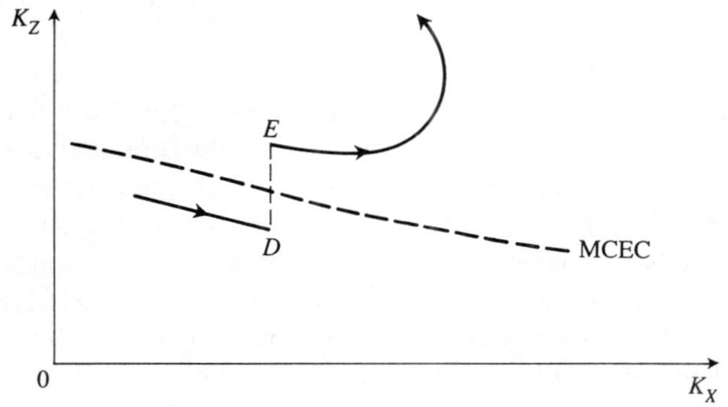

FIG. 4.7

also helps the economy to clear the minimum critical effort curve. Moreover, it is possible that this type of aid may convert an economy from a Fig. 4.4 situation to one of Fig. 4.3, such that take-off becomes possible. But even in this case, the economy will return to its original state if foreign aid ceases. That is, it will return to a state where the $\dot{K}_X = 0$ and $\dot{K}_Z = 0$ curves do not intersect, and the Z-sector will eventually disappear. In such cases, truly self-sustaining development will be impossible without reforming the structure of society. Moreover, foreign aid may sometimes have negative effects on a country's morale, fostering a mentality of dependence totally incompatible with a belief that Heaven helps those who help themselves.

4.6. Z-Sector Terms of Trade

We have so far assumed that the terms of trade of the Z-sector are constant. But if the Z-sector expands sufficiently to permit the export of a certain amount of Z-sector goods to acquire the foreign exchange needed to import food and raw materials such as coal, oil, iron ore, and metals, then the terms of trade of the Z-sector will deteriorate. Even if the prices of S-sector goods such as food and raw materials do not rise in absolute terms, the decline of the prices of Z-sector goods will make their price in terms of S-sector goods decline; that is, the terms of trade of Z-sector goods will deteriorate and the profit rate of the Z-sector will decline, with the effects set out in Section 4.3. But this will not significantly modify the general patterns of the dynamic paths. The upper branch of the $\dot{K}_Z = 0$ curve will shift downwards, and its lower branch will shift upwards; thus, the minimum critical effort curve will shift upwards, and take-off will become more difficult. Of course it is also possible that the situation may change from that of Fig. 4.3 to that of Fig. 4.4—for example, if the demand elasticity for exported Z-sector goods is very small, a small amount of exports of Z-sector goods will cause the terms of trade to deteriorate. But we do not know of any examples of economies whose development has been hampered by this.

The possibility of take-off may also depend on the worldwide supply elasticities of food and raw materials. Since expansion of the *Z*-sector will absorb labour from the *S*-sector and reduce the production of food and raw materials, it will become necessary to buy them from abroad. If their prices rise sharply the *Z*-sector profit rate will decline, making take-off difficult or even impossible. In the following section, we study the effects of a reduction of domestic food production. Where this happens, firm reform of the domestic agricultural sector is necessary to avoid a situation whereby the inavailability of food becomes a bottleneck that restricts the expansion of the *Z*-sector.

4.7. Food Prices

We have stated that as development proceeds, surplus labour in the *S*-sector decreases and hence domestic food production decreases. Implicitly we assumed that the necessary amount of food can be imported without causing a price rise. If there is surplus food on the world market or if the supply elasticity of food is very high, the reduction of the *S*-sector and thus the reduction of domestic food production will not hamper development, provided that the economy can produce plentiful export goods with which it can acquire sufficient foreign exchange. Theoretically, this is so. But food production depends on the climate, and the export and import of food is influenced by the international political situation. That is, food is sometimes used as a strategic device on the international political stage. It may be risky for a country to pursue a policy of heavy reliance on imports for its food supplies because poor harvests in food-producing countries or a change in the world political situation can make food imports difficult or impossible. Moreover, even if there are plentiful surplus foods at one time, there is no guarantee that the situation will last for a long time. No government, therefore, can adopt a policy of depending completely on imported food supplies.[8] We therefore assume that the price of foods will rise as development progresses. Then two effects are conceivable. First, the subsistence wage level will rise when it is measured in terms of *X*- or *Z*-sector goods. This effect is the same as that discussed in Section 4.4, i.e. the same as a general wage rise. The general patterns of the dynamic paths are not significantly affected by this, although it may hamper development in some cases. The second effect is to increase the *S*-sector profit rate. We study this effect more precisely in Chapter 5.

In fact, the effects discussed in this section and in the previous section will occur at the same time. The *Z*-sector expands, employment in the *S*-sector decreases, domestic food production decreases, and food imports increase. If the supply elasticity of food on the world market is small, food prices will rise; this is the effect discussed here. But suppose the *Z*-sector expands. Imports of food and raw materials will increase, and the export of *Z*-sector goods will increase to obtain the foreign exchange necessary to buy food and raw materials. If the elasticity of demand for Z-sector goods is small, the price of Z-sector goods will decline. This is the effect discussed in Sections 4.6 and 4.3. The effect discussed earlier in this section is the same as that of Section 4.4. Thus, the effect of S-sector goods becoming more costly and Z-sector goods becoming cheaper due to the expansion of the Z-sector is the sum of the effects discussed in Sections 4.3 and 4.4.

4.8. Investment of Non-wage Income in the *S*-Sector

Even in subsistence economies, there may be enormous inequalities in income and wealth distribution. Appropriate policies, whether introduced by the government or imposed by international agencies, may reduce extravagant consumption and divert the money saved to productive investment.[9] The heavy taxes imposed by the Japanese government on rental income in the *S*-sector is one example of this kind of device.

If a portion of the non-wage income of the *S*-sector is invested in the *X*- and *Z*-sectors, especially the latter, due to government policies or to the intentions of the landowners in the *S*-sector, our model must be modified somewhat. More precisely, we have to explain how non-wage income is divided among four types of uses—consumption, *S*-sector investment, *X*-sector investment, and *Z*-sector investment—and how the non-wage income level in the *S*-sector is determined. But this will severely complicate the analysis—for example, the level of non-wage income in the *S*-sector depends on the level of accumulated capital in the *S*-sector, requiring a differential equation which expresses the movement of the level of accumulated capital in the *S*-sector over time. We would then face the difficult problem of analyzing the characteristic features of the solution paths of a three-variable simultaneous differential equation system. Instead, the analysis that follows is rough and intuitive, but we are sure that greater precision would not significantly alter the general patterns of the dynamic paths. This generalization notwithstanding, the minimum critical effort curve will of course shift downwards because some portion of the non-wage income in the *S*-sector is invested in the *X*- and *Z*-sectors, especially the latter, instead of being spent on consumption, and hence the direction of the dynamic paths at each point shifts towards the north-east. Such policies have been adopted in Japan, since the Meiji restoration era beginning in 1868 and continuing until the end of World War II. The implications of this for investment in the agricultural sector will be discussed in Chapter 5.

4.9. Modification of Investment Distribution Functions

Our basic model assumes that the proportion of profits invested in each sector changes due to differences in the realized profit rates at the moment in time when the decision was made. It is quite possible, however, that the expected future levels of profit rates will not be the same as the present levels. That is, if the expectation is that the Z-sector will become more profitable, the investment distribution functions will be modified. Here we study the case where expectations are favourable to the Z-sector. The investment distribution functions are assumed to be as shown in Figs. 4.8 and 4.9.

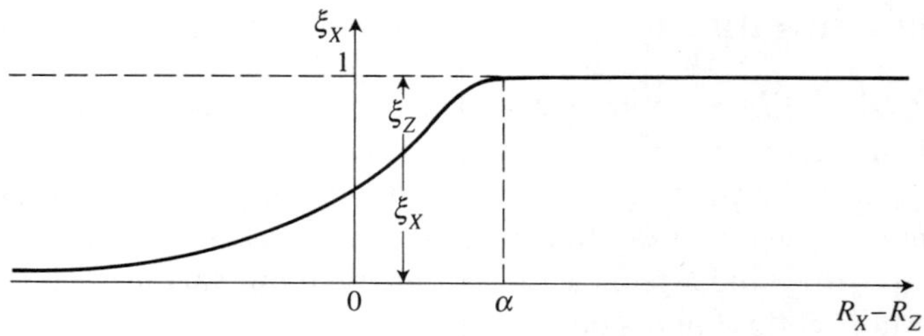

FIG. 4.8

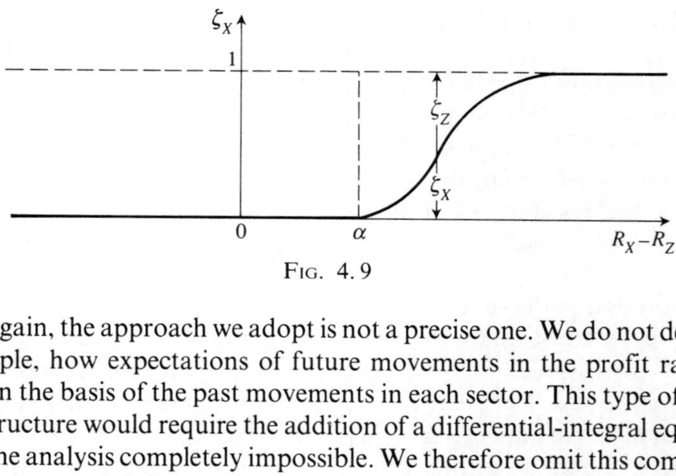

FIG. 4.9

Once again, the approach we adopt is not a precise one. We do not describe, for example, how expectations of future movements in the profit rates are formed on the basis of the past movements in each sector. This type of expectations structure would require the addition of a differential-integral equation, making the analysis completely impossible. We therefore omit this complexity and use only very simple investment distribution functions.

Now, the investment distribution function shown above can be seen to have shifted horizontally to the right relative to that of the basic model. When the difference between X- and Z-sector profit rates exceeds a certain level α, some portion of the profits obtained in the Z-sector, which would otherwise be wholly reinvested in that sector, will flow out of this sector and into the X-sector. Fig. 4.10 shows how the dynamic paths will change under these modified investment distribution functions.

In this case there may be more than one point of intersection between the $\dot{K}_X = 0$ and $\dot{K}_Z = 0$ curves. In general, the number of intersection points will be odd.[10] But there exists the case where P and Q coincide, in which instance the number becomes even. Such a case can be viewed as the one of multiple

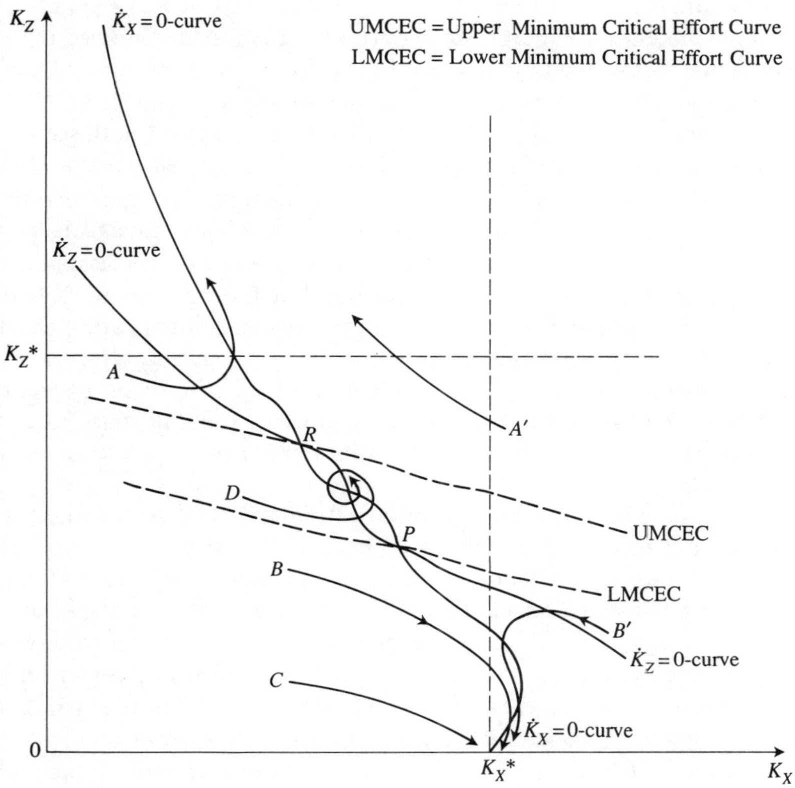

FIG. 4.10

roots of an algebraic equation. There is also the case where the two curves do not intersect. In this case take-off is impossible. (Otherwise, take-off is always possible.)

We now analyse the effects of the modification in the cases described in Figs. 4.3 and 4.4. The intersection points of the $\dot{K}_X = 0$ and $\dot{K}_Z = 0$ curves are located to the south-west of point (K_X^*, K_Z^*), where K_X^* and K_Z^* are values satisfying $R_X(K_X^*) = \mu$ and $R_Z(K_Z^*) = \mu$. The reason for this is as follows. Suppose $K_X \geqq K_X^*$ at the intersection point (K_X, K_Z). This point is different from (K_X^*, K_Z^*). Then $R_X(K_X) \leqq R_Z(K_Z^*) = \mu$. If some portion of profits flows out of the X-sector, $\dot{K}_X < 0$. This means that the point is not located on the $\dot{K}_X = 0$ curve. From this we see that for (K_X, K_Z) to be on the $\dot{K}_X = 0$ curve, the X-sector profit rate must exceed that in the Z-sector by more than α. In this case, the profit rate in the Z-sector becomes less than $\mu - \alpha$, and the transfer from X-sector to Z-sector ceases. Thus $\dot{K}_Z = 0$ cannot hold for this point, which means that it is not located on the $\dot{K}_Z = 0$ curve. The point (K_X, K_Z), therefore, cannot be the intersection point of the $\dot{K}_X = 0$ curve and $\dot{K}_Z = 0$ curves. Next, suppose $K_Z \geqq K_Z^*$. Then $R_Z(K_Z) \geqq R_Z(K_Z^*) = \mu$. The profit rate in the Z-sector is not less than μ. On the other hand, $K_X < K_X^*$. Thus, $R(K_X) > R_X(K_X^*) = \mu$. Then at the intersection point, some portion of the profits obtained in the X-sector is not re-invested in that sector but flows out of the sector and into the Z-sector, and the stationary state is sustained. But since $R_Z(K_Z) \geq R_Z(K_Z^*) = \mu$, the Z-sector expands. This contradicts the requirement that both sectors are in a stationary state at the intersection point. Thus it is impossible that $K_Z \geq K_Z^*$.

Fig 4.10 shows three cases of how the dynamic paths are modified in this case. There exist two minimum critical effort curves. The dynamic paths starting at B or C will eventually approach $(K_X^*, 0)$, and the Z-sector will disappear. The path starting at D is somewhat interesting, but not necessarily realistic: it shows a cyclical motion but eventually approaches a stationary point Q, where Q is one of the long-term stable equilibria. The path starting at A shows the eventual development case. The modification of the investment distribution functions makes the situation somewhat complicated, but the basic structure, or the general patterns of the dynamic paths, remain almost the same as in the basic model.

Here it should be noted that the modification makes take-off easier than in the basic case because even the upper minimum critical effort curve shifts downwards. This is seen from the fact that the intersection point of $\dot{K}_X = 0$ and $\dot{K}_Z = 0$ curves is located to the south-west of (K_X^*, K_Z^*). In the basic case, the $\dot{K}_X = 0$ curve and $\dot{K}_Z = 0$ curves intersect at (K_X^*, K_Z^*). But to show that the upper minimum critical effort curve shifts downward relative to that in Fig. 3.5 at every point on the curve, we may need some additional conditions. From the foregoing, we can see that a downwards shift occurs around the intersection point of the $\dot{K}_X = 0$ and $\dot{K}_Z = 0$ curves, but it cannot be shown that every point on the curve shifts downwards. A more precise explanation of this as well as of the case of an import tariff is provided in Appendix C.

We now consider the possibility of a shift in the investment distribution function caused not by government policies but by the initiative of the private capitalistic sector. Of course, the type of shift discussed above may not be the type of shift discussed in the following. But the shift favourable to the Z-sector in the following example will affect the possibility of development in ways similar to those discussed in this section. The example to be considered is the role played by *zaibatsu* (conglomerates) in pre-war Japanese economic development. The holding companies heading the conglomerates had complete control not only over the personnel administration of the affiliated enterprises but also over the funds available to them, and could direct them to investment in future prospective enterprises.[11] They established many heavy and chemical enterprises, siphoning off profits from affiliated enterprises in the light industry sector and transferring them to enterprises in the heavy and chemical industries.[12] Japan's policy of militarization also assisted this trend because the government purchase of weapons, ammunition, warships, and aircraft at the price proposed by producers gave the conglomerates very profitable opportunities—a trend familiar in other parts of the world too. But investment in weapons manufacturers does not directly contribute much to development, and indeed before the end of World War II, the growth rate of the Japanese economy was not exceptionally high. This is because despite the high saving ratio—especially high during the war period—investment was not directed to uses that would contribute to future development.

Nevertheless, the technologies fostered during the War, for example, making materials such as iron, steel, and cement, were put directly to practical use and contributed to the recovery and development of the post-war Japanese economy. The technologies for manufacturing aircraft were utilized, with some modifications, in the automobile industry. Likewise, know-how gained in the production of ammunition was applied to the chemical fertilizer industries. Similar correspondences with wartime production lines eased the transition to the manufacture of precision machinery and tankers, cargo-boats, and merchant ships—all fields in which Japan's representative industries now lead the world. Even the electronics industries can trace their roots to pre-war establishments. The firms that are involved are different, as the old Japanese conglomerates were dismantled by the allied powers because they had co-operated with Japan's militaristic pre-war government. But this corporate dismantling had no negative effects on the subsequent course of development; rather, it gave many able engineers an opportunity to demonstrate new ways to contribute to the economy. This fact emphasizes the importance of human capital for development.

Notes

1. When economies of scale in the *Z*-sector are exhausted because of the rise in the prices of productive factors other than the wage rate and the rental rate of capital (e.g. the land rent), the *Z*-sector satisfies rather the assumptions made concerning the *X*-sector and thus the separation of the *X*- and *Z*-sectors becomes meaningless. In fact, our model becomes very close to the one-sector growth model of Solow (1956). In the real world, unlimited growth of the heavy and chemical industries becomes impossible because economies of scale have been exhausted in these industries, or because, as will be shown later, the same effect is brought about by a decline in the prices of the products. In developed economies, however, new industries satisfying our assumptions concerning the *Z*-sector will emerge—e.g. electronics industries making ICs, computers, and communications devices, new material industries making fine chemicals and new ceramics, and biotechnological industries. Note also that firms in the traditional heavy and chemical industries endeavour to diversify into the new high technology fields. In Japan, these are styled 'light, thin, short, and small' industries, in contrast to the 'heavy, thick, long and large' heavy and chemical industries. Now, if we include the traditional heavy and chemical industries in the *X*-sector along with the traditional light industries, and include the above-mentioned newly emerging industries in the *Z*-sector, the

analysis of the model will apply with only one modification: if the economies reach this development stage, the S-sector will also have reached the stage where the assumptions made concerning the X-sector are valid, and thus we cannot assume that the S-sector is the source of an unlimited supply of labour at the subsistence wage level. Hence, the pattern of the dynamic paths will be one of those shown in Figs. 4.3, 4.5, and 4.10.

2. Note that the Keynesian concepts of the involuntary unemployment of labour and the shortage of effective demand are assumed not to exist in our development model. In this sense, we unintentionally accept Myrdal's opinion. See ch. 1. Growth models, especially neoclassical growth models, focus their concern on the supply side of the economy. The problem of the shortage of effective demand or the existence of involuntary unemployment of labour at each moment of time is somehow (for example, by appropriate monetary and/or fiscal policies) assumed to be avoided or solved, and all factors of production, including labour, are assumed to be always fully employed. Our development model also concentrates on the supply side. This basic approach is also used for the case where development has been successful. The path approaching the long-term equilibrium point Q in Fig. 4.3 shows the same features as the neoclassical growth model. There is also no involuntary unemployment. But we hasten to add, in order to avoid misunderstanding, that the existence of surplus labour or the disguised unemployment of labour in the S-sector is not excluded even for the case of successful development. In the case of Fig. 4.3, the saturation of economic development occurs for the technological reason of lost economies of scale. Thus, it is quite possible that surplus labour in the S-sector coexists with the saturation of development. But we shall see in the following two sections, the saturation of development may be caused by the shortage of skilled labour or of general labour. In these sections, however, we assume no involuntary unemployment of labour, in line with the neoclassical growth model.
3. Since the loss of economies of scale may be brought about through the rise of the rental rate caused by the shortage of land, we see that the shortage of productive factors other than capital may cause saturation. Capital, on the other hand, can be increased without limitation if other factors are plentiful. That is, capital is necessary for development, but not sufficient. The only way to escape from the trap of the long-term stationary state due to the shortage of productive factors is technological progress.
4. In this section, the wage rate in the Z-sector is not assumed to be fixed at $\overline{w}$, but to be an increasing function of N_Z, the amount of employment in the Z-sector (see n. 11 to ch. 2). However, we assume that the wage rate in the X-sector is fixed at $\overline{w}$. There are of course cases where there is no specific difference in the quality of labour required in the two sectors, but if the main X-sector industries are plantations, this assumption is natural. The case where a specific difference is not observed in the quality of labour in the two sectors will be studied in the next section.
5. The X-sector and Z-sector wage rates are the same, and are increasing in $(N_X + N_Z)$, the amount of labour in the capitalistic sectors.
6. Here we assume that employment in each sector will increase with the expansion of the sector, provided the wage rate is fixed. The scale of each sector is measured by the magnitude of K_X or K_Z. For the Z-sector, this assumption will hold as long as the terms of trade of Z-sector goods are fixed. For the X-sector, however, if a remarkable increase in the amount of production in the sector causes the terms of

trade to deteriorate, employment may decrease as the scale of the sector becomes large. That is, the substitution of capital for labour may occur without changing the output level significantly. In such cases, the arguments in the text do not hold. But whether Q, R, and T are located on a vertical straight line, a downwards-sloping line or an upwards-sloping line is irrelevant for the following discussions. Here, we want only to indicate that the points do not have to be on a vertical straight line.

7. Plantations require a vast area of land. If we take this land input into account, the X-sector production function cannot be assumed to exhibit constant returns to scale. Then, $R_X(K_X)$ will decrease very rapidly and saturation will soon be reached because land suitable for the particular crop is in limited supply. In addition, generally speaking, the overproduction of plantation products will entail a sharp decline in price. This is another cause of the saturation of plantation production.
8. But the existing policies of protecting domestic agriculture in the developed countries appear to have been adopted for the purpose of sustaining domestic farmers' incomes rather than for security purposes.
9. See e.g. Baran (1957), Dobb (1963), Myrdal (1968), or Lewis (1955). The latter wrote: 'No nation is so poor that it could not save 12 per cent of its national income if it wanted to; poverty has never prevented nations from launching upon wars, or from wasting their substance in other ways. Least of all can those nations plead poverty as an excuse for not saving, in which 40 per cent or so of the national income is squandered by the top 10 per cent of income receivers, living luxuriously on rents. In such countries productive investment is not small because there is no surplus; it is small because the surplus is used to maintain unproductive hordes of retainers, and to build pyramids, temples and other durable consumer goods, instead of to create productive capital. If this surplus were going instead as profits to capitalists, or as taxes to productivity-inclined governments, much higher levels of investment would be possible without inflation.'
10. Strictly speaking, the $\dot{K}_Z = 0$ curve may not be monotonically downward-sloping, but may exhibit some ups and downs as shown in Fig. 4.11, or may have some width as shown in Fig. B.2 in Appendix B. But the irregularity mentioned here will arise only in the region where $K_Z \leqq \overset{*}{K_Z}$. However, we omit the reason for this because the irregularities do not present any difficulties when examining the general patterns of dynamic paths. In particular, the irregularity will occur only in the case of a failure to achieve take-off. Here we will provide only a partial explanation. The

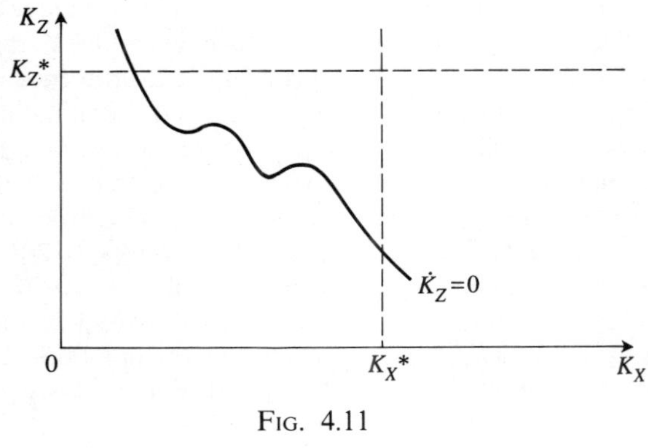

FIG. 4.11

proportion $\xi_Z(R_X - R_Z)$ of profits from the X-sector invested in the Z-sector is an increasing function of K_X, but the amount of investment from this source is $\xi_Z(R_X - R_Z) \cdot R_X \cdot K_X$. It may decrease, increase, or may not change as K_X increases, because $\xi_Z(R_X - R_Z)$ is increasing and R_X is decreasing in K_X, while K_X is, of course, increasing in K_X. This indeterminacy causes the irregularities in the shape of the $\dot{K}_Z = 0$ curve. As for the irregularity of the $\dot{K}_X = 0$ curve in the basic model, see Appendix B.

11. We so far have not evaluated positively the role of the profitability of the X-sector in the development process. However, if the profits of the X-sector can be directed to the Z-sector, whether through government policy or by conglomerates, then the more profitable the X-sector is, the better it is for development because the total investment funds in the country will be greater than would otherwise have been the case and thus investment in the Z-sector can be increased by such policies.
12. Korea seems to have conglomerates, some of which have begun to produce and export automobiles.

5

The Agricultural Sector

5.1. The Importance of Education and of Reform of the Social Structure

So far, we have treated the *S*-sector mainly as a passive sector that supplies a limitless amount of subsistence wage labour, but the positive role of this sector in development should not be underestimated. In this respect we have already mentioned the importance of improvement in the quality of labour for the achievement of take-off. Another role of the *S*-sector, as already mentioned, is that of supplier of food. The *S*-sector in our model does not coincide perfectly with the agricultural sector because we define the *S*-sector more broadly, as that in which the production system is non-capitalistic, but at least at the very earliest stage of development, the *S*-sector will largely coincide with the agricultural sector. That is, in the agricultural sector, the production system is not capitalistic but traditional and even feudalistic or sometimes close to slavery. Many economists, such as Lewis (1955), Rostow (1960), and Dobb (1963), insist that the productivity of the sector must be improved by capital accumulation and/or technological progress for the economy to develop. But this statement can be interpreted as stating that what is required is for the agricultural sector to transform itself into a sub-sector of the *X*-sector or the *Z*-sector rather than remaining in the *S*-sector because agricultural productivity will not improve if the sector stays as it is with conventional, feudalistic, or slavery production methods. Here we digress slightly.

Japan's experience in this respect may be somewhat peculiar because the agricultural sector remained feudalistic even after take-off, until the end of World War II. Why the Japanese economy succeeded in spite of this is not clear, but several factors can be cited. First, despite the feudalistic production methods and institutions, the sector was capitalistic in that landowners' behaviour was profit-orientated, in the sense that rent income was partly directed to investment in the *X*- and *Z*-sectors. This would certainly have compensated for the negative effects of feudalistic methods and institutions in food production. Secondly, the rapid increase in the demand for food caused by rapid population growth was met by imports from the newly acquired colonies of Taiwan and Korea. Japan thus did not face serious food shortages, although there were several riots over the high price of rice. A third factor was the quality of the workers even in the feudalistic agricultural sector: almost all people working in the *S*-sector would have been able to work in a modern factory with almost the same efficiency as those working in the factory for years.

They could read, write, and count, and could adjust to new circumstances with marvellous adaptability.[1] This may be the most important reason why Japan was able to develop successfully even though the agricultural sector remained feudalistic. Why Japan's traditional agriculturalists were so able and adaptable is beyond the scope of our discussion—some go back more than one hundred years in search of the cause, and attribute it to the educational policy adopted by the Meiji government. Suffice it to say that the Japanese consider education the most important value in an individual's life—a sense of values shared by the people in the Asian NIEs. As a matter of fact, it seems to be true that the workers in the agricultural sector in these countries have the same ability as those in Japan.

5.2. Modernization of the Agricultural Sector

As development proceeds, the capitalistic sector, especially the *Z*-sector, absorbs workers from the *S*-sector. In consequence, to supply the economy with necessary food and materials, the production technologies adopted in the *S*-sector have to become more intensive because the sector must produce more with less labour input.

At the initial stage of development, population usually increases. Development will bring about an increase in the labour productivity of the capitalistic sector, which in turn will cause wage income to exceed the subsistence wage level. This will, on one hand, encourage population to increase. On the other hand, development will increase the demand for labour in the capitalistic sector. This demand increase will further encourage population growth. Population growth necessitates an increase in food output. The material requirements of the capitalistic sector will also increase as development proceeds because the expansion of the scale of the sector cannot be maintained without an increase in material output.

As development proceeds, therefore, the *S*-sector must also expand. We can say this in a reverse way. If the *S*-sector cannot expand, the *X*- and *Z*-sectors cannot expand either, except where *S*-sector goods can be imported without causing a serious price rise. But how can the output of the *S*-sector be increased with less labour without causing a price rise? One way is to increase investment in that sector. Another way, which is related to increased investment, is technological progress. The third and most important way is to reform the production system and institutions. For example, the tenant or share-cropping system offers neither landowners nor tenants an incentive to introduce more productive technologies such as using machines and fertilizers, since the returns on the investment will have to be shared with people who had no part in the improvement. In fact, we shall see that development does not proceed satisfactorily in countries in which the tenant system remains: probably the single most important factor in the reform of the *S*-sector is the reform of the tenant system, which opens the way to investment and technological progress. If producers are assured of getting all the fruits of their endeavours, they will strive to do their best. The universal truth of this is revealed by the recent reforms in the socialist countries, where the introduction of free markets stimulated real improvements in output and quality. Where, on the other hand, the reform of the production system and institutions does not proceed, the process of development is obstructed, and the reform of the system is in turn impeded.

In short, the most important goal is to reform the *S*-sector so it becomes a subsector of the *X*- or *Z*-sectors.

Satisfactory development in the *Z*-sector will in turn stimulate the reform of the *S*-sector. First, many workers will abandon the *S*-sector to work in the *Z*-sector because of the wage inducements. This will bid up the wage in the *S*-sector such that in extreme circumstances the tenant system will become unworkable. Secondly, the attitudes of landowners—hitherto the highest stratum of society will be modified as they increasingly come into contact with the élite of the business world. They will be stimulated to change by the latter's way of life and attitudes, their apparent elegance and modernity; some of the *nouveaux riches* may actually be wealthier than the old land-owner class. This may encourage the landowners to imitate their way of earning income as well as their way of life. That is, the landowners will begin to be transformed into capitalists, and the *S*-sector will begin to be transformed from both the employee side and the employer side. In this way, it will become a subsector of the *X*- or *Z*-sectors, and the process of capital investment will be the same as in those sectors. Once wage income exceeds the subsistence level, it will no longer be spent exclusively on consumption; some portion will be saved or invested. Economies that have attained this level are no longer traditional economies, and their development would be better described by the so-called growth models. But this is not the task of this book.

Here, we shall comment on the desirability, or otherwise, of economic development. Development has many merits, but it also has many demerits. Working in a factory may diminish the condition of humanity: to work in a factory it is necessary to be punctual and disciplined, to lose freedom. Development may also destroy the natural environment. This may in turn destroy all life on earth, including human life. Even now, we do not know whether we are proceeding towards collapse or prosperity. We do not enter into a discussion of this serious problem here because at this stage of scientific research, the choice of whether or not to proceed with development must still be left to each nation to decide for itself. The only prescription we can provide for development is that every nation has the right to choose its own future, even if it may be the road to ruin.

But we would add one further point: the developed nations have no right to interfere with the developing nations' choice of the way to development unless they first throw out everything they have acquired through their own development.

Notes

1. Japan's *X*-sector started to expand just after the country opened up. Working conditions in the factories, however, were really harsh, especially in the fibre industries where 12- or 16-hour working days were not exceptional, even for female workers, and sometimes three workers took turns sleeping in the same bed in company dormitories. Many young female workers lost their lives due to tuberculosis caused by overwork. The situation was known widely as 'the pathetic life of mill girls'.

6

Implications and Concluding Remarks

We have stated the advantages of our theoretical model and analytical methods, and also pointed to some important aspects so far overlooked in explaining the process and mechanism of economic development. Our arguments are especially applicable to the late starters in industrialization, such as Japan, newly industrialized economies, and other so-called middle-income Asian countries. To foster better understanding and avoid misunderstanding, however, we sum up and emphasize a few major implications of our analysis.

1. *The advantages of our model.* What are the advantages of our model as compared with existing ones? First, our model consists of three sectors, whereas conventional development models have only two. It is obvious that a model consisting only of traditional and manufacturing sectors cannot highlight changes within the manufacturing sector, though these may be critical for development.

We argue for the three-sector model at the cost of complication because we believe that one of the most important tasks in clarifying the development processes is to analyse the mechanism of changing industrial composition, not just to assume a balanced growth path in which all industries expand proportionately. In reality, such, 'balanced' growth is rather rare. The so-called life-cycle theory of industrial development does try to focus on changes in industrial composition, but its proponents Akamatsu (1956) and Vernon (1966) do not explain the mechanism adequately. For explicit analysis, industries must be subdivided into at least two subsectors.

Thus, a development model must be able to explain how a subsector grows fast and then declines, with another subsector replacing it. New industries are usually slow to develop initially but gradually accelerate their expansion, until finally they catch up with the older industries and may eventually replace them. Our model is successful in explaining such a feature, whereas most existing models do not explicitly deal with the mechanism of accumulation and allocation of capital. It also clarifies interrelations between the X- and Z-sectors. A model that consists of only a single manufacturing sector *vis-à-vis* the traditional sector can never highlight this process. For reasons explained in the next section, the phase-diagram approach which we adopt as an analytical method is both appropriate and convenient for demonstrating dynamic interactions between two manufacturing subsectors.

2. *Our analytical approach.* Our analytical method, then, is based extensively on differential equation systems and phase diagrams. Some readers may have

an allergy to this method, but it is widely and effectively used in many fields of economics. First, the benefits gained through using such instruments in analysing the economic development process have significantly exceeded the costs of learning to apply them. Further, since the method is employed in broad areas of economics, having some knowledge on the method gives better access to the fruits of the human intellectual activity in this area. Roundabout production methods are often more productive than direct ones because of the unexpected spinoffs.

The phase-diagram method has another advantage, in that it avoids the ambiguity which is inevitably involved in using ordinary language. Without using phase diagrams it is impossible to visualize exactly the capital accumulation path, the minimum critical effort curve, and the stationary trap of poverty.

3. *Definition of success in economic development*. In our model, success in economic development is defined as the successful take-off of the Z-sector meaning that the Z-sector can expand on a self-sustaining growth path. However, this does not mean that the Z-sector is the only engine for development. We have already seen that a well-developed X-sector makes take-off easier. This can be seen from the fact that the minimum critical effort curve slopes rightwards and down, and hence, the hurdle to be cleared for take-off is lower for a larger value of K_X. We also indicated that take-off may be impossible without structural reforms to remove the barriers hindering rational resource allocation. Likewise, the minimum critical effort curve may be located too high for the economy to clear the hurdle with the policy instruments available to the government. This latter case suggests the usefulness of foreign assistance and/or capital importation.

We also indicated that government protection of the Z-sector with import tariffs can promote investment in the Z-sector, but that the final situation will be worse than under free trade should the economy fail to take off.

4. *Various obstacles to development*. It is true that the self-perpetuation of the Z-sector is necessary for take-off, but take-off may be impossible even if tremendous resources are invested in the Z-sector at the expense of the X-sector. The reasons for this are manifold. An extreme case is that the economy lacks a skilled labour force and the institutions that facilitate rational allocation of factors and resources. Religious barriers may likewise hinder the rational behaviour of workers and enterprises to the point that take-off is unattainable. Government apathy towards market mechanisms and bureaucratic mismanagement often result in enduring stagnation; and wasting resources in military build-up can also rob the economy of funds that would otherwise be directed towards getting the Z-sector over the minimum critical effort hurdle.

Thus, institution building, reform of the economic system, and the promotion of primary and secondary education are the prescription for industrialization. Economies at the lowest income level face these policy tasks even today.

One may question the relevance of our analysis to socialist economies that stagnate despite having given first priority to investment in heavy industries. Our answer is that the failure was due to negligence of market forces and waste of resources in military build-up.

5. *Development and the Z-sector*. It is true that the *X*-sector plays an important role in economic development, but in many of today's industrial countries the emphasis is now on the *Z*-sector rather than the *X*-sector. The point is that an economy which can adapt to the changing economic environment will be the most successful in improving the well-being of the nation. The two most important factors to which an economy must adapt are production technologies and terms of trade.

We deny that *X*-sector development automatically leads to *Z*-sector development. On the contrary, this is rather exceptional. Historically, the British economy is considered to be such a rare case. It was a pioneer economy, and hence its Z-sector was a monopolist supplier of its new products. Thus, it could sell the products at an early stage even though the cost was quite high. That is, it could charge the cost to customers. This does not apply in the case of latecomers such as Japan and other Asian countries. If they were or are unable to receive assistance from the government and/or foreign sources in the early stages of industrialization, their Z-sector had or has to desperately compete with the strong producers of the advanced industrial countries. The economic environment is tough for newcomers. This is one of the most important reasons why latecomer economies need government intervention to protect domestic production against competing imports.

Even in Germany, whose development came after that of England, government assistance was required in order for the economy to develop. If this was true for Germany, it will be even more true for latecomer economies, and particularly for economies which will pursue economic development in the future.

The US was a relative latecomer in industrialization before World War I, and the US government actually adopted protectionist policies for domestic manufacturing industries. However, the US had an advantage in developing its own heavy industries in certain respects. First, the US was located far from its competitors, and hence high transportation costs were a natural import tariff. Secondly, a rich endowment of natural resources reduced the dependence on imported raw materials. This meant that the US did not have to export in order to finance imports of industrial raw materials. Thirdly, there was a large domestic market. Together, these factors effectively bought time for the infant US *Z*-sector to grow in an isolated market without overwhelming competitive pressure from the industrial countries of Europe. In this way it was able to grow to the point of self-sustaining growth.

6. *The necessity of the policy fostering the Z-sector.* The UK and the US cases, however, are exceptional. Most developing economies cannot follow these models. The world market has been much more integrated since World War II, and most developing countries depend heavily on imports of industrial

material as well as capital equipment. To finance these imports, they must export their products. Since they have comparative advantages in light industries—for the X-sector is generally more labour-intensive than the Z-sector, and wage rates in developing countries are lower than those in the developed countries—it is easier for them to start with the X-sector. However, developing countries are not able to expand their production and export their products without limitation, for if they continue to specialize in light industries, the share of their X-sector goods in the world market will become sufficiently large to cause a sharp deterioration of the terms of trade. To make matters worse, the production costs of their exportable will not decline, due to the lack of economies of scale in the X-sector. They therefore have to diversify their exports, to include heavy industrial goods which they have so far imported: only by raising the Z-sector can a developing economy benefit from economies of scale and avoid deterioration of the terms of trade. But this requires that the economy can afford to invest sufficiently in the heavy industries to achieve such economies of scale, and that it is successful in sustaining take-off.

Although there is a danger of failure and of enduring distortions due to vested interests in protection, we none the less interpret the cases of Japan and South Korea as examples of success in establishing heavy industries. As the exposition in Parts II and III deals only with the mechanism underlying the development of the Japanese economy, brief mention is made here of the case of South Korea from our theoretical viewpoint. The government of South Korea intervened actively in resource allocation among industries by means of protection and subsidies, especially at the start of the development period. In consequence, the shipbuilding, steel, automobile, and electronic industries have developed so rapidly that these industries are now important export sectors. Many other so-called Newly Industrializing Economies (NIEs) are on a similar track. Some middle-income countries (using the World Bank classification) are now following NIEs, and some of ASEAN (Association of South-East Asian Nations) countries are likewise catching up with NIEs.

We conclude that the keys to successful industrialization for a country are as follows: success is not achieved completely by free trade with other countries, nor by increasing specialization in producing X-sector goods, nor by extreme interventionist policies with regulatory rigidity. In other words, the key to successful industrialization for a latecomer economy is to supplement market mechanisms by guiding resource allocation indirectly so as to avoid unfavourable specialization and to reap future gains from industrial diversification. The prospects for terms of trade and economies of scale are two key factors to consider. We believe neither in government dictatorship nor in a pure laissez-faire approach.

Appendix A

Proposition 1. Suppose

$$F_Z(\alpha K_Z, \alpha N_Z) > \alpha F_Z(K_Z, N_Z)$$

for $\alpha > 1$. Then

$$\frac{\partial F_Z}{\partial K_Z} K_Z + \frac{\partial F_Z}{\partial N_Z} N_Z - F_Z \geqq 0.$$

Proof. The following expression

$$F_Z(\alpha K_Z, \alpha N_Z) - \alpha F_Z(K_Z, N_Z)$$

is positive for $\alpha > 1$. This is easily seen from the assumption. However, it is negative for $0 < \alpha < 1$. Let $K_Z' = \alpha K_Z$ and $N_Z' = \alpha N_Z$. Then $\frac{1}{\alpha} K_Z' = K_Z$ and $\frac{1}{\alpha} N_Z' = N_Z$.

Let $\beta = \frac{1}{\alpha}$. Then $\beta > 1$. Applying the inequality in the assumption, $F_Z(\beta K_Z', \beta N_Z') > \beta F_Z(K_Z', N_Z')$. Then, dividing both sides by β, $\frac{1}{\beta} F_Z(\beta K_Z', \beta N_Z') > F_Z(K_Z', N_Z')$.

Since $\alpha = \frac{1}{\beta}$, $\alpha F_Z(K_Z, N_Z) > F_Z(\alpha K_Z, \alpha N_Z)$. Thus, we can view the expression as increasing in the vicinity of $\alpha = 1$ (see Fig. A.1). Thus, the derivative of this expression cannot be negative. Differentiate it with respect to α and let $\alpha = 1$. Then we obtain

$$\frac{\partial F_Z}{\partial K_Z} K_Z + \frac{\partial F_Z}{\partial N_Z} N_Z - F_Z \geqq 0$$

If F_Z is a homogeneous function, the above proposition is easily obtained from a general proposition concerning homogeneous functions, such as Euler's theorem. However, we do not assume homogeneity, which is why we have given the proof. It should be noted that the equal sign cannot be dropped. The case shown in Fig. A.1 is that in which the above expression is increasing in α but the derivative at $\alpha = 1$ is zero.

Proposition 2. The profit rate in the Z-sector

$$R_Z = \frac{1}{K_Z}\left[F_Z(K_Z, N_Z) - \bar{w} N_Z\right]$$

is strictly increasing in K_Z.

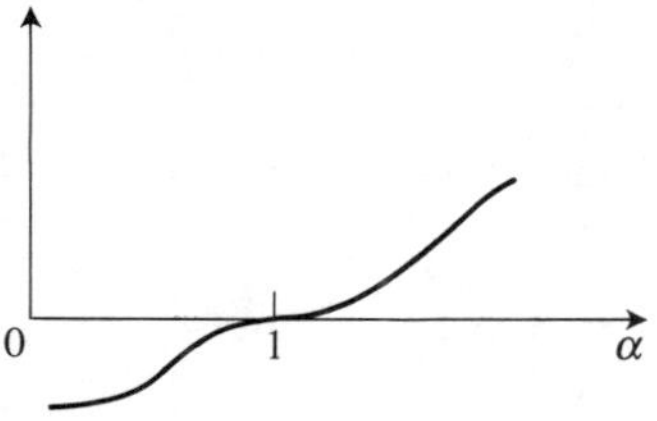

FIG. A.1

Proof. We have shown in the text that

$$\frac{dR_Z}{dK_Z} \geqq 0$$

Suppose that the relation

$$\frac{dR_Z}{dK_Z} = 0$$

holds for all values of K_Z in an interval of positive length. That is, $R_Z = a$ (constant) for all values of K_Z such that

$$K^* < K_Z < \bar{K}.$$

Then, for all values of K_Z in this interval,

$$F_Z(K_Z, N_Z) - \bar{w}\, N_Z = aK_Z$$

holds as an identity. Differentiate this expression with respect to K_Z and N_Z. We then obtain in this interval,

$$\frac{\partial F_Z}{\partial K_Z} + \frac{\partial F_Z}{\partial N_Z}\frac{dN_Z}{dK_Z} - \bar{w}\frac{dN_Z}{dK_Z} = a,$$

$$\frac{\partial F_Z}{\partial N_Z} = \bar{w}.$$

This last relation is (3.9) in Chapter 3 of the text. It then follows that

$$\frac{\partial F_Z}{\partial K_Z} = a.$$

Thus,

$$\frac{\partial^2 F_Z}{\partial K_Z^2} = 0.$$

But we assumed that

$$\frac{\partial^2 F_Z}{\partial K_Z^2} < 0.$$

This means that $\frac{dR_Z}{dK_Z}$ can be zero for scattered values but not for all values in an interval of positive length. Fig. A.2 shows this situation, but the case in Fig. A.3 is impossible. Thus, R_Z is an increasing function of K_Z.

The text mentions the case shown in Figs. 3.5 or 4.4 as a possibility. This case will arise if $R_Z(K_Z) < \mu$ for all values of K_Z. But we assumed that the Z-sector exhibits increasing returns to scale, which would seem to preclude this possibility. In fact, if we

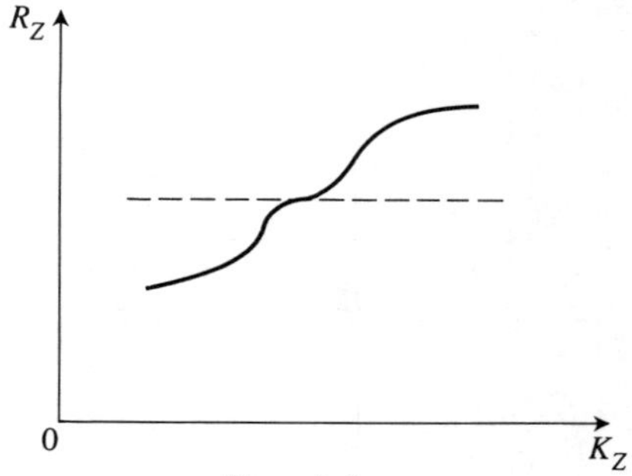

FIG. A.2

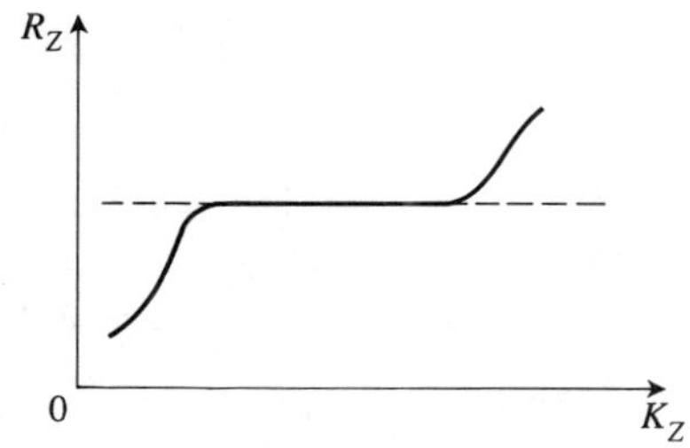

FIG. A.3

assume that the production function in the Z-sector is homogeneous, the possibility is excluded because under the assumptions of homogeneity and economies of scale, R_Z can take any large value, provided we take a very large value of K_Z.[1] The case of Figs. 3.5 or 4.4 would thus indeed be impossible. But in our model we do not assume a homogeneous production function in the Z-sector, in which case R_Z can be smaller than μ for any value of K_Z. For example, suppose the production function is

$$K_Z^{\alpha(1+\gamma(K_Z))} N_Z^{1-\alpha} \quad (0 < \alpha < 1) \text{ for } K_Z > 1,$$

and define the function $\gamma(K_Z)$ so it satisfies the following conditions.

$$\frac{d\gamma(K_Z)}{dK_Z} < 0, \lim_{K_Z \to \infty} \gamma(K_Z) = 0, \lim_{K_Z \to \infty} K_Z^{\gamma(K_Z)} = C \text{ (= constant)},$$

$$\frac{dK_Z^{\gamma(K_Z)}}{dK_Z} > 0.$$

Since we are here concerned with a property of the production function for large values of K_Z, we only define it for values of K_Z larger than 1.

Through a simple calculation, the profit rate can be shown to equal

$$\alpha\, \overline{w}^{\frac{-1+\alpha}{\alpha}} (1-\alpha)^{\frac{1-\alpha}{\alpha}} K_Z^{\gamma(K_Z)}.$$

Thus, if the above conditions are satisfied, the profit rate increases as the value of K_Z increases, but if we take a sufficiently small value of C, it will not exceed a certain value. On the other hand, provided that the last of the four conditions is satisfied, the production function exhibits increasing returns to scale. A $\gamma(K_Z)$ function satisfying the above conditions can be obtained as follows. Let $G(K_Z)$ be the function shown in Fig. A.4, and define $\gamma(K_Z)$ as follows:

$$K_Z^{\gamma(K_Z)} \equiv G(K_Z).$$

Then

$$\gamma(K_Z) = \log G(K_Z)/\log K_Z.$$

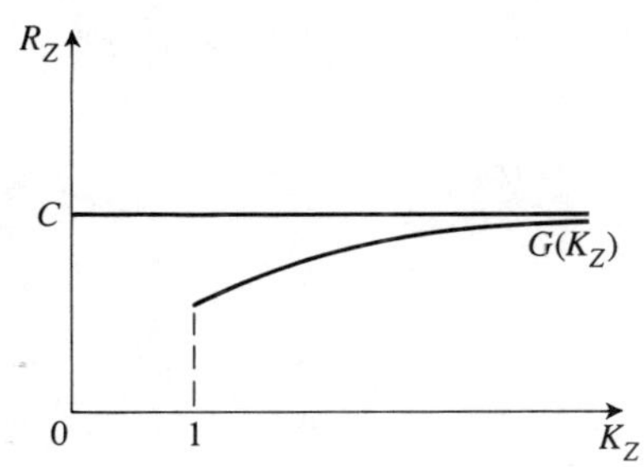

FIG. A.4

Appendix B

In this appendix we obtain the phase diagram of the following differential equation system:

$$\dot{K}_X = \varphi_X(K_X, K_Z) - \mu K_X, \tag{B.1}$$

$$\dot{K}_Z = \varphi_Z(K_X, K_Z) - \mu K_Z. \tag{B.2}$$

Here

$$\varphi_X(K_X, K_Z) \equiv \xi_X(R_X(K_X) - R_Z(K_Z)) \cdot R_X(K_X) \cdot K_X + \zeta_X(R_X(K_X) - R_Z(K_Z)) \cdot R_Z(K_Z)K_Z.$$

$$\varphi_Z(K_X, K_Z) \equiv \xi_Z(R_X(K_X) - R_Z(K_Z)) \cdot R_X(K_X) \cdot K_X + \zeta_Z(R_X(K_X) - R_Z(K_Z)) \cdot R_Z(K_Z) \cdot K_Z.$$

Denote the values of K_X and K_Z satisfying $R_X(K_X) = \mu$ and $R_Z(K_Z) = \mu$ as K_X^* and K_Z^*, respectively. Divide the first quadrant of the $K_X - K_Z$ plane into the four regions A, B, C, and D as shown in Fig. B.1. Let the locus of points on the $K_X - K_Z$ plane satisfying each of the following relations be known as the $\dot{K}_X = 0$ curve and the $\dot{K}_Z = 0$ curve, respectively. Then

$$\varphi_X(K_X, K_Z) - \mu K_X = 0 \tag{B.3}$$

and

$$\varphi_Z(K_X, K_Z) - \mu K_Z = 0. \tag{B.4}$$

Our first aim is to determine the location of these two curves on the $K_X - K_Z$ plane. That is, in which of the four regions A, B, C, and D are these curves located, and where are the curves located to one other? Take a point in A. For this point,

$$K_X < K_X^* \text{ and } K_Z < K_Z^*.$$

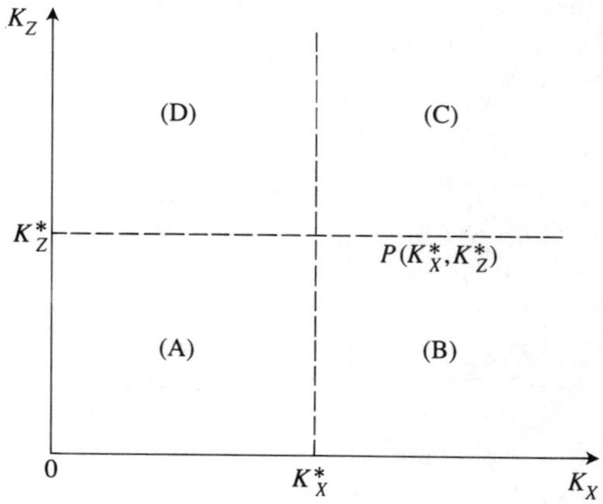

Fig. B.1

Since $R_X(K_X)$ is decreasing in K_X and $R_Z(K_Z)$ is increasing in K_Z,

$$R_X(K_X) > R_X(K_X^*) = \mu, \; R_Z(K_Z) < R_Z(K_Z^*) = \mu.$$

Thus,

$$R_X(K_X) - R_Z(K_Z) > 0.$$

Then,

$$\xi_X = 1, \;\; \zeta_X > 0, \;\; \xi_Z = 0, \;\; \zeta_Z < 1$$

and

$$\dot{K}_X = \varphi_X(K_X, K_Z) - \mu K_X = \xi_X \cdot R_X \cdot K_X + \zeta_X \cdot R_Z \cdot K_Z - \mu K_X \quad R_X \cdot K_X - \mu K_X > 0,$$
$$\dot{K}_Z = \varphi_Z(K_X, K_Z) - \mu K_Z = \xi_Z \cdot R_X \cdot K_X + \zeta_Z \cdot R_Z \cdot K_Z - \mu K_Z \geq R_Z \cdot K_Z - \mu K_Z < 0$$

From the above, we see that $\dot{K}_X > 0$ and $\dot{K}_X < 0$ for all points in A. By similar reasoning, we see that $\dot{K}_X < 0$ and $\dot{K}_Z > 0$ for all points in C. From these results, we see that both the $\dot{K}_X = 0$ curve and the $\dot{K}_Z = 0$ curve are not located in regions A and C. It is also easy to see that the $\dot{K}_X = 0$ and $\dot{K}_Z = 0$ curves intersect at (K_X^*, K_Z^*).

Suppose K_Z is fixed at a value greater than K_Z^*. Then $R_Z(K_Z) > \mu$. Now increase the value of K_X from 0 to K_X^*. Then, the value of $R_X(K_X)$ decreases from a certain value $R_X(0)$—possibly including infinity—to μ. As long as $R_X(K_X) \geq R_Z(K_Z)$ $(> \mu)$, $\dot{K}_X > 0$. This can be seen as follows. Since $R_X(K_X) \geq R_Z(K_Z) > \mu$, $\xi_X = 1$, and $\zeta_X \geq 0$. Then

$$\dot{K}_X = \xi_X \cdot R_X \cdot K_X + \zeta_X \cdot R_Z \cdot K_Z - \mu K_Z \geq R_X \cdot K_X - \mu K_X > 0.$$

Next, suppose $\mu < R_X(K_X) < R_Z(K_Z)$ for a value of K_X. Then $\xi_X < 1$, $\zeta_X = 0$. In this case,

$$\dot{K}_X = \xi_X \cdot R_X \cdot K_X - \mu K_X = (\xi_X \cdot R_X - \mu) K_X. \tag{B.5}$$

If the value of K_X increases and eventually approaches K_X^*, R_X decreases and eventually approaches μ, and ξ_X also decreases or at least does not increase. Then the relation $\xi_X \cdot R_X - \mu = 0$, and thus $\dot{K}_X = 0$ holds for one and only one value of K_X. Furthermore, we can easily see that the value of K_X for which $\dot{K}_X = 0$ holds depends negatively, or at least not positively, on the value of K_Z. This is seen from the fact that the value of ξ_X, and hence the right-hand side of relation (B.5), is decreasing, or at least not increasing, in K_Z.

From the above we can see that the $\dot{K}_X = 0$ curve is downwards sloping for $K_X \leq K_X^*$ and is located in region D. Since the right-hand side of (B.5) is a continuous function in both K_X and K_Z, the $\dot{K}_X = 0$ curve is also continuous.

Next, fix K_Z at a value such that $K_Z < K_Z^*$. If $K_X = K_X^*$, then $R_X(K_X) = \mu$. Since $K_Z < K_Z^*$, $R_Z(K_Z) < \mu$. Then, $\dot{K}_X > 0$. This may be explained as follows. Since $R_X(K_X) = \mu$ and $R_Z(K_Z) < \mu$, $R_X(K_X) > R_Z(K_Z)$. Thus, $\xi_X = 1$ and $\zeta_X > 0$. Then, from (B.1),

$$\dot{K}_X = \varphi_X(K_X, K_Z) - \mu K_K = \xi_X \cdot R_X \cdot K_X + \zeta_X \cdot R_Z \cdot K_Z - \mu K_X = \zeta_X \cdot R_Z \cdot K_Z > 0.$$

Now, fix the value of K_Z, and increase the value of K_X from K_X^* to infinity. Then, on the right-hand side of (B.1), R_X decreases monotonically from μ to zero,[2] and $\xi_X(R_X - R_Z) = 1$ while $R_X \leq R_Z$, but thereafter it also decreases or at least does not increase. ζ_X also decreases or at least does not increase. Since the value of K_Z is fixed, the right-hand side of (B.1), that is $\varphi_X(K_X, K_Z) - \mu K_X$ decreases monotonically. If K_X increases sufficiently, then $\varphi_X(K_X, K_Z) - \mu K_X$ becomes negative. From this we see that $\dot{K}_X = 0$ for one and only one value of K_X. At this point, $K_X > K_X^*$ and $K_Z < K_Z^*$. We also obtain $R_X(K_X) < \mu$ and $R_Z(K_Z) < \mu$. From this, we see that the profit rate in each sector is not enough to compensate for the depreciation of capital, even if the profits in each sector are wholly reinvested in the sector. Since $\dot{K}_X = 0$, this means that the profits of the X-sector are wholly reinvested in the X-sector, and some portion of the profits of the Z-sector flows out of that sector and is invested in the X-sector. This also means that $R_X > R_Z$. From the above we see that the $\dot{K}_X = 0$ curve is located in region B for values of K_X greater than K_X^*.

The shape of the $\dot{K}_X = 0$ curve in this region is as shown in Fig. B.2. To see the reason for this, suppose $K_X = K_X^*$ and $K_Z < K_Z^*$. Since $\xi_X = 1$ and $R_X = \mu$, we obtain

$$\dot{K}_X = \xi_X \cdot R_X \cdot K_X + \zeta_X \cdot R_Z \cdot K_Z - \mu K_X = \zeta_X \cdot R_Z \cdot K_Z.$$

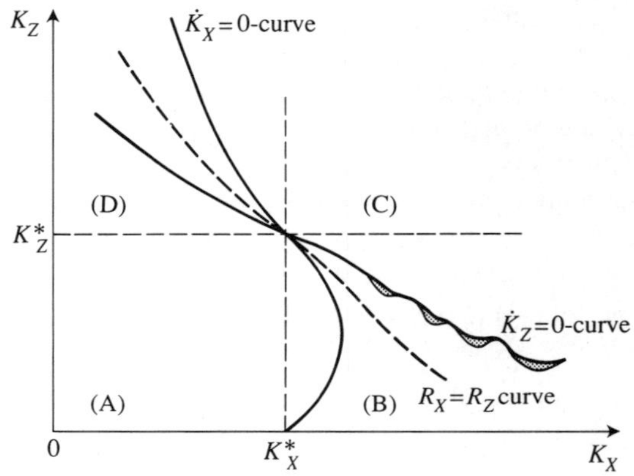

FIG. B.2

Now suppose K_Z decreases from K_Z^* to zero. Then both R_Z and K_Z decrease. On the other hand, ζ_X increases as K_Z decreases. Thus their product—that is, $\zeta_X \cdot R_Z \cdot K_Z$—is not necessarily monotonically decreasing as K_Z decreases. But we can see that it eventually approaches zero.[3] For $0 \leqq \zeta_X \leqq 1$, R_Z decreases as K_Z decreases, and K_Z decreases to zero. This means that $\dot{K}_X = 0$ at point $(K_X^*, 0)$, and that the terminal point of the $\dot{K}_X = 0$ curve is $(K_X^*, 0)$.

By similar logic, we see that the $\dot{K}_Z = 0$ curve is downwards sloping in D, but in B it can take the irregular shape shown in Fig. B.2. Why the shape may become irregular in this region will not be explained in detail because our main concern is the pattern of paths in A and D. However, an approximate explanation would be as follows. In region B, $R_X(K_X) < \mu$, and $R_Z(K_Z) < \mu$. Thus $\dot{K}_Z = 0$ is possible only for the case in which $R_Z > R_X$, $\xi_Z > 0$, and $\zeta_Z = 1$. The portion of the profits of the X-sector invested in the Z-sector is $\xi_Z \cdot R_X \cdot K_X$. ξ_Z is increasing in K_X, R_X is decreasing in K_X, and of course K_X is increasing in K_X. Thus the product of ξ_Z, R_X, and K_X may not necessarily be monotonic. Roughly speaking, this is why the $\dot{K}_Z = 0$ curve may take the irregular shape shown in Fig. B.2.[4]

Next, we show why the $\dot{K}_X = 0$ curve is located above the $\dot{K}_Z = 0$ curve in region D, and why the opposite holds in region B.

The dotted line curve in Fig. B.2 is the locus of points for which $R_X(K_X) = R_Z(K_Z)$ is satisfied. Call it the $R_X = R_Z$ curve. The reason for the downwards slope is easily seen. Now in region D, $R_X > \mu$, and $R_Z > \mu$. Then on the $R_X = R_Z$ curve, $R_X = R_Z > \mu$. Then,

$$\dot{K}_X = R_X \cdot K_X - \mu K_X > 0,$$

and

$$\dot{K}_Z = R_Z \cdot K_z - \mu K_z > 0$$

on this curve. Since $\dot{K}_X > 0$ on the left side of the $\dot{K}_X = 0$ curve and $\dot{K}_X < 0$ on the right side, the $R_X = R_Z$ curve is located to the left of the $\dot{K}_X = 0$ curve. Similarly, we see that the $R_X = R_Z$ curve is located above the $\dot{K}_Z = 0$ curve. From this we see that the $\dot{K}_X = 0$ curve is located above the $\dot{K}_Z = 0$ curve. By similar reasoning, we see that in region B the $\dot{K}_X = 0$ curve is located below the $\dot{K}_Z = 0$ curve.

We implicitly assumed in the text that the profit-optimizing problem in the Z-sector has a solution for which profits are non-negative for any value of K_Z. But in reality it is possible that the optimal solution is to cease production for small values of K_Z. In this case, the shape of the $\dot{K}_X = 0$ curve should be modified as shown in Fig. B.3. The reason is easily seen. The profit rate R_Z in the Z-sector is zero for small values of K_Z. Then the profits from the Z-sector that are invested in the X-sector are also zero because $\zeta_X \cdot R_Z \cdot K_Z = 0$. Thus, the $\dot{K}_X = 0$ curve becomes a vertical straight line for small values of K_Z. The terminal point of this line is point $(K_X^*, 0)$.

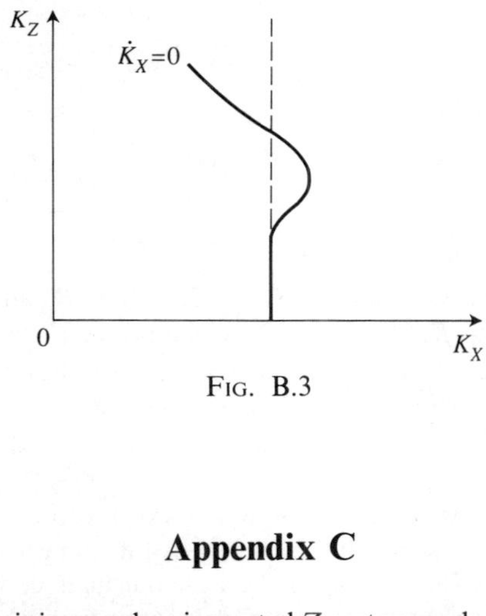

FIG. B.3

Appendix C

If an import tariff of π is imposed on imported Z-sector goods, the domestic price of Z-sector goods also becomes $1 + \pi$. Suppose the ratio of Z-sector goods in the wage-earners' market basket is a constant $\alpha(< 1)$. Then the subsistence wage level goes up to

$$(1 - \alpha)\overline{w} + \alpha(1 + \pi)\overline{w} = (1 + \alpha\pi)\overline{w},$$

and the profit rate in the X-sector becomes

$$R_X = \frac{1}{K_X}[p(X)\ X - (1 + \alpha\pi)\overline{w}N_Z].$$

Corresponding to (3.2) in the text,

$$\frac{\partial F_X}{\partial N_X} = \frac{(1 + \alpha\pi)\overline{w}}{p'(X)\ X + p(X)} \tag{C.1}$$

(3.1) in the text does not change. Here, calculate $\dfrac{dR_X}{d\pi}$.

$$\frac{dR_X}{d\pi} = \frac{1}{K_X}\left\{(p'(X)X + p(X))\frac{dX}{d\pi} - \alpha\overline{w}\,N_Z - (1+\alpha\pi)\overline{w}\frac{dN_X}{d\pi}\right\}$$

$$= \frac{1}{K_X}\left[\left\{\frac{(1+\alpha\pi)\overline{w}}{\dfrac{\partial F_X}{\partial N_X}}\right\}\frac{dX}{d\pi} - \alpha\overline{w}N_Z - (1+\alpha\pi)\overline{w}\frac{dN_X}{d\pi}\right].$$

Since

$$\frac{dX}{d\pi} = \frac{\partial F_X}{\partial N_X}\cdot\frac{dN_X}{d\pi},$$

we obtain

$$\frac{dR_X}{d\pi} = \frac{1}{K_X}(-\alpha\overline{w}\,N_X) < 0.$$

Next,

$$R_Z = \frac{1}{K_Z}\,[(1+\pi)Z - (1+\alpha\pi)\overline{w}N_Z].$$

This relation holds under the assumption that Z-sector goods are not exported, or that they are exported with a subsidy making up the difference between the domestic price and the export price.

Corresponding to (3.9) in the text, we obtain

$$\frac{\partial F_Z}{\partial N_Z} = \frac{(1+\alpha\pi)\overline{w}}{1+\pi} \tag{C.2}$$

Calculate $\dfrac{dR_Z}{d\pi}$.

$$\frac{dR_Z}{d\pi} = \frac{1}{K_Z}\left[Z + (1+\pi)\frac{dZ}{d\pi} - \overline{w}\alpha N_Z - (1+\alpha\pi)\overline{w}\frac{dN_Z}{d\pi}\right].$$

Then, from (C.2) and $\dfrac{dZ}{d\pi} = \dfrac{\partial F_Z}{\partial N_Z}\dfrac{dN_Z}{d\pi}$,

$$\frac{dR_Z}{d\pi} = \frac{1}{K_Z}\,[Z - \overline{w}\,\alpha\,N_Z] \geq \frac{1}{(1+\pi)K_Z}\,[(1+\pi)Z - (1+\alpha\pi)wN_Z\overline{w}].$$

The value $(1+\pi)Z$ is the output level and $(1+\alpha\pi)\overline{w}N_Z$ is the share of labour in it. If $Z > 0$ and $N_Z > 0$ at the optimal production level, then $(1+\pi)Z \geq (1+\alpha\pi)\overline{w}N_Z$, and thus $\dfrac{dR_Z}{d\pi} \geq 0$. If $Z = 0$ and $N_Z = 0$ at the optimal production level in the absence of an import tariff, then the import tariff contributed to making production profitable. In either case, the profit rate in the Z-sector will rise due to the import tariff, except in the trivial case where zero production is the most profitable for the Z-sector, both with and without the import tariff.

In the above, we examined the increase or decrease in the profit rates in terms of S-sector goods. But if investment goods are Z-sector goods, imported or domestic (and this case may be usual), we have to see how the profit in terms of Z-sector goods in each sector changes. The change in the profit in the X-sector due to the import tariff is

$$\frac{d\left(\dfrac{R_X\,K_X}{1+\pi}\right)}{d\pi} = -\frac{\alpha\overline{w}N_X}{1+\pi} - \frac{R_X\,K_X}{(1+\pi)^2} < 0,$$

and that in the Z-sector is

$$\frac{d\left(\dfrac{R_Z K_Z}{1+\pi}\right)}{d\pi} = \frac{Z - \alpha\overline{w}N_Z}{1+\pi} - \frac{R_Z K_Z}{(1+\pi)^2}$$

$$= \frac{1}{(1+\pi)^2}\left[(1+\pi)Z - (1+\pi)\alpha\overline{w}N_Z - (1+\pi)Z + (1+\alpha\pi)\overline{w}N_Z\right]$$

$$= \frac{1}{(1+\pi)^2}(1-\alpha)\overline{w}N_Z > 0.$$

From the above, we can see that the profit in the Z-sector rises even in terms of Z-sector goods.

From the results obtained above, we see that the minimum critical effort curve shifts downwards. Fig. C.1 shows the reverse case. The solid lines show the situation without the import tariff, and the dotted lines the situation with the import tariff. But we shall show that this case does not arise. The profit rates in the X-sector and the Z-sector are functions of K_X and π and of K_Z and π and of K_Z and π, respectively, so they can be written as $R_X(K_X, \pi)$, and $R_Z(K_Z, \pi)$, respectively. R_X is decreasing in both K_X and π. R_Z is increasing in both K_Z and π.

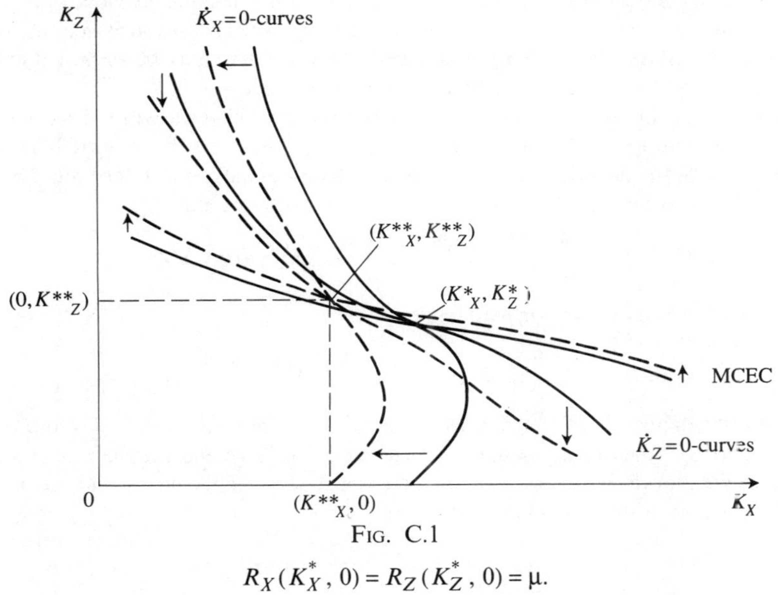

FIG. C.1

$$R_X(K_X^*, 0) = R_Z(K_Z^*, 0) = \mu.$$

Suppose

$$R_X(K_X^{**}, \pi) = R_Z(K_Z^{**}, \pi) = \mu \quad (\pi > 0).$$

Point (K_X^{**}, K_Z^{**}) is the intersection of the $\dot{K}_X = 0$ and $\dot{K}_Z = 0$ curves with the import tariff. We easily see that $K_X^{**} < K_X^*$ and that $K_Z^{**} < K_Z^*$. Then (K_X^{**}, K_Z^{**}) is located to the south-west of (K_X^*, K_X^*). This means that Fig. C.1 is impossible. But the case shown in Fig. C.2 is possible. That is, one or more parts of the MCE curve with import tariff may locate above that without tariff. The former is shown as the dotted-line curve in Fig. C.2, and the latter as a solid line. In the text, we mentioned that the minimum critical effort curve

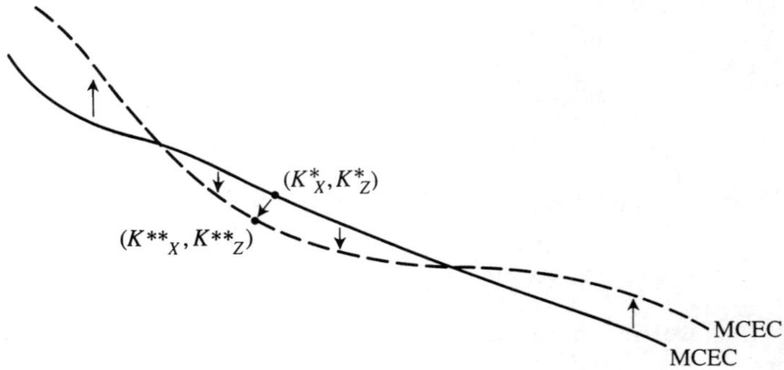

FIG. C.2

will not shift wholly downwards. This statement is viewed as referring to the case of Fig. C.2. Possibility, however, does not imply inevitability.

Next, consider technological progress. Technological progress in the X-sector keeps the value of $R_Z(K_Z^*)$ constant and increases the value of $R_X(K_X^*)$, and thus the minimum critical effort curve will shift upwards, making take-off more difficult. But it can easily be seen that technological progress in the Z-sector will make the value of $R_Z(K_Z^*)$ increase and thus the minimum critical effort curve will shift downwards, making take-off easier. But the shift of the minimum critical effort curve may be as shown in Fig. C.2; we discussed only the movement around point (K_X^*, K_Z^*).

Suppose that some portion of Z-sector goods is exported at the world price without subsidies and suppose also that the import tariff is imposed on imported Z-sector goods. Denote the portion as γ. For simplicity, assume that γ is not dependent on the level of the import tariff. Then, by a calculation similar to the above,

$$R_Z = \frac{1}{K_Z}\,[(1+\pi)(1-\gamma)Z + \gamma Z - \overline{w}(1+\alpha\pi)\ N_Z].$$

Differentiate this expression with respect to π.

$$\frac{dR_Z}{d\pi} = \frac{1}{K_Z}\,[(1-\gamma)Z - \overline{w}\alpha N_Z]. \tag{C.3}$$

The right-hand side of this expression may take a negative value, if γ and α are large—that is, if the export share of Z-sector goods is large and the share of Z-sector goods in the market basket of wage-earners is also large. This means that the import tariff may become unfavourable for take-off.

Notes

1. The essential points of our analysis will not be affected even if we assume the homogeneity of the $F_Z(K_Z, N_Z)$ function. As mentioned in the text, the case in Figs. 3.5 or 4.4 will be excluded. But even in this case, the value of K_Z^* that satisfies $R_Z(K_Z^*) = \mu$ may be so large that in the real world it may be practically impossible to elevate the starting point above the minimum critical effort curve. In this respect,

the case of Figs. 3.5 or 4.4 should be viewed as the case where the value of $\overset{*}{K_Z}$ is infinitely large.

2. It is not necessary that R_X eventually approaches zero as K_X approaches infinity. We need to make a slight modification in the general case, but the essential feature is the same as under the restrictive assumption.
3. Because $\zeta_X \cdot R_Z \cdot K_Z$ is not necessarily monotonically decreasing for a fixed value of K_X such that $K_X \geqq \overset{*}{K_X}$, when K_Z decreases from $\overset{*}{K_Z}$ to 0, the shape of the $\dot{K}_X = 0$ curve in region B might be the one shown in Fig. B.4. But $\zeta_X \cdot R_Z \cdot K_Z$ is monotonically decreasing in K_X. The $\dot{K}_X = 0$ curve, therefore cannot take the shape shown in Fig. B.5 or Fig. B.6.
4. The logic employed here to show the irregularity of the $\dot{K}_Z = 0$ curve is the same as that employed in n. 10 in Ch. 4.

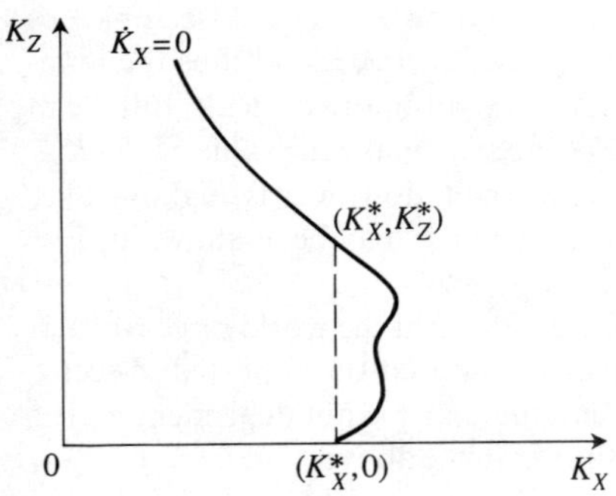

Fig. B.4

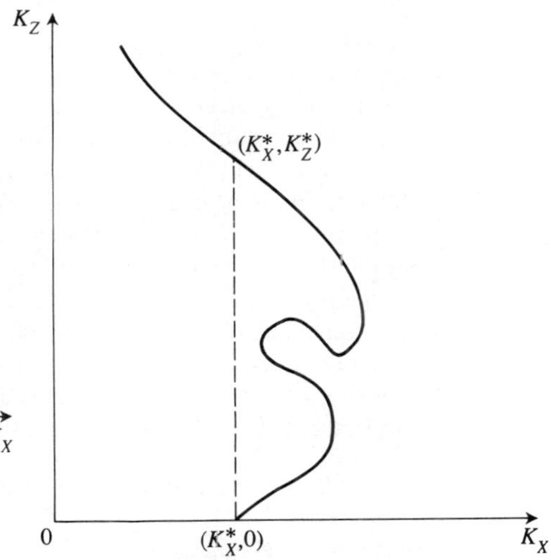

Fig. B.5

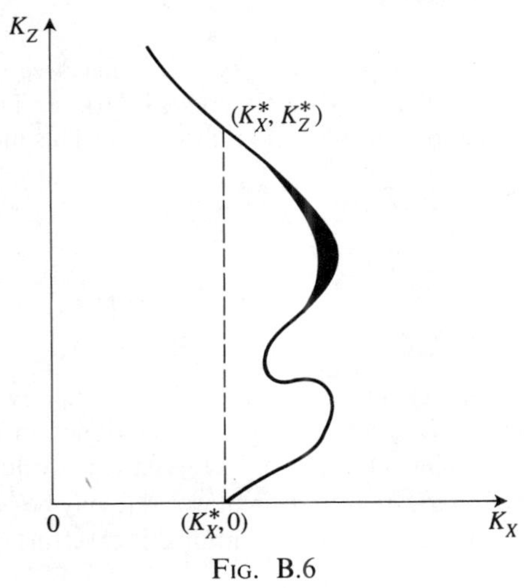

Fig. B.6

PART II

The Economic Development of Japan Pre-World War II

7

Introduction and Summary

7.1. Introduction

Part I developed a theoretical model which clarifies the courses of capital accumulation of a national economy. The model consists of three sectors: the subsistence sector (S-sector), the light industry sector, including plantations (X-sector), and the heavy industry sector (Z-sector). We now consider its implications for testing our theory against Japanese experiences.

Let us start with the development of the Japanese economy before World War II. At that time Japan was a monocultural economy as Gunnar Myrdahl (1960) once characterized the Indian economy, with only a few industrial sectors, such as silk and tea. Even so, it successfully developed modern industries which are classified as Z-sector. If we pick up an economy which remains at pre-take-off stage, the only thing we can test is a case of falling into the 'trap' of poverty or of monocultural stagnation. Although we wish some day to test a case of take-off failure, we concentrate here on testing of a success case.

A second implication of testing the model on the Japanese economy is that an economy which may be classified as pre-modern and has not fully subscribed to the modern Western culture or institutions, or a monocultural economy, can take-off toward industrialization. The Japanese economy had not had contact with Western culture for a few centuries, yet it managed to industrialize in the relatively short period of one hundred years from the Meiji restoration of 1868.

To paraphrase the above observations using the terminology of Part I, the curves for $\dot{K}_x = 0$ and $\dot{K}_z = 0$ might have not intersected, or the actual point in the $K_X - K_Z$ diagram might have been so close to the K_X axis, and thus so far below the minimum critical effort curve, that the economy might not have been able to take-off. In actuality, however, the economy did succeed in taking off, thanks to government support which played a critical role by implementing various minimum critical efforts. We concluded that it was in the latter part of 1920s that the Japanese economy reached the point of Z-sector take-off. The significance of this is that even an economy which apparently had no common features with Western capitalism at the start of the development process succeeded in achieving self-sustaining growth, and this development process can be explained by a theory predicated on the economic rationalism of Western society.

A third observation is more appropriately characterized as a rationale for testing the Japanese economy rather than an implication. The availability of statistical data of the type required for testing the model is frequently limited,—data on capital stock are scarce in most developing countries even for recent years—but this is not the case in Japan. It is true that these data are not always consistently prepared even in Japan, but, none the less, we were able to acquire enough data to test our theory.

One thing should be mentioned here, and that is that although there are quite a number of empirical studies on the development of Japanese economy—e.g. Lockwood (1954), Shinohara (1961), Fujino (1965), Ohkawa (1965*a*), and Minami (1970)—we are not aware of any study that sheds light on the mechanism of capital allocation and accumulation in a dynamic model. Minami's study, which is based on Lewis's (1954) model focuses on the course of development, but it is not dynamic. Our study, in contrast, covers changes in the terms of trade, the classification of industrial sectors based on our own theoretical model, the mechanism of capital accumulation and allocation between sectors, and various functions undertaken by the government in making the minimum critical efforts. Thus, we believe it to be the first work that comprehensively describes the development of modern Japan in the framework of a theory with empirical tests.

7.2. Correspondence between Theory and Empirical Test

A most important assumption that must be subjected to empirical testing is whether industries can in fact be classified into three sectors with the characteristics assumed in the theoretical model.

The *S*-sector, the subsistence sector which consists of agriculture and related sectors, is not explicitly described in the model but plays an important role in industrial development, in that it affects the course of economic development through the supply of food as well as of labour.

The *X*-sector, the light industry sector, must fulfil the following two premisses. First, the relative price of *X*-sector products against those of the *S*-sector is a decreasing function of *X*-sector output; secondly, its production function is characterized by constant returns to scale. These conditions, however, are not necessary but sufficient conditions for the important interim result that the *X*-sector profit rate in terms of *S*-sector products is negatively related to *X*-sector capital stock.

The *Z*-sector, the heavy industry sector, must be compatible with the following: the terms of trade of the *Z*-sector against the *S*-sector must be independent of *Z*-sector output, and its production function must be characterized by increasing return to scale. These points are tested empirically in Chapter 8. To test our assumptions on the relationship between the terms of trade and the level of production of the *X*- and *Z*-sectors, we do not have to estimate demand functions *per se*—merely whether the relative prices of *X*-sector products *vis-à-vis* *S*-sector products decrease as *X*-sector production expands, and whether *Z*-sector terms of trade are independent of the level of production of the *Z*-sector itself.

The point is that the *X*-sector of a developing country often occupies such a large share of the world market that it faces unfavourable terms of trade if it expands production of its traditional *X*-sector. This is true particularly for the primary products in which many developing countries specialize. Thus, we must test whether the prices of *X*-sector goods declined as Japan expanded its *X*-sector production. We first apply this test on individual sectors to find the relationship between prices and production. Since the above assumption implies that the *X*-sector holds a significantly large share of the world market, like a monopolistic producer, we examine Japan's market share in major possible *X*-sector goods.

Conversely, if *Z*-sector terms of trade against *S*-sector goods are independent of *Z*-sector production, the implication is that Japan has a small share of the world market. Once again, we test for this without estimating demand functions but by examining the relationship between *Z*-sector prices and production, and then supplement this data by investigating Japan's world market share in the tested *Z*-sector industries.

One advantage of this approach is that it avoids the controversial issue of whether terms of trade have deteriorated over time for developing countries: test merely whether the terms of trade are independent of production in the sector in question, or worsen as production increases. Changes in the terms of trade over time, *per se*, are irrelevant.

Although our method of testing is naïve in that no strict statistical tests are applied, testing was none the less carried out in two stages: first on individual sectors and second on aggregate data. The price data used in our examination are those of gross output, for both the pre-war and the post-war period.

Another important assumption is related to production technology, i.e. that *X*-sector and *Z*-sector production functions are characterized by constant and increasing returns to scale, respectively. The problem in testing these assumptions is that actual data reflect both scale factors and technological development, and that as in other countries, data for capital stock in Japan are not well developed. Even if production functions are estimated in a sophisticated way, therefore, they are not free from criticism.

For our theory, however, technological progress, if it exists, does not have to be separated from scale factors. All we need to show is that *Z*-sector profits increase as capital stock expands, and that *X*-sector profits decline as production increases. Technical progress can theoretically be discussed separately as was done in Part I, but in empirical examination we do not intend to differentiate them. The inadequacy of available capital data makes it meaningless to produce sophisticated production functions. This is well known to all who are engaged in long-term development studies.

In Chapter 9 we consider two important interim results that play a critical role in deducing the paths of capital accumulation. One is that the profit rate (profit over capital stocks) of the *X*-sector declines in terms of *S*-sector products while that of the *Z*-sector increases as capital accumulation of the respective sectors proceeds. Therefore, the profit rate of the *Z*-sector, which starts from a low level, will rise as capital is accumulated and will finally exceed that of the *X*-sector. We next consider whether a growing proportion of *X*-sector profit is directed to investment in the *Z*-sector, as described in the model.

In making an empirical test, we face a serious dilemma: it is desirable to make industry coverage as broad as possible in order to cover all sources of investment funds, as the theory assumes that the proportion of total resources available to invest in the *X*- and *Z*-sectors increases with respect to $R_x - R_z$ and $R_z - R_x$ respectively. However, the consistency of data declines as coverage expands.

For the pre-war period the only reliable data we have for calculating the profit rate are in any case the profit and capital stock data shown on the balance sheets of the major companies. As data consistency is thereby maintained, estimation results are much better than for the post-war period even though the coverage of industries and enterprises is substantially reduced. For the post-war period we used aggregate profit data estimated from macro production (value added) minus wage bills. Industry and firm coverage thereby improved significantly, but estimation results deteriorated.

The data on capital accumulation which we have considered so far are the products of a combination of government policies and market forces. In development economics, these policy efforts are called 'minimum critical efforts' which push industries above a curve which divides the $K_x - K_z$ space into 'successful take-off' on the one hand and poverty or a 'stagnation trap' on the other. It is therefore necessary to investigate what specific policy actions have been taken for the 'big push' towards self-sustained development. These features will be examined in Chapter 10. The government efforts include:

1. Roles of foreign capital
2. Establishment of 'state enterprises' and their transfer to private sectors
3. Industry subsidies, taxation, and investment in social overhead capital
4. Roles of import tariffs
5. Supply of foods and labour and wage rates.

Chapter 10 of this part corresponds to Chapter 4 of Part I, 'Modification of the Model'. The period covered by the empirical test is from 1868 through 1940, that is from the Meiji restoration to the beginning of World War II.

7.3. Summary of Findings

Due to data constraints it is almost impossible to include all industries in the classification of industrial sectors from 1868 to 1940 since the data on production, prices, capital stock, and labour force must be mutually consistent. We have therefore concentrated on selected manufacturing sectors for making a quantitative test of the theory, as set on below:

X-*sector*: food-processing, textiles, lumber and related products, printing and publishing, miscellaneous manufacturing.

Z-*sector*: machinery, chemical, cement, metal.

The classification is based on two conditions, i.e. terms of trade and production technology.

Not unexpectedly, the *X*- and *Z*-sectors do not completely fulfil the two sets of conditions at certain points. For instance, in the period 1884–1910, the terms of trade of the *Z*-sector were negatively correlated to its output level, whereas our assumption was that they should be independent of output. For the rest of the period, however, and especially for the 1920s and 1930s, the data fit our assumption very well. It therefore seems safe to say that the classification of industries into *X*- and *Z*-sectors is generally relevant. In fact, we applied the test of classification in two stages: first individual sectors were tested for classification, and then the two sectors thus aggregated were tested against the two sets of conditions.

After the test of basic assumptions, we then applied the test of the interim results—that is, $\frac{dR_x}{dK_x} < 0$, and $\frac{dR_z}{dK_z} \geq 0$, because these interim results have critical importance for the phase diagram of capital accumulation. Since the data required for testing these results need to be much more detailed and at the same time consistent with each other, however, our test is limited to the period between 1920 and 1940. Another reason for concentrating on this period is that this is when R_z exceeded R_x for the first time since 1868—that is, this is the period in which capital accumulation in the *Z*-sector first accelerated.

If individual industries in the *X*- and *Z*-sectors satisfy all the conditions we assumed, testing of $\frac{dR_x}{dK_x} < 0$, and $\frac{dR_z}{dK_z} \geq 0$ is only an indirect way of testing that enterprises maximize their short-run profits. When classification into the two sectors is difficult due to shortage of data, however, this test can assure that our aggregation and classification are supported by the data.

After testing the above interim results, we observed that R_z exceeded R_x in the first half of the 1930s. It was extremely difficult, however, to make a strict test of our hypothesis about investment allocation between the *X*- and *Z*-sectors because of limited data availability. We therefore had to confine our efforts to an indirect test of whether profits tend to be invested more in the sector where profit rate is higher, with some 'stickiness'. The model assumes that when the Z-sector profit rate exceeds that of the *X*-sector, all the profits earned in the *Z*-sector will be re-invested in the same sector, that some portion of the profits from the *X*-sector be invested in the *Z*-sector, and that this portion is an increasing function of $R_z - R_x$. We have tested whether the ratio of investment in the Z-sector to that in the X-sector increases as the difference, $R_z - R_x$, expands.

Our investigation of minimum critical efforts by the government and the examination of the initial condition of food and labour supply in Chapter 4 led us to several conclusions. First, we affirm the results of other empirical studies: foreign capital did not play a critical role in development. Major borrowing from abroad before 1940 was restricted to foreign loans incurred at the very beginning of the Meiji period, including debt issues by such major municipal governments as Tokyo and Osaka for city construction; borrowing to finance the Russo-Japanese War in the early 1900s; and bond issues for the reconstruction of Tokyo after the devastation by the Kanto earthquake (1923). The amount of borrowing was small relative to the foreign debts of today's developing countries. It is to be noted that Japan became a net capital exporter after World War I, when European nations issued bonds on the Tokyo market.

Secondly, we attach importance to the roles of the government enterprises which laid the foundations for the industrial development of the later period. By establishing these companies, the government pushed the K_x, K_z point north, above the minimum critical effort curve. Although state enterprises are often criticized for their inefficiency, we believe that they can play important roles in a country where modern institutions have not yet been established, and that they can prepare industrial bases for future development. In the case of Japan, mining sectors run by the government played a much more important role than might be expected in earning invaluable foreign currency. In the manufacturing sectors, state enterprises became a base for accelerating the development of the Z-sector. The evidence for this is that when we add the K_z of the government sector to that of the private sector, it shifts the total K_z substantially northwards in the phase diagram. When these state enterprises were later sold off to the private sector their prices were kept low, thereby subsidizing further Z-sector development. Furthermore, the government had played another important role in taking the risk of operating the new industries at the start-up stage.

Thirdly, it can be argued that the subsidization of industries and allocation of social overhead capital were also concentrated in the Z-sector. A first channel of assistance was the preferential treatment in taxation from which Z-sector industries, including the steel sector, benefited. Subsidies were not large, however, because direct government operation of the new industries made heavy subsidies unnecessary. As mentioned earlier, the subsequent sale of these enterprises to the private sector was also a form of subsidization. The current study tested only the manufacturing sectors, but had we included such sectors as transportation, communications, and public utilities, we would have found significant subsidies, and this would have shifted the production function of the Z-sector even higher than that of the X-sector. Thus, the government promoted investment in the Z-sector through subsidies, both explicit and implicit, as well as through the allocation of infrastructure investment. Transportation, both inland and marine, and energy industries received huge subsidies. This laid the groundwork for the acceleration of *Z*-sector development and also created domestic demand for *Z*-sector products.

Fourthly, tariff policy did not especially favour the *Z*-sector. First of all, the Meiji government had to accept an unfavourable treaty *vis-à-vis* the Western imperialistic powers, and it was not until 1899 that the government obtained the sovereign right to decide import tariffs. In 1911 the government succeeded in making the Western countries agree to revise the treaty again, but as many tariff agreements remained in place, rates were in general low. A third amendment was achieved in 1926, but overall we can say that the government did not use tariffs to protect the *Z*-sector. Rather, tariffs were used to restrict imports of consumption goods.

Japan's inability to set its own tariff level in the Meiji period ironically kept its foreign trade freer. As the theory in the Part I suggests, the government could avoid the worst case in which a failure of industrialization with protective tariffs would make the economy worse off than under free trade. In the 1920s, the government could have exercised protective tariffs to promote the *Z*-sector, but it did not do so, probably because it was more concerned about the increased costs for production in the *X*-sector due to higher prices of *Z*-sector goods.

Thus, tariffs were not used specifically to benefit the *Z*-sector prior to 1945, but rather subsidies, preferential taxes, and infrastructure construction; instead, tariffs were utilized to encourage domestic savings by discouraging the consumption of luxury goods.

Finally, let us review the conditions of food and labour supply. Technological progress in the subsistence sector plays a dual role in the industrialization of the *Z*-sector. On one hand, it improves the terms of trade for the *Z*-sector because it ensures an abundant food supply. At the same time, it contributes to the labour supply, in that labour can be transferred from the *S*-sector to the *Z*-sector at a constant wage rate. Labour productivity in agriculture increased magnificently after the Meiji restoration. Technical improvements played an important role in this, together with an increase in inputs of intermediate goods such as fertilizer and the promotion of fixed investment. In the later period, particularly from the 1920s on, the availability of food from overseas colonies in Asia contributed to keeping the terms of trade of the *Z*-sector favourable.

It is to be noted that even though international prices of foodstuffs may be stable because of the high elasticity of supply, a developing country cannot always import as much as it needs because it lacks foreign exchange. In this case, the domestic price of *Z*-sector products falls relative to *S*-sector products and profit rates in the *Z*-sector deteriorate, which in turn slows *Z*-sector expansion. This is a problem which many developing nations face today.

With respect to labour supply, Minami (1970) argued that labour supply in the pre-World War II era had always been unlimited, in the sense that the marginal productivity of labour in agriculture remained lower than the institutional subsistence wage rate. This means that the economy never had to face a general shortage of labour until 1940. But what about the shortage of skilled labour for the *Z*-sector, the subject on which our model focuses? Our conclusion is that the supply of *Z*-sector labour was also unlimited.

This is because the *Z*-sector wage rate had never risen substantially relative to that of the *X*-sector. Why, then, did the *Z*-sector not suffer from a labour shortage, like the shortage of skilled labour in many developing countries today? The reason is that, from the beginning, the Meiji government put the utmost priority on improving primary and secondary education, and this ensured an adequate supply of skilled labour. This will be discussed more thoroughly in Chapter 10.

7.4. Some Peculiar Conditions

While it is true that Japan achieved the self-sustaining development of heavy industries—that is, the take-off of the *Z*-sector—in about sixty years from the start of the new government of 1868, there were several peculiar conditions which characterized the environment of development in Japan. These factors must be taken into consideration when we compare the development performance of Japan with that of today's developing countries.

First was the recurrence of wars. The best thing that happened to the Japanese economy prior to World War II was World War I, which devastated the industrial production of most of the European countries and thereby accelerated the development of the *Z*-sector in Japan between 1914 and 1920. In contrast to this, the effects of the wars in which Japan was directly involved were ambiguous. For instance, although the ending of the Sino-Japanese War in 1895 certainly brought enormous reparations, increasing Japan's foreign reserves and thereby facilitating Z-sector expansion, it siphoned off manpower and economic resources from industries to military sectors.

A similar argument can be developed about the Russo-Japanese War, which ended in 1905. The victory brought Japan additional land, but it also placed an enormous burden on the economy. More significantly, incessant battles in China, such as the Manchuria incident (1931) and the Japan–China War, which started in 1937, pulled the Japanese economy into war-time control.

Direct control by the militaristic government killed the market mechanism, making it difficult for us to test the theory using the actual data. However, it would seem that repeated warfare created demand for Japanese *Z*-sector goods and enabled it to avoid serious recessions. It is on this point that today's LDCs find themselves in a very different situation; the external environment is substantially different.

To point out another difference, the acquisition of colonies is something that today's developing countries could not dream of. Yet after every war, Japan obtained foreign lands which became important sources of food. Korea, Taiwan, and Manchuria functioned as such at a time of domestic industrial development. This enabled the *Z*-sector to maintain favourable terms of trade. Investment in the *Z*-sector had to subject to the constraint of foreign exchanges. Needless to say, this does not imply that colonization is justifiable but, it did give the developing Japanese economy an expanding agricultural frontier, a situation crucially different from that of LDCs. These differences in the given environments somewhat limit the relevance of the Japanese experiences to such countries.

None the less, both the theory and empirical test suggest that there are various alternatives which contemporary LDCs can appropriately pursue. Expansion of the agricultural frontier is easily translated into technological progress in that sector, the importance of which is often neglected. It cannot be overemphasized that strengthening of primary and secondary education is of critical importance. The government interventions, though often criticized severely by the proponents of a free market, can and must play constructive roles while the market mechanism is unfavourable to new industrial sectors.

8

Classifying Industrial Sectors

As explained in Part I, our model classifies the national economy into three sectors; in the first dimension it is divided into two sectors, namely, the capitalist sector and the traditional sector, with the former being subdivided into two sectors, i.e. the *X*-sector and the *Z*-sector. These two subsectors of the capitalist sector have their own characteristics with regards to production technology and demand.

If we apply this criterion for sector classification to the Japanese economy, the subsistence sector (hereafter the '*S*-sector') seems to correspond roughly to the agricultural sector, and the capitalist sector to the industrial sector. Before World War II, there was a huge amount of disguised unemployment in the rural sector, and wage rates were low—around subsistence level. Thus, some empirical studies argue that before 1945 there was no 'turning point' in the Japanese economy as normally defined by development economists.[1]

While there is a good reason to include forestry, and fishery along with agrculture in the *S*-sector,[2] in what follows forestry and fishery are often excluded for lack of data. Since the share of these two subsectors was small throughout the period studied, this exclusion does not cause any serious problem.

The capitalist sector is divided into the *X*-sector and the *Z*-sector, both of which fall into the category of secondary industries.[3] Our task in this chapter is to apply our theory to the classification of the *X*- and *Z*-sectors. In other words, we test the assumptions about the production technology and demand characterizing individual industries against actual data in order to classify them into the two sectors. Whether the terms of trade that an industry faces are independent of the production level and whether production is subject to constant return to scale are critical factors in capital accumulation.

Our results show, first, that our tripartite classification of industries is relevant to the Japanese economy in the period 1868–1945; and secondly, that the two subsectors in the capitalist sector, namely, the *X*- and *Z*-sectors, are consistent with the two sets of assumptions we have prepared, i.e. industries can be classified according to production technology and demand conditions. If the model can thus account for the classification of Japanese industries prior to World War II, this may in turn suggest that the course of capital accumulation may be as the model predicts.

8.1. Terms of Trade

8.1.1. Terms of trade in the model

As already stated, the terms of trade of *X*- and *Z*-sector industries are the prices of the products of each sector relative to that of the *S*-sector. The most basic assumption of the model is that the terms of trade of the *X*-sector *vis-à-vis* the *S*-sector are a declining function of *X*-sector output, but that *Z*-sector terms of trade are independent of *Z*-sector output. These conditions were therefore the basis of our classification. In reality, however, a country will have numerous industrial groupings, so that to make this classification we have to disintegrate the conventional groupings and reaggregate those industries having the defined characteristics. A further problem is that when considering the relative prices of goods, we must also consider what prices we should pick up when there are numerous market places. Here we investigate the behaviour of prices both in the domestic and the international market places.

There are a few reasons we do so. First, since in reality trade barriers prevent the integration of markets, the demand curve that producers face does not necessarily represent the world demand; some producers sell only to the domestic market. While we found no contradiction between price behaviour in the two markets, the assumptions had to be made on both.

Secondly, the data for the prices of outputs of disaggregated industries generally did not differentiate between import and export prices in the pre-World War II period. This may have been because there was no significant discrepancy between domestic and world prices before Japan won the right to impose import tariffs in 1911, so that trade conditions would have been close to free trade, and domestic prices close to world prices. After 1911, however, it is suspected that the price discrepancy between overseas and domestic markets became more marked. As the import tariffs in these days were high for *X*-sector goods, domestic prices of *X*-sector goods are expected to have been significantly higher than those of the world market. Therefore, it is not sufficient to compare the production costs of *X*-sector goods with the domestic prices, because a large part of those products were sold at international price. Thus, when the export–output ratio is high, it is more appropriate to compare the production of *X*-sector goods to international terms of trade rather than to domestic terms of trade. In other words, when export accounts for a significantly large share of output, the international terms of trade should be adopted in classifying industries on the basis of a terms of trade criterion.

Thirdly, a separate investigation of domestic and international markets seems to be more relevant when we discuss the implications for contemporary LDCs although this is not a theoretical necessity. Many such countries may be termed 'monocultural economies', in the sense that they heavily depend on a few goods for their exportables. As the following analyses will clarify, for a long period Japan's exports were also restricted to a few *X*-sector goods exports, and the *X*-sector was a major component of Japanese industry. In this sense, Japan's position in international trade before 1945 was similar to that of today's developing countries, in that the economy depended on the export of *X*-sector goods. It will thus be useful to scrutinize the behaviour of Japan's export market, in particular when we examine the implications of Japan's experiences for today's LDCs.

To summarize, for the goods whose export–output ratios are high, the overseas terms of trade, or international relative prices, are appropriate variables to adopt when applying the terms of trade criterion to classifying industries into *Z*- and *X*-sectors. For other goods, domestic terms of trade will have to be considered too.

We accordingly adopted both domestic and overseas prices as variables in applying the terms of trade criterion. Our major findings are that the classification of industries following the terms of trade criterion leads to classification of the so-called light industries, such as textile and food processing industries, the *X*-sector; and classification of the so-called heavy industries, including metal and chemical industries, in the *Z*-sector. This applies for most of the period between 1868 and 1945, though there were a few short periods of exception, which will be discussed below.

8.1.2. Overseas terms of trade

In this section, we investigate the following questions that are closely related to the price behaviours of industries:

1. What were the market shares of Japanese exports in the world market by industry?
2. What were the market shares of Japanese exports in specific important overseas markets?
3. What was the relationship between terms of trade and export quantity for representative goods?
4. What were the ratios of export to output by industry?

The classification of industries into X- and Z-sectors following the results of the above procedures and the testing of the relations between terms of trade and export quantities of the aggregated *X*- and *Z*-sectors will also be discussed.

The answers to questions 1 and 2 are expected to provide indirect evidence of the relationship between the terms of trade and export quantity. If the market share is significantly large, the demand curve for the good in question is likely to be downward sloping but will be horizontal otherwise. The reason question 2 is considered is that the world market was not so integrated as it is today, and thus specific markets which were important to Japan's exports merit special attention.

Needless to say, question 3 directly tests our assumptions on individual industries. Question 4 will help us to understand which markets, domestic or overseas, were more important for specific goods, which will in turn suggest which relative price should be taken as the terms of trade variable. When the export–output ratio is low, it would seem appropriate to take relative domestic prices for testing the assumption. In fact, for most industries whose export–output ratio was low, relative domestic prices proved to be independent of production, so these industries were classified as *Z*-sector. As a final step in the analysis, we also made a direct test of the *X*- and *Z*-sectors as newly aggregated.

Table 8.1 shows the market shares of Japanese manufactured exports by industry. Japanese data were available for only four years before 1945, but Maizels (1963) gives consistent data sets for world trade in manufactured goods for these time points, and comparison of Japanese trade data with Maizels's data showed high consistency. The table gives data for four main commodity classifications—metals and machinery, chemical industries, textiles and clothes, and miscellaneous manufacturing—broken down by subsectors as appropriate.

It appears that the share of textile and clothes in total exports was significantly higher than that of other sectors, rising steadily from 3% at current prices in 1899 to about 19% in 1937. The trend in miscellaneous manufacturing followed the textile and clothes sector, but exports of heavy industrial products and chemical goods were almost negligible in the world market.

The export shares of metals, machinery, and chemical products in total Japanese exports were also small, as is apparent from the coefficients of export specialization in Table 8.2.[4] Not unnaturally, Japan was specialized in labour-intensive industries and not in the capital-intensive heavy goods and chemical industries.

As the world market was far from fully integrated before 1945, world market indicators might not be entirely relevant however. Accordingly Table 8.3 gives the market shares of Japanese industrial exports by major market. It likewise shows that textile and other light manufacturing industries accounted for relatively large shares, whereas heavy industries did not. Thus, there were no substantial discrepancies between the world market and regional markets for Japanese exports.

So far we have focused on the performance of Japanese industries in the overseas market. Several notes are in order here. First, as these data exclude raw materials and food, tea, which was a very important exportable especially in the early Meiji period, was not covered. Secondly, trade statistics of Japan before 1945 did not separate transactions with colonies from trade with foreign countries. This aspect will be discussed in the Statistical Appendix (Section 8.3). But in general it can be stated that adjustment for trade with colonies strongly supports our assumptions.

Let us now consider the relations between terms of trade (in terms of food prices) and export quantities under the above market share conditions. The assumption was that terms of trade of light industries would decline as exports

CARDIFF
UWCC LIBRARY

TABLE 8.1. Japan's share of world exports (%, f.o.b)

	At current prices							At 1913 prices			At 1955 prices				
	1899	1913	1929	1937	1950	1955	1957	1899	1913	1929	1929	1937	1950	1955	1957
Metals and engineering	0.6	0.8	0.7	2.6	2.1	3.2	3.7	0.8	0.8	0.6	0.5	3.6	2.5	3.2	3.8
Metals	1.7	1.7	0.6	2.5	4.9	6.4	3.7	2.1	1.7	0.6	0.5	3.3	5.9	6.4	3.2
Machinery	—	1.5	0.6	2.4	0.9	1.6	2.0	—	0.1	0.7	0.5	3.5	0.9	1.6	2.2
Passenger road-vehicles[a]	—	0.3	0.5	2.6	1.5	2.4	6.1	—	0.3	0.4	0.4	3.5	1.4	2.4	6.7
Other metal goods	—	0.5	1.4	4.0	2.1	3.7	3.8	—	0.5	1.3	1.0	5.0	2.4	3.7	3.2
Chemicals	0.4	1.0	1.9	3.0	0.8	2.4	2.6	0.7	1.0	2.2	1.6	4.1	0.8	2.4	2.8
Intermediates	0.5	0.7	1.8	3.1	0.6	1.5	1.4	0.5	0.7	2.0	1.6	4.4	0.6	1.5	1.5
Finished chemical	—	1.6	2.0	2.7	1.0	3.8	4.8	1.4	1.6	2.4	1.6	3.7	1.1	3.8	5.1
Textiles and clothing	2.4	4.2	9.8	18.9	8.5	15.1	18.7	2.5	4.3	9.7	8.8	21.8	8.0	15.1	18.7
Yarns	5.4	8.9	2.7	10.7	2.6	8.7	10.1	7.3	8.9	1.8	2.4	11.4	2.8	8.7	10.2
Fabrics	1.5	3.0	12.9	22.1	11.4	17.7	23.1	1.0	3.0	12.8	12.0	26.4	11.0	17.7	23.1
Made-up goods	1.9	4.0	6.7	17.5	5.3	13.8	14.8	1.2	4.0	7.9	5.3	17.7	4.4	13.8	14.8
Other manufactures	1.5	2.5	3.5	6.6	2.7	5.3	6.7	1.4	2.5	4.1	3.7	11.7	2.7	5.3	7.1
Intermediates	0.7	0.6	1.5	2.5	0.7	1.4	4.1	0.6	0.6	1.9	1.3	3.1	0.6	1.4	4.3
Finished goods	1.8	3.1	4.6	8.6	3.8	7.3	8.3	1.6	3.1	5.1	4.9	16.2	3.7	7.3	8.7
TOTAL (manufactures)	1.5	2.4	4.0	7.0	3.4	5.1	5.8	1.7	2.4	3.7	3.2	9.5	3.3	5.1	6.1

[a] includes other transport equipment.

Source: Maizels (1963).

TABLE 8.2. Japanese coefficients of export specialization

	At current prices							At 1913 prices			At 1955 prices				
	1899	1913	1929	1937	1950	1955	1957	1899	1913	1929	1929	1937	1950	1955	1957
Metals and engineering	0.413	0.343	0.177	0.380	0.638	0.635	0.629	0.482	0.343	0.171	0.172	0.378	0.751	0.635	0.617
Metals	1.087	0.715	0.157	0.351	1.450	1.253	0.639	1.272	0.715	0.165	0.164	0.343	1.799	1.253	0.523
Machinery	0	0.067	0.151	0.338	0.266	0.314	0.338	0	0.067	0.176	0.152	0.363	0.258	0.314	0.355
Passenger road-vehicles[a]	0	0.130	0.129	0.371	0.444	0.464	1.050	0	0.130	0.107	0.117	0.361	0.430	0.464	1.092
Other metal goods	0	0.200	0.356	0.569	0.653	0.729	0.636	0	0.200	0.356	0.301	0.524	0.709	0.729	0.512
Chemicals	0.253	0.429	0.476	0.425	0.229	0.474	0.455	0.440	0.429	0.585	0.500	0.434	0.247	0.474	0.458
Intermediates	0.339	0.317	0.460	0.452	0.179	0.294	0.239	0.291	0.317	0.551	0.513	0.455	0.196	0.294	0.250
Finished chemical	0	0.690	0.500	0.386	0.292	0.756	0.821	0.842	0.690	0.636	0.483	0.406	0.317	0.756	0.833
Textiles and clothing	1.591	1.821	2.439	2.726	2.538	2.955	3.195	1.480	1.821	2.631	2.773	2.289	2.444	2.955	3.081
Yarns	3.467	5.344	0.680	1.561	0.784	1.680	1.762	4.430	3.781	0.467	0.744	1.205	0.871	1.680	1.682
Fabrics	0.987	1.264	3.206	3.163	3.402	3.481	3.926	0.628	1.264	3.456	3.759	2.766	3.360	3.481	3.803
Made-up goods	1.238	1.708	1.667	2.500	1.590	2.688	2.517	0.710	1.708	2.145	1.660	1.860	1.316	2.688	2.467
Other manufactures	1.000	1.043	0.886	0.944	0.812	1.039	1.158	0.820	1.043	1.097	1.159	1.228	0.811	1.039	1.166
Intermediates	0.447	0.250	0.362	0.359	0.190	0.279	0.698	0.348	0.250	0.500	0.403	0.323	0.193	0.279	0.719
Finished goods	1.161	1.318	1.141	1.229	1.125	1.432	1.426	0.961	1.318	1.370	1.540	1.702	1.125	1.432	1.445

[a] includes other transport equipment.

Source: Maizels (1963).

TABLE 8.3. Japan's industrial exports to major markets (%)

			Industrial countries				Semi-industrial countries				USSR	Rest of world	World total
			United Kingdom	Continental Western Europe	North America	Total (including Japan)	Southern dominions	India (including Pakistan)	Other semi-industrial countries	Total			
Metals and engineering		1899	2.2	0.6	0	0.9	—	—	—	—	—	9.4	2.1
	a	1913	2.8	0.6	2.4	1.3	—	2.0	—	0.6	—	4.4	1.7
		1929	0.8	0	0	0.1	0	1.8	—	0.5	0	2.1	0.6
		1929	0.6	0	0.2	0.2	0	1.8	—	0.5	1.9	1.5	0.5
	b	1937	0.2	0.2	0.4	0.2	0.7	13.6	1.0	2.8	1.7	10.2	3.3
		1950	0	2.0	6.7	3.7	11.6	9.4	7.2	9.0	—	7.9	5.9
		1955	2.7	0.9	3.6	2.0	8.5	29.1	18.3	17.4	0	9.8	6.4
Chemicals	a	1899	0	0	0	0	0	—	—	0	0	3.4	0.7
		1913	1.6	0.6	1.2	0.9	0	5.9	—	1.3	0	1.4	1.0
		1929	0	1.6	3.3	1.6	0	3.2	0	0.9	0	4.0	2.2
		1929	0.7	0.7	2.7	1.1	1.5	3.3	0	1.2	0	2.8	1.6
	b	1937	1.7	1.8	4.1	2.2	1.3	7.5	1.5	2.9	0	7.1	4.1
		1950	0	0.4	0.7	0.4	0	1.0	0.6	0.6	—	1.5	0.8
		1955	1.4	0.4	1.3	0.8	1.5	2.8	1.2	1.6	—	5.0	2.4

Textiles and clothing		1899	0.4	1.0	2.9	1.2	0	0.6	—	0.3	0	6.7	2.5
	a	1913	2.7	1.8	3.7	2.5	1.6	1.8	0.4	1.2	3.2	10.3	4.3
		1929	3.5	1.1	5.0	2.8	9.4	29.4	1.8	12.4	20.0	16.8	9.7
		1929	3.5	1.1	4.4	2.6	7.0	22.8	1.4	10.2	30.0	14.9	8.8
	b	1937	13.7	4.2	14.6	9.2	14.6	57.7	11.0	22.9	—	30.1	21.8
		1950	8.5	2.1	9.3	5.1	5.1	34.0	1.2	9.3	—	9.9	8.0
		1955	6.2	2.3	23.2	10.2	8.3	29.5	6.1	10.2	—	21.5	15.1
Other manufactures		1899	0.5	0	2.1	0.6	0	3.0	0	0.6	0	2.7	1.0
	a	1913	1.1	0.9	3.5	1.5	0.8	3.7	0	0.8	0	4.7	2.0
		1929	1.1	0.8	3.8	1.9	1.3	16.3	0.5	2.6	4.8	6.8	3.4
		1929	1.1	0.7	3.1	1.6	1.3	10.2	0.5	2.3	3.4	5.8	2.9
	b	1937	5.3	2.3	10.3	6.0	7.3	24.7	3.6	9.1	9.4	16.5	10.0
		1950	0	0.5	4.3	2.6	1.0	3.1	1.4	1.5	0	3.2	2.6
		1955	1.9	1.0	8.1	4.7	2.8	6.2	2.6	3.2	0	6.3	4.9

Note: Southern dominions consist of Australia, New Zealand, and Union of South Africa.

[a] at 1913 constant prices.
[b] at 1955 constant prices.

Source: Calculated from Maizels (1963).

increased, whereas those of heavy industries would remain constant, or independent of exports. The prices of food, more precisely, the *S*-sector or prices used below are indices of prices of 'cultivated agricultural products' (*koshu nosanbutu kakaku shisu*), the major components of which are cereal crops and vegetables.

Figs. 8.1–8.5 plot the relationship between the terms of trade against the *S*-sector and the quantity of export for some representative goods as listed in *Nihon Boeki Seiran* (Detailed Statistics of Japan's Foreign Trade). The products considered are: refined sugar, silk, cotton yarn, cotton fabric, silk products, hosiery products, coal and steel. The terms of trade were calculated using the unit price index of export items (year 1913 = 100), and the export quantity was estimated by dividing the value of export for each product by the price index and converting the result into indices using 1913 as the base year. The figures indicate that the terms of trade for raw silk, silk products, and cotton yarn declined as their exports expanded. On the other hand, the terms of trade for cotton fabrics and steel remained independent of export trends. This supports the results of the investigation into market shares, namely that most textile industries show *X*-sector characteristics, and their terms of trade behaved as such except for cotton fabrics. In the late nineteenth century the United

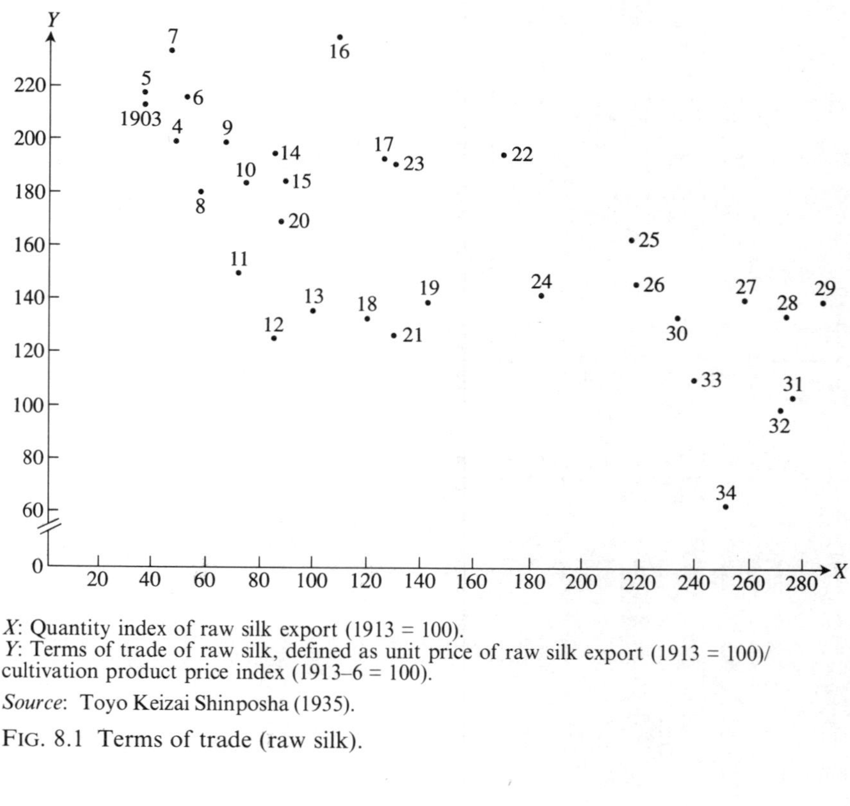

X: Quantity index of raw silk export (1913 = 100).
Y: Terms of trade of raw silk, defined as unit price of raw silk export (1913 = 100)/ cultivation product price index (1913–6 = 100).

Source: Toyo Keizai Shinposha (1935).

FIG. 8.1 Terms of trade (raw silk).

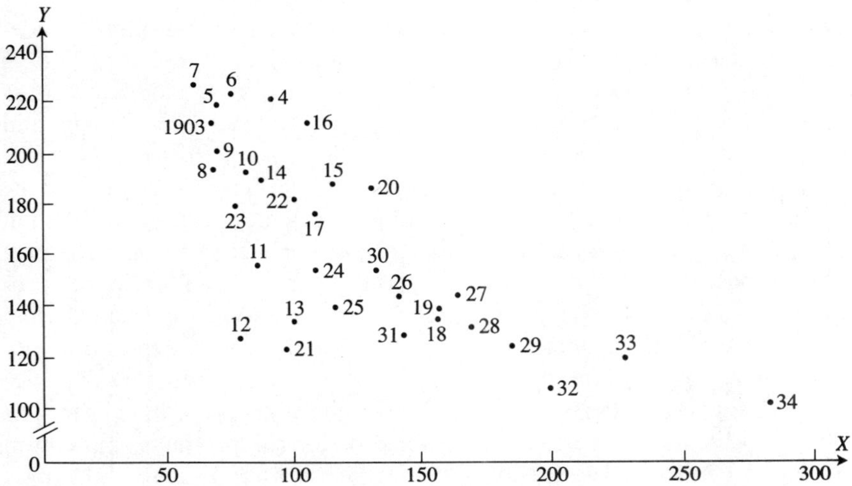

X: Quantity index of silk product export (1913 = 100).
Y: Terms of trade of silk products, defined as unit price of silk products export (1913 = 100) / cultivation product price index (1934–6 = 100).

Source: Toyo Keizai Shinposha (1935).

FIG. 8.2 Terms of trade (silk products).

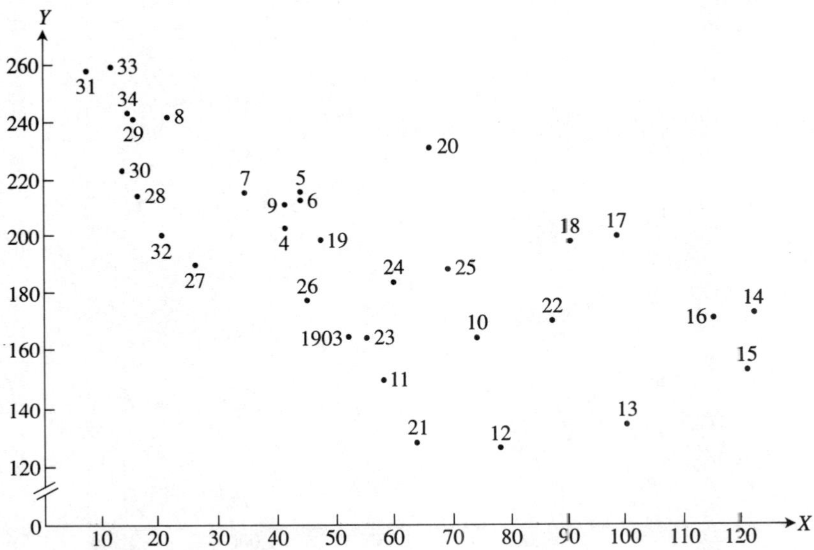

X: Quantity index of cotton yarn export (1913 = 100).
Y: Terms of trade of cotton yarns defined as unit price of cotton yarns export (1913 = 100) / cultivation product price index (1934–6 = 100).

Source: Toyo Keizai Shinposha (1935).

FIG. 8.3 Terms of trade (cotton yarns).

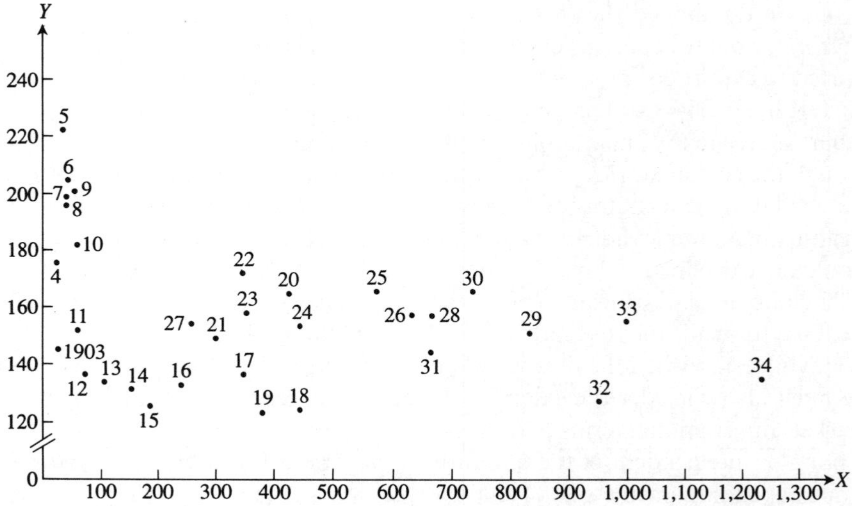

X: Quantity index of cotton fabric export (1913 = 100).
Y: Terms of trade of cotton fabric, defined as unit price of cotton fabric export (1913 = 100) / cultivation product price index (1934-36 = 100).

Source: Toyo Keizai Shinposa (1935).

FIG. 8.4 Terms of trade (cotton fabric).

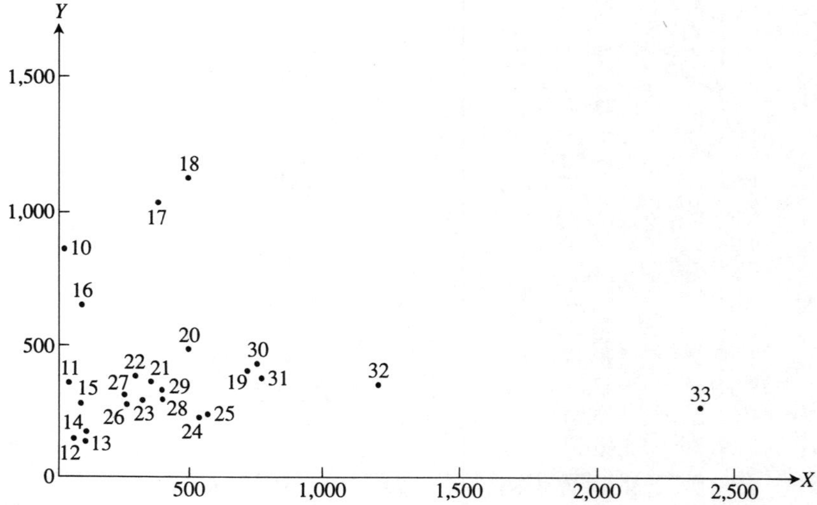

X: Quantity index of iron export (1913 = 100).
Y: Terms of trade of iron defined as unit price of iron export (1913 = 100) / cultivation product price index (1934–6 = 100).

Source: Toyo Keizai Shinposha (1935).

FIG. 8.5 Terms of trade (iron).

Kingdom dominated the world market for cotton fabrics, so Japan had no alternative but to reduce prices in order to penetrate the market. In fact, Japan started to export cotton fabrics in 1910, to India, and subsequently overtook the UK in South-East Asian markets by price reduction. Thereafter, however, Japan succeeded in maintaining its prices through grade improvement. Although the cotton fabrics sector was thus somewhat peculiar, we nevertheless included it in the X-sector together with the other textile subsectors because if an adjustment were made for quality improvement, the terms of trade here too would have declined as Japan's export share expanded.

The above analysis leads us to a tentative classification: textile industries are X-sector while the steel industry is Z-sector.

Having established the overseas demand structure for Japanese industries, we must also consider the extent to which Japan depended on the overseas market. International terms of trade may be negative with respect to Japanese exports or independent of them, but the implications for domestic producers differ depending on the size of the overseas market relative to the domestic market. Domestic demand will be investigated in Section 8.1.2.

Table 8.4 shows the relative importance of domestic and overseas markets for selected products. The export–output ratio was high for tea (green tea), raw silk, cotton, and silk fabrics—except for the latter, around 50%. In contrast, the ratio for steel was negligible. Indeed, Japan started to export steel

TABLE 8.4(*a*). Japan's and world production of cotton yarn (tons)

	(1) Japan's production	(2) Japan's exports	(3) World production	(2)/(1)%	(2)/(3)%	(1)/(3)%
1884	16,471					
1890	21,822	6		—		
1895	69,139	2,120		3.1		
1900	121,574	37,572		30.9		
1905	165,517	48,129		29.1		
1910	—	60,701		—		
1914	329,312	99,174		30.1		
1920	299,414	53,890		18.0		
1926	412,212	55,944		13.6		
1930	451,912	10731		2.4		
1935	651,861	17,385		2.7		
1937	736,140	23,351	2,302,680	3.2	1.0	32.0
1940	439,419	27,646	1,230,720	6.3	2.2	35.7

Note: Figure for 1937 includes UK and Germany and that for 1940 excludes the two countries.

Sources: Columns (1), (2) Nippon Ginko (1966); (3) Toyo Keizai Shinposha, 3 (1950): 741 (annual data converted from monthly averages).

TABLE 8.4(*b*). Japan's production and export of selected commodities (¥m.)

	Cotton fabrics			Silk and silk fabrics			Tea		
	(1) production	(2) exports	(2)/(1)% %	(3) production	(4) exports	(4)/(3)% %	(5) production	(6) exports	(6)/(5)% %
1910	122	22	18.0	142	33	23.2	14	15	107.1
1915	182	41	22.5	149	43	28.9	16	15	93.8
1920	694	340	49.0	560	158	28.2	34	17	50.0
1925	772	437	56.6	491	117	23.8	36	15	41.7
1930	493	285	57.8	425	66	15.5	24	8	33.3
1935	810	503	62.1	633	77	12.2	23	11	47.8
1940	595	407	68.4	635	38	6.0	72	25	34.7

Note: Exports of tea include re-exports from Taiwan.
Sources: Shinohara (1967): 220, 228–9

TABLE 8.4(*c*). Japan's production and export of iron ('000 tons)

	(1) Japan's production	(2) Japan's exports	(3) World production	(2)/(1)%	(2)/(3)%
1885	0		n.a.		
1890	2		n.a.		
1895	2		n.a.		
1900	2		27,183		
1905	72		43,906		
1910	168		58,260		
1915	343	29	66,613	8.5	0.04
1920	533	97	72,349	18.2	0.10
1925	1,037	108	90,374	10.4	0.10

Sources: (1), (2), (3) Asahishimbun (1930): 1200, 1259.

TABLE 8.4(*d*). Japan's and world production of raw silk (tons)

	(1) Japan's production	(2) Japan's exports	(3) World production	(2)/(1)%	(2)/(3)%
1868		727	6,427		11.3
1870		408	7,662		5.3
1875		697	9,592		7.3
1880	1,999	1,145	10,577	57.3	10.8
1885	1,905	1,379	9,043	72.4	15.2
1890	3,458	2,018	11,411	58.4	17.7

Table 8.4d (*cont.*)

	(1) Japan's production	(2) Japan's exports	(3) World production	(2)/(1)%	(2)/(3)%
1895	6,410	3,486		54.4	—
1901	7,069	5,219	19,229	73.8	27.1
1905	7,309	4,345	18,830	59.4	23.1
1910	11,904	8,802	24,495	73.9	35.9
1915	15,172	10,647	23,665	70.2	45.0
1920	21,877	10,437	20,845	47.7	50.1
1925	31,066	26,171	40,675	84.2	64.3
1930	42,619	28,194	58,890	66.2	47.9
1935	43,733	33,189	64,130	75.9	51.8
1940	42,768	17,621	58,890	41.2	29.9

Note: World production data are summation of those of Italy, Austria, France, Spain, Central Asia, Middle East, India, and China plus Japanese exports.

Sources: Column (1) Nippon Ginko (1966); (2) Matsui (1963): 174; (3) Asahishimbun (1958): 1265; Toyo Keizai Shinposa, 3 (1950): 295.

TABLE 8.4(*e*). Japan's production and export of vessels (¥'000)

	Production	Exports	Exports/ Production %
1909	16,788	333	2.0
1914	18,327	734	4.0
1919	333,917	3,628	1.1
1920	301,239	15,829	5.3
1921	217,268	3,492	1.6
1925	96,274	2,040	2.1
1926	99,755	389	0.4
1927	96,830	1,404	1.5
1928	84,830	2,831	3.3
1929	126,297	6,390	5.1
1930	174,153	5,451	3.1
1931	87,901	3,960	4.5
1932	92.709	7,487	8.1
1933	118,404	1,724	1.5
1934	153,379	1,112	0.7
1935	205,869	1,289	0.6

Sources: MITI (1962); Nippon Ginko (1966).

TABLE 8.4(*f*). Japan's production and exports of earthenware and porcelain (¥'000)

	(A) Production	(*A'*) Production Exports	(B)	(B)/(A)%	(B)/(*A'*)%
1926	215,186	73,970	33,182	15.4	44.9
1927	196,937	74,363	30,491	15.5	41.0
1928	207,158	76,726	34,642	16.7	45.2
1929	229,984	74,767	36,962	16.1	49.4
1930	174,149	62,419	27,171	15.6	43.5
1931	155,452	54,198	19,307	12.4	35.6
1932	176,566	65,263	22,937	13.0	35.1
1933	237,543	85,247	35,634	15.0	41.8
1934	275,544	92,364	41,877	15.2	45.3
1935	312,705	99,368	42,736	13.7	43.0
1936	357,358	108,172	243,19	212.1	39.9
1937	443,887	115,191	55,660	12.5	48.3
1938	481,252	110,008	36,880	7.7	33.5
1939	648,821	164,703	46,233	7.1	28.1
1940	777,573	196,144	66,578	8.6	33.9

A: ceramic, stone and clay products.
A': earthenware and porcelain.

Sources: Toyo Keizai Shinposha (1950): 539; MITI (1962).

TABLE 8.4(*g*). Japan's production and export of textile machines (¥'000)

	Production	Exports	Imports	Exports/ Production %
1909	1,050		—	—
1914	879	421	6,256	47.9
1919	21,914	3,393	16,185	15.5
1920	21,371	3,389	20,570	15.9
1921	21,010	4,431	34,604	21.1
1922	11,481	5,035	33,612	43.9
1923	16,924	3,801	25,364	22.5
1924	17,519	3,590	14,294	20.5
1925	19,360	3,454	9,437	17.8
1926	18,883	2,624	9,645	13.9
1930	21,222	3,852	7,824	18.2
1935	86,016	12,547	4,613	14.6
1940	51,988	23,000	—	44.2
1945	29,283	9,217	—	31.5

Source: Arisawa (1959).

only in the late 1910s, some ten years after Yawata Steel (the government steel mill that was one of the predecessors of today's Nippon Steel) commenced operation. In the field of machinery manufacture, only textile machinery developed rapidly because this sector had two large demand sources: the domestic textile sector, and textile sectors in South and East Asia. The ship-building industry was the next to develop, and these two subsectors expanded their shares in the machinery sector; however, their export–output ratios never became as high as that of the textile industry.

To sum up our findings, textile and miscellaneous manufacturing sectors (mostly sundry goods) enjoyed a fairly large world market share, while Japan's share of the world market for metals, machinery and the chemical industries was negligible. The terms of trade of the textile sectors were negative with respect to their export quantity, whereas those of the steel sector were independent of exports. The export dependency of tea and textile products was significantly high while that of steel and machinery was extremely low.

As regards overseas terms of trade, we can therefore classify industries into *X*- and *Z*-sectors in the following way:[5]

X-*sector*: food-processing and tobacco, textile yarn, fabrics and clothes of various materials, leather and related products, lumber and lumber products, paper and paper manufactures, miscellaneous manufactures.

Z-*sector*: oil and fat products, chemistry, mining, metal manufactures, machinery.

The industrial classification was based on data compiled by Ohkawa *et al.* (1967) which, in turn, were based on *Nihon Boeki Seiran*. The relationship between terms of trade and exports for *X*- and *Z*-sectors are plotted in figs. 8.6 and 8.7. It is clear that the terms of trade (measured along the vertical axis) decline as real exports (measured along the horizontal axis) expand. For the *Z*-sector, there were some subperiods when terms of trade declined in the short run as export quantities increased, but only around 1900: after that, an expansion of exports did not lead to a decline in the terms of trade. We can therefore conclude that the demand curve is horizontal.

8.1.3. Domestic terms of trade

So far we have studied the relationship between relative prices in the international market and export quantity by industry and classified industries into *X*- and *Z*-sectors. In this section, we compare the domestic terms of trade between industries with the volume of production, using the industrial classification made in the preceding section. We observe that the domestic terms of trade of the *X*-sector *vis-à-vis* the *S*-sector are negatively related to production, while those of the *Z*-sector are independent of production. Details of the analysis are given below.

In order to relate terms of trade to production, we need statistics on production

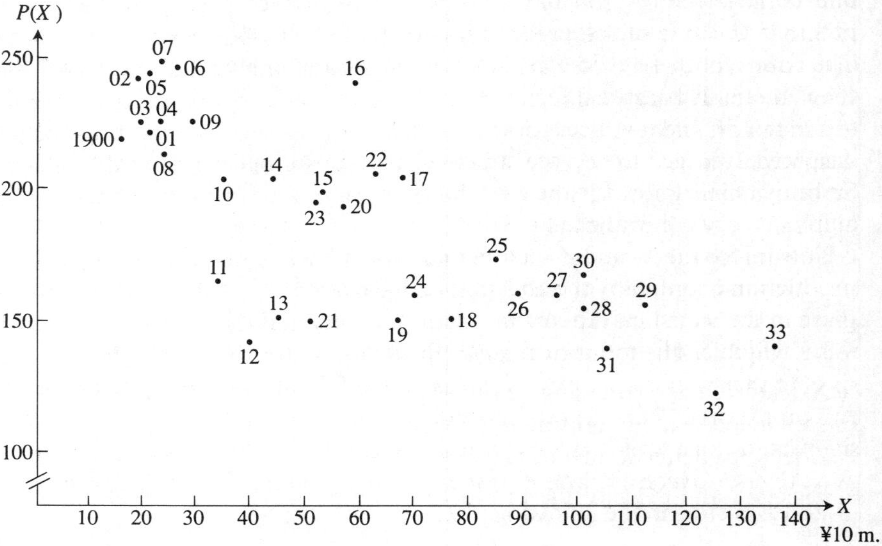

X: Export value of *X*-sector (¥.m. at 1934–6 constant prices).
P(*X*): Index of export price of *X*-sector / cultivation product price index.

FIG. 8.6 Terms of trade of *X*-sector measured against exports.

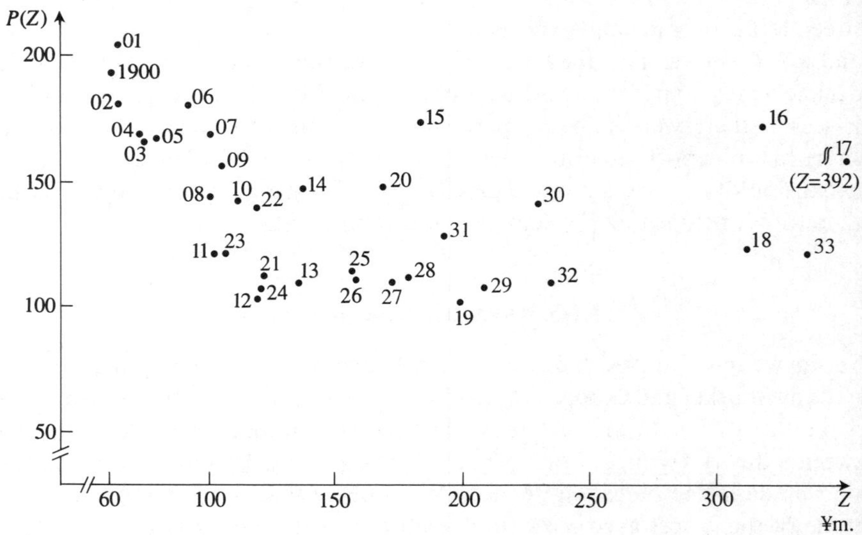

Z: Export value of *Z*-sector (¥m. at 1934–6 constant prices).
P(*Z*): Index of export prices of *Z*-sector / cultivation product price index.

FIG. 8.7 Terms of trade of *Z*-sector measured against exports.

and corresponding price data by industry which are consistent with each other. As the basic statistics for this work, we use Shionoya's (1966) estimates. The nature of these estimates and the way we aggregate them is explained in the Statistical Appendix (Section 8.3).

First of all, price indices for the *X*- and *Z*-sectors were calculated from the weighted average of the price indices of individual industries as estimated by Shionoya. The index for the *S*-sector was calculated from the price index of cultivated agricultural products, as described in the Section 8.1.1. Production quantities for the *X*- and *Z*-sectors were estimated by deflating the aggregated production quantity by the above price deflators. Details of the procedure are given in the Statistical Appendix (Section 8.3).

Individual industrial sectors were classified into the two sectors as below:

X-*sector*: food-processing, textiles, lumber and lumber products, ceramics, miscellaneous manufacturing.

Z-sector: chemical, metal, machinery industries.

Figs. 8.8 and 8.9 plot the production and terms of trade for each sector.

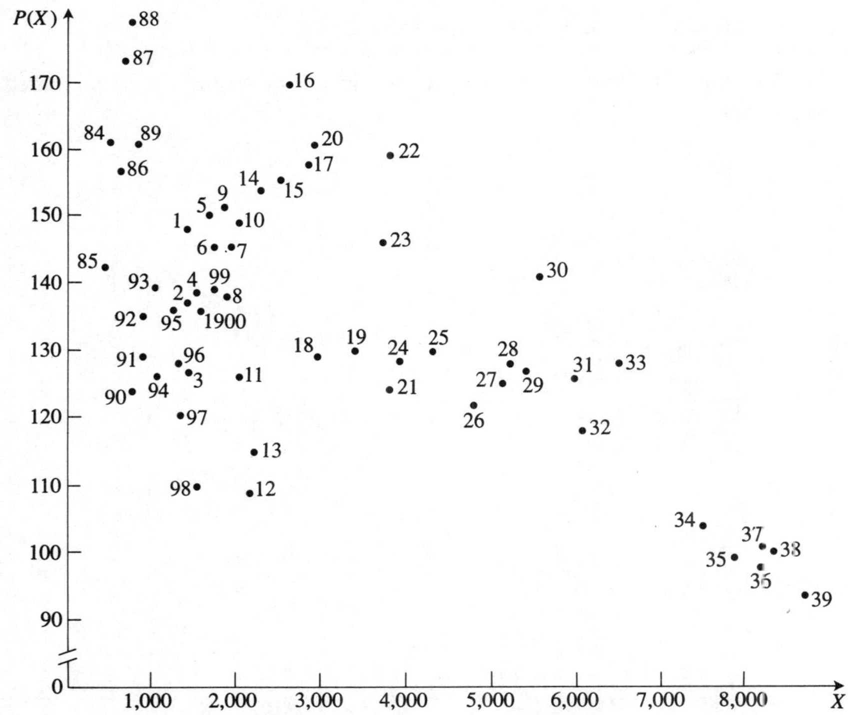

X: Production value of *X*-sector (¥m. at 1934–6 constant prices).
P(*X*): Index of production price of *X*-sector / cultivation product price index.

FIG. 8.8(i) Terms of *X*-sector measured against domestic production (based on original series).

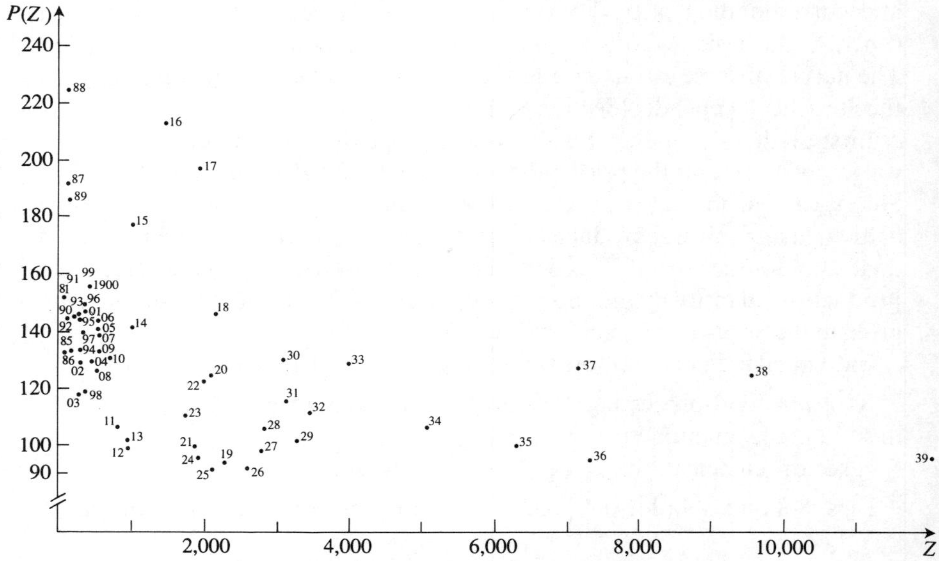

Z: Production value of Z-sector (¥m. at 1934–6 constant prices).
$P(Z)$: Index of production price of Z-sector / cultivation product price index.

FIG. 8.8(ii) Terms of trade of Z-sector measured against domestic production (based on original series).

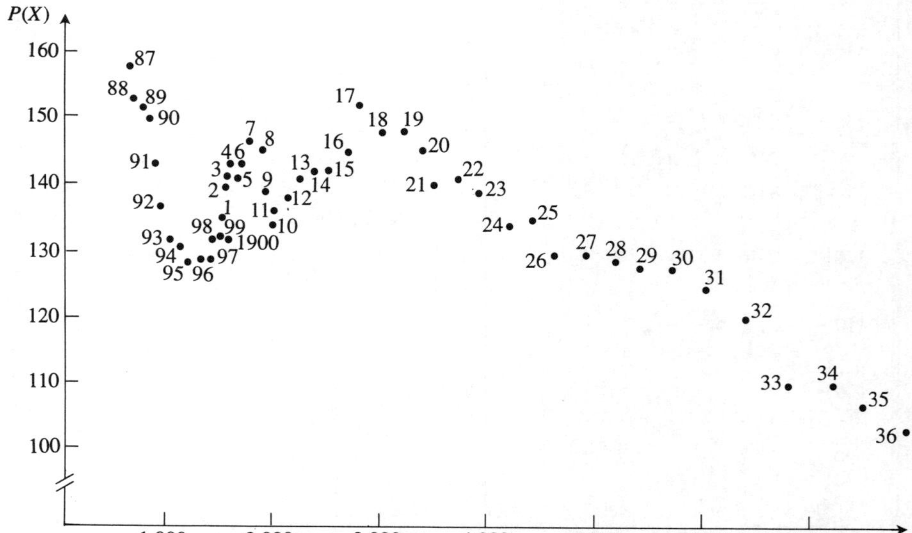

X: Production value of X-sector (¥m. at 1934–6 constant prices).
$P(X)$: Index of production price of X-sector / cultivation product price index.

FIG. 8.9(i) Terms of trade of X-sector measured against domestic production (based on 7-year moving average series).

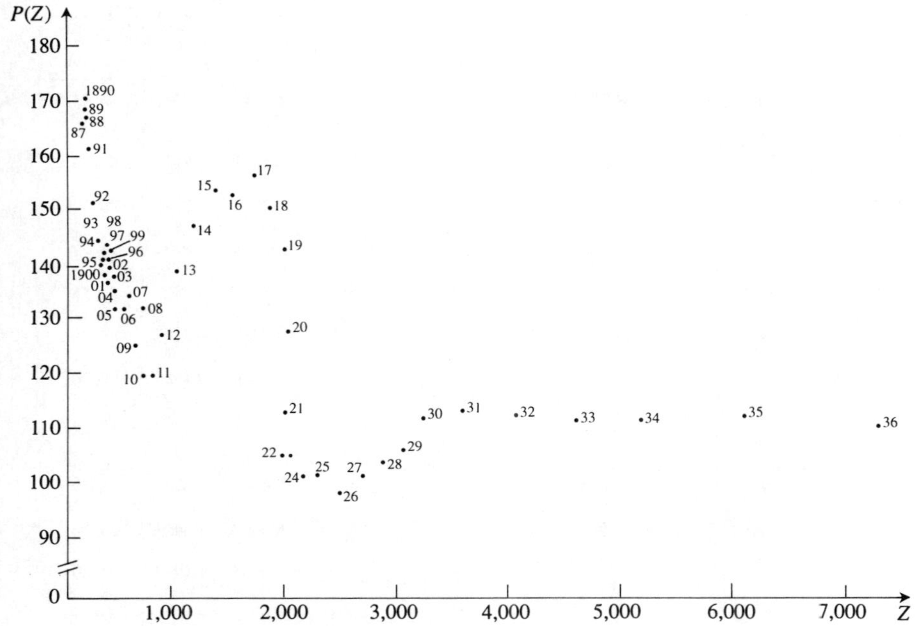

Z: Production value of *Z*-sector (¥m. at 1934–6 constant prices).
P(*Z*): Index of production price of *Z*-sector / cultivation product price index.

FIG. 8.9(ii) Terms of trade of *Z*-sector measured against domestic production (based on 7-year moving average series).

They show clearly that in the *X*-sector, terms of trade declined steadily as production expanded. For the *Z*-sector, the relationship is not so clear until 1920, but thereafter terms of trade are independent of production, although production reached a level several times higher than in the 1920s.

To test more strictly, we regressed the terms of trade variable on production. The result was that *X*-sector terms of trade were negatively related to production for the period 1915–39, and that the parameter of the production variable for the Z-sector was not significantly different from zero for the period 1921–39. We can therefore conclude that the assumptions about terms of trade in our theoretical model are supported at least for the years 1921–39. Let us look at the details of statistical analysis.

Estimation results of various forms during various periods are summarized in Table 8.5. For the *X*-sector, terms of trade are definitely negatively related to production for the entire period (1884–1939). The parameter of production variable is negative and statistically significant at a 5% level. For the *Z*-sector, on the other hand, we cannot reject statistically the hypothesis that the parameter for that period is negative. Intuitively this can be understood by looking at the preceding figures. For some fifty years, *Z*-sector terms of trade certainly

TABLE 8.5 Regression Analysis of terms of trade

	Constant term	Coefficient of production	$\bar{R}$	*S*	*d.w.*	Period
X-sector	150.77	–0.0053 (–6.0617)*	0.6276	15.4294	1.0240	1884–1939
Linear-regression original series	169.15	–0.0277 (–2.6881)*	0.5292	16.2317	1.2196	84–00
	123.68	–0.0075 (1.4480)	0.2050	15.2672	1.2156	95–20
	177.08	–0.0095 (–8.9000)*	0.8747	10.4783	2.0709	15–39
Moving average (7 years)	146.75	–0.0040 (–7.2052)*	0.7138	7.8062	0.1550	87–36
	170.20	–0.0296 (–5.9486)*	0.8518	5.4382	0.3103	87–00
	123.09	0.0077 (5.1668)*	0.7119	4.5165	0.4686	95–20
	168.57	–0.0079 (–2.0614)	0.9761	2.9248	0.8460	15–36
Log-linear regression original series	2.4934	–0.1101 (–5.232)*	0.5693	0.0540	0.8757	84–39
	2.6963	–0.1852 (–2.704)*	0.5320	0.0489	1.3422	84–00
	1.7638	0.1143 (1.426)	90.1993	0.0488	1.1944	95–20
	3.5012	–0.3801 (–9.043)	0.8780	0.0355	1.6471	15–39
Moving average (7 years)	2.3943	–0.0788 (–5.547)*	0.6147	0.0295	0.1187	87–36
	2.8446	–0.2337 (–7.441)*	0.8983	0.0140	0.3731	87–00
	1.7390	0.1224 (5.568)*	0.7386	0.0136	0.4735	95–20
	3.1572	–0.2854 (–12.965)*	0.9425	0.0154	0.4019	15–36
Z-sector	142.07	–0.0052 (–3.393)*	0.400	27.77	0.6550	1884–1939
Linear regression original series	169.74	–0.0661 (–2.656)*	0.434	20.52	1.1123	84–10
	121.61	0.0125 (0.555)	0.0	42.40	0.7235	10–20
	106.76	0.00026 (0.232)	0.0	13.64	1.5567	21–39
Moving average (7 years)	144.40	–0.0078 (–5.771)*	0.630	15.54	0.1399	87–36
	173.52	–0.0766 (–10.060)*	0.901	5.89	0.3836	87–10
	117.78	0.0160 (1.998)	0.480	12.19	0.4306	10–20
	101.79	0.0018 (2.536)*	0.515	4.39	0.5878	21–36

Notes: $\bar{R}$: coefficient of correlation modified by degree of freedom; S: standard error of estimates.

t-values are in parentheses.

*statistically significant at 5% level.

declined as production increased significantly, but after 1920 they became independent of production.

In other words, in the period between 1884 and 1920, the terms of trade changed for reasons other than the level of production. The *Z*-sector in the earlier period was not competitive in the international market-place, and was subsidized by the government. Therefore, the terms of trade deteriorated as production expanded. This was behind the apparent negative relation between *Z* and P_Z/P_S. This in turn brought about price decreases, while efficiency improved as the result of a learning process. To some extent, a similar argument holds for the *X*-sector. We therefore tried to divide the entire period into three subperiods and applied the same test.

With respect to the X-sector, the three subperiods designated were 1884–1900, 1895–1920, and 1915–39. Three wars (Japan–China (Ming Dynasty) War (1894–5), Japan–Russo War (1904–5), and World War I (1914–18)) might have affected the terms of trade so significantly as to make the regression analysis for the entire period 1884–1939 ambiguous. As the preceding figures show, the prices of industrial goods rose significantly during war periods. So we divided the period into three subperiods to test our assumptions on the terms of trade: 1884–1900, 1895–1920, and 1915–39. As three wars are included during the second period, the movements of the price variables were much influenced by them. The overlapping of some years was to increase the number of the sample. The parameter of the production variable used in regressing the terms of trade on production was negative, with a significance level of 5% level a *t*-test. The *F*-test for correlation between terms of trade and production shows that the relation is significant at a 5% level.[6] For the merged period of 1884–1920, the parameter was positive, but the *t*-test indicated that it was not significantly different from zero at a 5% significance level.

For the Z-sector, three different subperiods were used: 1884–1910, 1910–20, and 1921–39. For the period 1884–1910 it turned out that the hypothesis of negative relations between terms of trade and production could not be re jected. For the periods 1910–20 and 1921–39 the parameter was positive, but the *t*-test showed that it did not differ from zero at a 5% significance level, and the *F*-test showed that the positive relation was not significant. Thus we can conclude that *Z*-sector terms of trade are independent of production for the period 1910–39. Concerning the period 1910–20, it is to be noted that the First World War raised the terms of trade for the *Z*-sector drastically; in other words, exogenous factors dominated the relations.

From the above investigation, we can conclude that *X*-sector terms of trade declined (rose) as production increased (decreased) in the period 1915–39, and that *Z*-sector terms of trade were independent of production in the period 1921–39. Therefore the period which satisfies our terms of trade assumptions completely is 1921–39.

Since the terms of trade assumptions were satisfied even when a strict criterion was applied for the period 1921–39, the classification into *X*- and *Z*-sector industries was completely supported for this period. As this is the critical period in the take-off, we shall investigate the process of capital allocation and accumulation during this period more rigorously.

8.2. Production Technology

8.2.1. Production technology in the model

Our development model assumes that the X- and Z-sectors have distinct production technologies, although both use capital and labour as production factors. In the theoretical model, the production function is general rather than the more specific CES– or Cobb–Douglas functions, which would be more restricting and therefore inferior for our purposes. Instead, our only assumptions are that the X-sector is subject to constant return to scale and the Z-sector to increasing return to scale. These assumptions, combined with the assumptions about terms of trade for the respective sector, are critical in determining changes in the profit rate. To paraphrase, the relationship between profit rate and capital stock in each sector depends not only on production technology but also on the terms of trade assumptions.

As mentioned in Part I, the assumptions that the X-sector's terms of trade decline as production expands and that its production function is one of homogeneity of degree are only sufficient conditions to generate a declining profit rate as capital accumulation proceeds. In other words, the profit rate can decline if X-sector terms of trade deteriorate as production expands, despite increasing returns to scale in production. However, we shall test our assumptions in a very strict way, i.e. whether returns to scale are constant in the X-sector and increasing in the Z-sector.

8.2.2. Methods of testing

It is difficult to test these assumptions without specifying production functions. A sector has technology giving constant returns to scale if, for a given positive figure λ, production expands by λ, when the input of all production factors increases by λ. However, as it is difficult strictly to test this in a general form on observable data only, we adopt the Cobb–Douglas function for testing our assumptions.

The steps we take can be summarized as follows. The production function is expressed as:

$$Y = CK^{\alpha} L^{\beta}, \tag{8.1}$$

where Y, K, and L stand for production, capital stock, and labour input, respectively. The superscripts α and β are the parameters which express production technology, and C represents a parameter for unit conversion.[7]

In order to avoid multicollinearity in measuring the function, we transform (8.1) into:

$$Y/L = C(K/L)^{\alpha} L^{\alpha+\beta-1}. \tag{8.2}$$

We make time-series data for Y/L, K/L, and L, and apply a simple least-square estimation of the logarithm of (8.2). Let us express the result as:

$$y = C\, k^{\alpha}\, L^{\lambda},$$

where $y = Y/K$, $k = K/L$, respectively, and α, λ, and C are estimated parameters. As $\lambda = \alpha + \beta - 1$, testing of $\lambda = 0$ is equivalent to testing $\alpha + \beta = 1$, which implies constant returns to scale. Here, however, α and β, which are calculated from α and λ later, must satisfy $0 < \alpha < 1$ and $0 < \beta < 1$. That is, the marginal productivity of the respective production factor is diminishing.

8.2.3. Estimated production functions

The above method was used to test our technological assumptions in the respective industrial sectors. Ideally, the test should be applied to disaggregated data, and the X- and Z-sectors aggregated only after testing. In practice this is extremely difficult, because the industrial classification is not consistent among all the sectors for which we obtain data on production, capital stock, and labour input. Data on capital stock are rare in any country, especially for the early period of economic development. We therefore, tested on the X- and Z-sectors as classified in the preceding section based on terms of trade conditions. In reality, some sectors may not satisfy the terms of trade conditions and the technology conditions at the same time. If an individual sector does not satisfy the conditions and the effect of this is significant, it will affect the aggregated production function and the relationship between profit rate and capital stock. For this reason we take a simpler approach to testing technology assumptions, using sectors aggregated according to the terms of trade criteria.

First, we estimate production functions for the X- and Z-sectors for the period 1919–40, with the following results:

$$\log \frac{Y_{X_1}}{L_{X_1}} = -\,2{,}8987 + \underset{(6.3948^{*})}{0.8128} \log \frac{K_{X_1}}{L_{X_1}} + \underset{(0.6300)}{0.1794} \log L_{X_1}$$

$$\bar{R} = 0.9278 \qquad S = 0.0391 \qquad d.w. = 0.849$$

$$\log \frac{Y_{X_1}}{L_{Z_1}} = -\,2.77325 + \underset{(7.4858^{*})}{0.8206} \log \frac{K_{Z_1}}{L_{Z_1}} + \underset{(4.5453^{*})}{0.1706} \log L_{Z_1}$$

$$\bar{R} = 0.8497 \qquad S = 0.0385 \qquad d.w. = 1.254.$$

Here, Y_i, K_i, and L_i are real production, capital stock, and labour of the i-th sector ($i = x, z$), respectively. $\bar{R}$, S, and $d.w.$ are respectively, the correlation coefficient adjusted for the degree of freedom, the standard error of estimates, and the Durbin–Watson ratio, which indicates the degree of auto-correlation of the residual of estimated equations. The figures in parentheses under the estimated parameters are the t-values obtained in testing significance; an asterisk indicates that they are significant at the 5% level. This statistical notation applies to all the expressions below. Last, the subscript 1 indicates that in

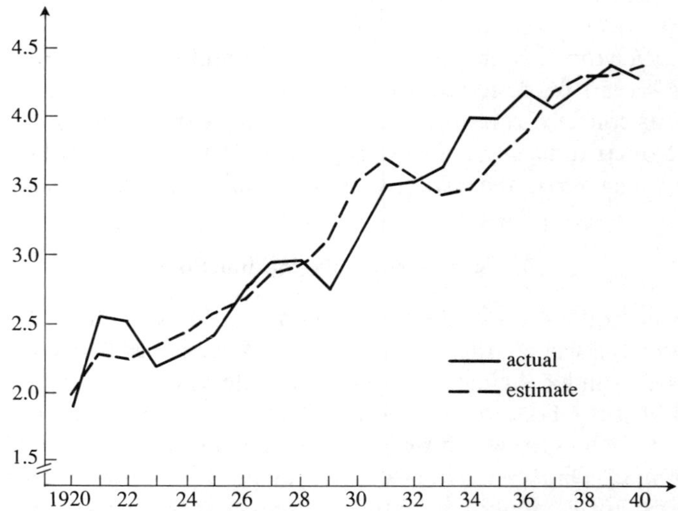

Note: Unit = real production per employee (1,000 yen at 1934–6 constant prices/person).

FIG. 8.10 Estimated performance of production function of X-sector (pre-war period).

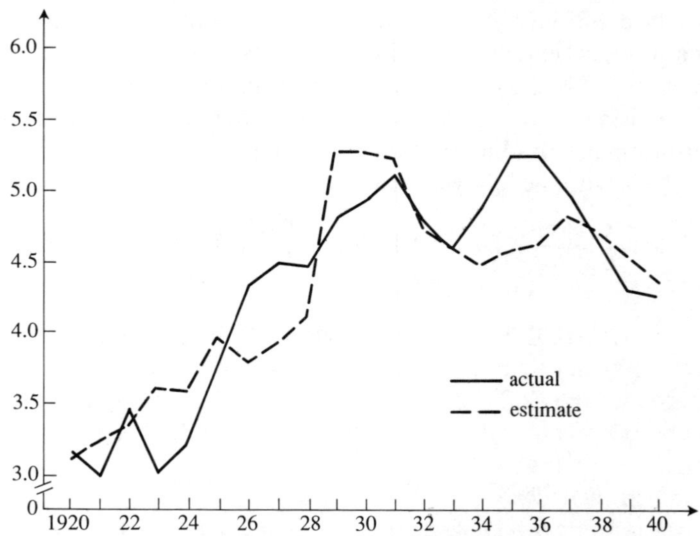

Note: Unit = real production per employee (1,000 yen at 1934–6 constant prices/person).

FIG. 8.11 Estimated performance of production function of Z-sector (pre-war period).

the absence of more reliable data, statistics of factory buildings have been taken as representative of capital stock.

According to our estimates, the production elasticity of capital in the X-sector is 0.813, significantly different from zero at the 5% level, and $\lambda = \alpha - \beta - 1 = 0.18$, which, however, is not significantly different from zero. We therefore, cannot reject the hypothesis that the X-sector production function is subject to constant returns to scale. As production technology is of homogeneity of degree one, if factor markets are in perfect competition then the production elasticity of factors expresses the distribution shares of respective factors—that is, in our case, the capital share of income and the labour share of income in the X-sector are 81.3% and 18.7% respectively. Needless to say, the law of decreasing marginal productivity of the respective factors is satisfied.

The estimation results with regards to the Z-sector reveal that the production elasticity of capital and the economies of scale parameter are both statistically significant. That is, economies of scale do exist in the Z-sector. The production elasticity of capital turns out to be 0.82, and the production function is of homogeneity of degree 1.17, which implies that the production elasticity of labour equals 0.35. Thus, the assumption of diminishing marginal productivity of each factor is satisfied.

To make sure, we estimate production functions using the same data sets but assuming constant returns to scale, and compare the results with those estimated above. It turns out that the correlation coefficients and the t-statistic both improve in the X-sector when constant returns to scale are assumed, and deteriorate in the Z-sector:

X-*sector*:

$$\log \frac{Y_{X_1}}{L_{X_1}} = -\ 1.7550 + \underset{(11.6478^*)}{0.8768} \log \frac{K_{X_1}}{L_{X_1}}$$

$$\bar{R} = 0.9301 \qquad S = 0.03508 \qquad d.w. = 0.879.$$

Z-*sector*:

$$\log \frac{Y_{Z_1}}{L_{Z_1}} = -\ 1.9264 + 0.5575 \log \frac{K_{Z_1}}{L_{Z_1}}$$

$$\bar{R} = 0.6698 \qquad S = 0.05247 \qquad d.w. = 0.537.$$

We therefore conclude that during the period 1919–40, production technology was subject to constant returns to scale in the X-sector and increasing returns to scale in the Z-sector.

Data availability causes some problems. The data on real production (Y_i) are Shionoya's estimates, which cover all production by family firms and private and government factories. On the other hand, the available data for labour (L_i) cover only private factories employing five persons, or more; government factories, private factories with less than five employees, and family firms are excluded. Furthermore, data on capital stock, K_i, cover factory buildings only; equipment and machines are excluded. Thus, the data for Y_i, K_i, and L_i are not consistent. Further, we hope that the relationships among the data components have not varied drastically in the estimation period so that our first approximation can still be justified. Needless to say, more rigorous testing will require more consistent data.

The use of factory building data for an approximation of capital stock, as discussed in the Statistical Appendix (Section 8.3) is obviously a compromise. The use of power data as an alternative (horsepower of the prime mover or the generating capacity) means we can adopt a consistent set of data for production, capital stock, and labour input for private and government factories, though still excluding family firms, with the following results:

X-*sector*:

$$\log \frac{Y_{X_2}}{L_{X_2}} = -1.60092 + \underset{(7.4148^*)}{0.655808} \log \frac{K_{X_2}}{L_{X_2}} + \underset{(1.4819)}{0.346087} \log L_{X_2}$$

$$\bar{R} = 0.9051 \qquad S = 0.03405 \qquad d.w. = 1.0934.$$

Z-*sector*:

$$\log \frac{Y_{Z_2}}{L_{Z_2}} = -0.263189 + \underset{(6.8250^*)}{0.615079} \log \frac{Y_{Z_2}}{L_{Z_2}} + \underset{(2.5026^*)}{0.0973719} \log L_{Z_2}$$

$$\bar{R} = 0.8574 \qquad S = 0.04053 \qquad d.w. = 1.0374.$$

Here, the data for Yi, Ki, and Li ($i = x, z$) are production, capital stock measured in power of generator, and those engaged in all factories except family firms. The sample period is from 1920 to 1940 when this set of data is available.

The above results confirm our previous conclusion: the scale parameter of the X-sector is not significantly different from zero, and that of the Z-sector is positive and statistically significant. In other words, the production technology of the X-sector is subject to constant returns to scale and that of the Z-sector to increasing returns to scale. Assuming the scale parameter of the X-sector to be zero, the production elasticity of capital and labour in the X-sector are 0.656 and 0.344, respectively. The corresponding figures for the Z-sector are 0.615 and 0.482, respectively. All these estimates satisfy the law of diminishing marginal productivity. As we have seen, estimating production functions on the assumption of constant returns to scale does not change estimation performance significantly in the case of the X-sector but increases statistical errors for the Z-sector. (For details, see Section 8.3.6).

It should be noted that in the mid-1930s Japan entered into a war-time economy, with increasing government control over resource allocation. Thus, production functions estimated using observable data and assuming market mechanisms are influenced by government intervention. When the sample period is divided into two, i.e. 1920–30 and 1930–40, estimation performance improves for the former period and deteriorates for the latter. This reflects the fact that government control in a war-time economy represents a significant distortion of economic activities. (Details of the analysis are given in the Statistical Appendix in Section 8.3.)

From all the above, we conclude that the production technology of both the *X*- and *Z*-sectors satisfies the assumptions of our model for the period 1920–40.

8.3. Statistical Appendix

This section shows sources of data, characteristics of data, and possible modifications of the procedure for estimating production functions for future empirical studies. The data discussed are prices, production, capital stock, labour force, and external trade.

8.3.1. Price indices

The data we used in estimating terms of trade and production at constant prices were taken from Ohkawa *et al.* (1967).

This publication offers three alternatives for the deflator for *S*-sector goods: (*a*) rice; (*b*) cultivated agricultural products, and (*c*) all products (see Table 8.6). The difference between (*b*) and (*c*) is that the former excludes livestock and silkworm production. As in our classification the latter properly belongs to the textile subsector of our *X*-sector and (*a*) is too narrowly defined, we consider an index based on (*b*) to be most suitable for our *S*-sector deflator. However, (*b*) also includes tea, which we consider representative of the *X*-sector; but as it comprises only 5%–6% of all agricultural products, distortion may be considered negligible. The base period for the index is set at 1934–6.

Several price indices exist for the capitalist sectors in the period we are considering, but the only ones consistently available by detailed industrial sectors for a long period are the so-called Shionoya estimates, estimated by Shionoya (1966). A futher advantage is that the data on production and prices are mutually consistent because they have been estimated by the same author. The weights used in compiling the deflators for our *X*- and *Z*-sectors are shown in Table 8.7 as they reflect changes in industrial composition, which is in itself informative for our subject.[8]

The terms of trade index was calculated by dividing the *X*- and *Z*-sector deflators thus estimated by the *S*-sector deflator. All indices have 1934–6 as their base period. Export and import price indices were constructed by taking a weighted average of the price indices for categorized trade prices published in Ohkawa et al. (1967). These indices, compiled by Yamada Katsumi from Toyo Keizai Shinposha (1935) and other sources, have 1928–30 as the base; we therefore had to modify them as appropriate to a base period of 1934–6, but could not recompile them since the Toyo Keizai Shinposha (1935) data are

TABLE 8.6 Weights of commodities used in preparing agricultural price index (%)

	1874–76	1904–06	1934–36	1954–56
Rice	60.00	51.99	52.72	49.94
Mugi (wheat)	9.55	10.62	9.17	9.37
Miscellaneous cereals	3.04	2.53	0.78	0.66
Potatoes	2.82	3.20	3.48	5.05
Pulses	4.57	4.16	2.34	3.75
Vegetables	3.79	6.38	7.20	8.76
Fruit and nuts	0.47	1.86	2.71	3.83
Industrial crops	10.18	6.46	5.57	5.71
SUBTOTAL	94.42	87.20	83.97	87.07
Silk cocoons	4.75	9.19	9.83	2.91
Livestock and poultry products	0.83	3.61	6.20	10.02
TOTAL	100.00	100.00	100.00	100.00

Source: Ohkawa *et al.* (1967): 207.

TABLE 8.7 Weights of commodities used in preparing price indices of manufactured goods (%)

Commodity	1873–89	1887–1903	1901–15	1913–33	1931–45
Foods	36.2	43.4	31.6	21.7	15.9
Textiles	24.5	23.2	32.5	34.7	31.2
Wood and woodproducts	12.3	9.1	11.3	6.8	3.8
Chemical products	13.6	8.1	9.2	10.1	15.5
Stone, clay and glass products	2.2	1.8	2.0	2.6	2.9
Metals	3.6	2.8	4.2	6.7	13.5
Machinery	1.6	7.5	4.5	12.9	12.1
Miscellaneous	6.0	4.1	4.7	4.5	5.1
Total *X*-sector	79.0	79.8	80.1	67.7	56.0
Total *Z*-sector	21.0	20.2	19.9	32.3	44.0
TOTAL	100.00	100.0	100.0	100.0	100.0

Note: X sector: foods, textiles, wood and woodproducts, printing and binding, and miscellaneous; Z sector: metal and metal products, machinery, ceramics, and chemicals.

Source: Ohkawa *et al.* (1967): 194–207.

available up to 1933 only, so adjustment beyond changing the base period would have been extremely cumbersome. The weights we used in changing the base period are shown in Table 8.8.

TABLE 8.8 Weights of commodities used in preparing export and import price indices
(%)

Commodity	Exports	Imports
Animals and plants	0.22	0.06
Foods, beverages, and tobacco	8.20	14.09
Yarns, textile fabrics and their products	71.31	34.59
Skins, bones and hair	0.75	1.47
Oils, waxes and chemicals	4.01	14.70
Crude minerals, ores, metals and their products	6.94	14.70
Machinery	1.89	8.07
Wood and wood products	1.31	4.48
Paper, paper products and Miscellaneous	5.37	7.13
X-sector	87.16	61.82
Z-sector	12.84	37.47
Total	100.00	100.00

Note: *X* and *Z* sector, as Chapter 8.
Source: Ohkawa *et al.* (1967).

8.3.2. Production series

Production series at constant prices for the *X*- and *Z*-sectors were calculated by dividing production at current prices by the sectoral deflators as estimated in Section 8.3.1. The Shionoya estimates were used as the basis, but since factor inputs differed in coverage, they had to be modified. The Shionoya estimates, production at current prices, are compiled from factory statistics which cover both private and government factories regardless of size. We classified these estimates into the *X*- and *Z*-sector. Constant price production is that which is deflated by the price indices of respective sectors. These series are used as production for testing terms of trade assumptions.

For estimating the production function, however, a further modification was required because employment statistics are not available for firms employing less than five persons. Official production statistics for these family firms are available only from 1939—in *Shokosho* (1944)—but Umemura Mataji has estimated the production of family firms for the period 1878–1942. (See Ohkawa 1965). Assuming that the production of family firms accounted for a constant share of total production, we constructed production series for the *X*- and *Z*-sectors.

By taking this step, production series of family firms is converged to Umemura estimates and the shares of family firms' production in *X*- and Z-sectors are fixed at the base year 1939–40. This procedure is applied to the period from 1920 through 1940. This is how we derived production series consistent with our data on labour input.

Table 8.9 compares production by family firms with the Shionoya estimates, which include family firms. The share of family firms in production, naturally, is much higher in the *X*-sector than in the *Z*-sector. It is by these procedures that we have obtained consistent production data for estimating production functions.

8.3.3. Capital stock

Capital stock in the production function normally consists of durable production goods (including machines, transport equipment, and power generators and transmitters), ships, and buildings and other structures. Statistics of capital stock for the pre–World War II period is available in Ohkawa *et al.* (1966) where separate time-series for primary and non-primary industries for the period 1877–1940 are given.

For this study, however, we need data on capital stock by more detailed industrial sectors, and moreover we need data on net capital stock and not gross stock. Since no such data are available, we use generator capacity by industry as a proxy variable. These data are available in Minami (1965) which gives the total horsepower of factory generators, in private factories, excluding family firms, for the period 1884–1940, for eight industrial sectors: textile-spinning and weaving, metal, machines, ceramic, chemical, lumber and related products, printing and publishing, and food-processing. These eight sectors and other industries (excluding gas and electricity) make the total. For the miscellaneous manufacturing sector, power data are estimated by subtracting that of eight sectors from the total. Strangely, for some years this procedure yielded negative figures; in these cases we substituted the arithmetical average of the data of the years before and after the year in question.

The Minami estimates are not always consistent; data for government factories are included for the period 1922–8, for example, but not otherwise. We have therefore taken power data from 1920 through 1940 for government factories from *Factory Statistics* and added them to the Minami estimates for private factories. Shokosho (1944) in fact includes data for local government factories too, but as these are small, however, and are included only from 1928, we have ignored them.

Another proxy for capital stock are data on factory buildings. Such data are available for ten industrial sectors, i.e. the gas and electricity sector and miscellaneous manufacturing in addition to the eight sectors covered in the Minami estimates, in the Ohkawa *et al.* (1966), estimated by Ishiwata Sigeru. This series covers both factory buildings and warehouses, but it excludes those of government factories and includes those of family firms. To overcome this problem we obtained an estimate for family firms by applying a ratio of building value per factory by industry and then subtracting this figure from the data, and used a similar method to obtain an estimate for government factories which was then added to that for private factories. A most serious defect

TABLE 8.9. Estimates of production share of family firms (¥m.)

		Foods	Textiles	Wood products	Chemicals	Ceramics	Metals	Machinery	Others	Total
Factory statistics	1939	1,786	4,904	743	4,191	649	5,835	6,334	452	24,792
5 employees and over	1940	2,334	4,976	1,022	4,623	777	5,900	8,443	542	27,092
Less than 5 employees (A)	1939	433	378	269	148	66	120	155	305	1,874
	1940	550	401	333	156	73	117	175	339	2,146
TOTALS	1939	2,219	5,282	1,012	4,339	715	5,955	6,489	757	26,666
	1940	2,884	5,377	1,355	4,779	850	6,017	8,618	881	29,238
Shionoya estimates (B)	1920	2,186	3,136	791	1,070	287	741	1,566	435	10,212
	1925	2,452	3,891	599	1,103	311	631	644	531	10,162
	1930	2,019	2,715	393	1,146	248	780	781	500	8,582
	1935	2,203	4,328	524	2,188	418	1,912	1,717	707	13,997
	1939	3,558	6,435	1,412	5,181	782	5,437	7,113	1,668	31,586
	1940	3,580	6,711	1,967	5,781	920	5,482	8,740	1,942	35,123
(A)/(B)	1939	0.1217	0.0587	0.1905	0.0286	0.0844	0.0221	0.0218	0.1829	0.0593
	1940	0.1536	0.0598	0.1693	0.0270	0.0793	0.0213	0.0200	0.1746	0.0611
Average		0.1377	0.0593	0.1799	0.0278	0.0819	0.0217	0.0209	0.1788	0.0602

Note: Average of (A)/(B) gives average of share for 1939 and 1940.
A: less than 5 employees.

Source (for Shionoya estimates): Shionoya (1966).

of this method is that we could not distinguish between the value of buildings of a large government factory and those of small factories; moreover building values are in fact based on the industrial census which had a sampling of only 12.6%. Since there is no alternative source of data on building values, it would seem that generator capacity provides a more accurate proxy for capital equipment than do data on buildings.

8.3.4. Labour force

Shokosho (1944) provides data on the number of employees by industry in both government and private factories employing five persons or more, excluding family firms. It is for this reason that we exclude family firms from our production function. The basic data by industry are obtained from MITI (1962), but some figures in Shokosho (1944), which form the basis of the data in this book, do not differentiate the government factory and private factory for some years. What is even worse is that there were no reports for some years. We therefore take data on government factories to make a consistent data series.

Having constructed labour data for the *X*- and *Z*-sectors in this way, however, we obtained an excellent series of data—the so-called Akasaka (1964) estimates, covering government and private factories as well as family firms for the periods 1899–1919 and 1919–42. Using data from this source for future estimates would clearly improve the results.[9]

8.3.5. Value of exports and imports

The value of foreign trade for *X*- and *Z*-sectors was calculated from trade statistics for seventeen categories in the Toyo Keizai Shinposha (1935) in order to make it consistent with our production and price data.

It should be noted that trade statistics for the pre-World War II period do not cover trade between Japan and, its colonies, so these data are not included in our analysis. Furthermore, the Toyo Keizai Shinposha (1935) gives data on the value of trade in current prices for the period 1868–1933, but indices of trade prices are available only for the period 1900–39. Our investigation therefore is limited to the period 1900–33, for which both series are available.

8.3.6. Estimated production functions

The major results obtained from estimating the production functions are listed below. The notation is explained in Section 8.2.3.

X-*sector*

$$\log \frac{Y_{X_1}}{L_{X_1}} = -2.89866 + \underset{(6.3948^*)}{0.812842} \log \frac{K_{X_1}}{L_{X_1}} + \underset{(0.63000)}{0.179374} \log L_{X_1}$$

$$\bar{R} = 0.9278 \qquad S = 0.03909 \qquad d.w. = 0.8490 \ (1919\text{–}40)$$

$$\log \frac{Y_{X_1}}{L_{X_1}} = -\ 1.75496 + \underset{(11.647^*)}{0.876821} \log \frac{K_{X_1}}{L_{X_1}}$$

$\bar{R} = 0.9031 \qquad S = 0.03850 \qquad d.w. = 0.9301\ (1919–40)$

$$\log \frac{Y_{X_2}}{L_{X_2}} = -\ 1.60092 + \underset{(7.4148^*)}{0.655808} \log \frac{K_{X_2}}{L_{X_2}} + \underset{(1.4819)}{0.346087} \log L_{X_2}$$

$\bar{R} = 0.9501 \qquad S = 0.034054 \qquad d.w. = 1.0934\ (1920–40)$

$$\log \frac{Y_{X_2}}{L_{X_2}} = 0.539087 + \underset{(13.206^*)}{0.757685} \log \frac{K_{X_2}}{L_{X_2}}$$

$\bar{R} = 0.9468 \qquad S = 0.03510 \qquad d.w. = 0.9255\ (1920–40)$

$$\log \frac{Y_{X_2}}{L_{X_2}} = 0.785351 + \underset{(4.1369^*)}{0.592855} \log \frac{K_{X_2}}{L_{X_2}} - \underset{(-0.12200)}{0.0460443} \log L_{X_2}$$

$\bar{R} = 0.8411 \qquad S = 0.03413 \qquad d.w. = 1.4612\ (1920–30)$

$$\log \frac{Y_{X_2}}{L_{X_2}} = 0.501120 + \underset{(5.4214^*)}{0.582229} \log \frac{K_{X_2}}{L_{X_2}}$$

$\bar{R} = 0.8599 \qquad S = 0.03221 \qquad d.w. = 1.4910\ (1920–30)$

$$\log \frac{Y_{X_2}}{L_{X_2}} = -\ 5.62586 - \underset{(-\ 8.4454^*)}{0.144743} \log \frac{K_{X_2}}{L_{X_2}} + \underset{(6.4520^*)}{1.00370} \log L_{X_2}$$

$\bar{R} = 0.9243 \qquad S = 0.01775 \qquad d.w. = 2.0983\ (1930–40)$

$$\log \frac{Y_{X_2}}{L_{X_2}} = 0.551173 + \underset{(1.8557)}{0.570299} \log \frac{K_{X_2}}{L_{X_2}}$$

$\bar{R} = 0.4436 \qquad S = 0.04170 \qquad d.w. = 0.4593\ (1930–40).$

Z-sector:

$$\log \frac{Y_{Z_1}}{L_{Z_1}} = -\ 2.77325 + \underset{(7.4858^*)}{0.820563} \log \frac{K_{Z_1}}{L_{Z_1}} + \underset{(4.5453^*)}{0.1706456} \log L_{Z_1}$$

$\bar{R} = 0.8496 \qquad S = 0.03726 \qquad d.w. = 1.2541\ (1919–40)$

$$\log \frac{Y_{Z_2}}{L_{Z_2}} = -\ 1.92643 + \underset{(4.2528^*)}{0.557525} \log \frac{K_{Z_2}}{L_{Z_2}}$$

$\bar{R} = 0.6697 \qquad S = 0.05247 \qquad d.w. = 0.5371\ (1919–40)$

$$\log \frac{Y_{Z_2}}{L_{Z_2}} = -\ 0.263189 + \underset{(6.8250^*)}{0.615079} \log \frac{Y_{Z_2}}{L_{Z_2}} + \underset{(2.5026^*)}{0.0973719} \log L_{Z_2}$$

$\bar{R} = 0.8573 \qquad S = 04052 \qquad d.w. = 1.0374\ (1920–40)$

$$\log \frac{Y_{Z_2}}{L_{Z_2}} = 0.300674 + 0.640895 \log \frac{K_{Z_2}}{L_{Z_2}}$$
$$(6.3347^*)$$

$\bar{R} = 0.8134 \qquad S = 0.04580 \qquad d.w. = 0.7906\ (1920–40)$

$$\log \frac{Y_{Z_2}}{L_{Z_2}} = -\ 2.13085 + 0.538078 \log \frac{K_{Z_2}}{L_{Z_2}} + 0.426497 \log L_{Z_2}$$
$$(4.0672^*) \qquad\qquad (0.78585)$$

$\bar{R} = 0.8204 \qquad S = 0.04676 \qquad d.w. = 1.3879\ (1920–30)$

$$\log \frac{Y_{Z_2}}{L_{Z_2}} = 0.309923 + 0.576403 \log \frac{K_{Z_2}}{L_{Z_2}}$$
$$(4.7899^*)$$

$\bar{R} = 0.8288 \qquad S = 0.04575 \qquad d.w. = 1.2759\ (1920–30).$

8.3.7. Trade with Japanese colonies

Since statistics of Japan's trade with Sakhalin and the South Pacific islands which used to be Japanese colonies before World War II are not available, the data on trade with colonies shown in Table 8.10 refer only to Taiwan and Korea. Exports to these two colonies steadily expanded, accounting for about 20% of Japanese exports in 1933.

Even though the aggregate figure is small, it is important to note that some of the products that figured significantly in Japan's trade with these colonies may not have been 'exportables' in the wider sense. If an industry takes a large market share in colonial markets, but only a small share in world markets, it is classified in the *Z*-sector as far as terms of trade are concerned. On the other hand, it would be classified as the *X*-sector if it had a large share in both world and colonial markets. Table 8.11 considers this with regard to the trade in cotton and silk textiles and machinery. It can be seen that the colonies accounted for only 10% of cotton and silk cloth exports but 30% of machine exports; in other words, textiles were 'exportable' whereas machines were less 'exportable', taking a relatively small share in the overseas markets.

TABLE 8.10. Japanese trade with colonies before World War II

	Value of exports (¥'000)				Percentage distribution			
	To foreign countries	To Taiwan	To Korea	Total	To foreign countries	To Taiwan	To Korea	Total
1897	163,135	3,724		166,859	97.8	2.2		100.0
1900	204,430	8,439		212,869	96.0	4.0		100.0
1905	321,534	13,484		335,018	96.0	4.0		100.0
1910	458,429	29,070	14,415	501,914	91.3	5.8	2.9	100.0
1915	708,307	40,587	43,708	792,602	89.4	5.1	5.5	100.0
1920	1,948,395	112,041	139,819	2,200,255	88.6	5.1	6.4	100.0
1925	2,305,590	129,906	234,624	2,670,120	86.4	4.9	8.8	100.0
1930	1,469,852	123,127	278,194	1,871,173	78.6	6.6	14.9	100.0
1933	1,861,046	149,912	339,817	2,350,775	79.2	6.4	14.5	100.0

Source: Toyou Keizai Shinposha (1935): 663.

TABLE 8.11. Japan's colonial trade in textiles and machinery (¥'000)

		(1) Exports from Japan proper and Sakhalin	(2) Exports to Korea	(3) Exports to Taiwan	(2)+(3)=*A*	(1)+(2)+(3)=*B*	*A*/*B*%	(4) Total Exports	*B*/(4)%	(5) Production	*B*/(5)%	Imports Colonies' from the world
Cotton and silk fabrics	1905	41,751		2,245	2,245	43,996	5.1	335,018	13.1	521.5	8.4	3,415
	1910	53,260		3,418	3,418	56,678	6.0	501,914	11.3	622.0	9.1	4,813
	1915	81,730	9,993	4,653	14,646	96,376	15.2	792,602	12.2	922.0	10.5	17,878
	1920	493,682	28,199	6,039	34,238	527,920	6.5	2,200,255	24.0	3,136.0	16.8	38,198
	1925	549,835	52,446	15,708	68,154	617,989	11.0	2,670,120	23.1	3,891.0	15.9	70,870
	1930	338,033	44,990	13,394	58,384	396,417	14.7	1,871,173	21.2	2,715.0	14.6	59,319
	1933	573,537	62,227	15,106	77,333	650,870	11.9	2,350,775	27.7	3,652.0	17.8	77,379
Machinery	1905	3,097		400	400	3,497	11.4	335,018	1.0	117.8	3.0	789
	1910	3,357		1,502	1,502	4,859	30.9	501,914	1.0	133.0	3.7	6,672
	1915	10,031	1,841	1,733	3,574	13,605	26.3	792,602	1.7	272.0	5.0	4,877
	1920	50,011	10,858	11,744	22,602	72,613	31.1	2,200,255	3.3	1,566.0	4.6	29,196
	1925	29,656	10,684	6,137	16,821	46,477	36.2	2,670,120	1.7	644.0	7.2	19,132
	1930	35,266	25,017	11,621	36,638	71,864	60.0	1,871,173	3.8	781.0	9.2	42,115
	1933	67,622	29,905	10,985	40,890	108,512	37.7	2,350,775	4.6	1,065.0	10.2	41,971

Sources: Columns (1)–(3) Toyo Keizai Shinposha (1935); (5) Shionoya (1966).

Notes

1. The turning point is defined as when the labour supply is transformed from 'unlimited' to 'limited' in the sense of Arthur Lewis, i.e. when industrial sectors cannot increase their employment without paying higher wage rates.
2. Another way to classify the *S*-sector is to include family firms in manufacturing in this sector in the pre-World War II period, as they constituted an important source of labour supply to industrial sectors especially in urban areas.
3. We included the mining sector in the secondary industries.
4. This coefficient is defined as (the Japanese export of good *i* / Japan's total export) divided by (world export of good *i* / world total export).
5. In reality, the classifications of goods and industries do not always coincide. Our classification of industries is therefore approximate.
6. The statistic $(n - P - 1)R^2/P(1 - R^2)$ follows an *F*-distribution, where n, P, and R are sample size, number of independent variables, and correlation coefficient, respectively.
7. We did not specify technological progress in the theoretical model. If there exists neutral technological progress, the effect is reflected on $\alpha + \beta$. As technological progress has the same effect as increasing returns to scale in our analysis, however, we do not have to separate these factors.
8. The weights used by Ohkawa (1967) did not coincide with those used for period II in Table 8.7. We therefore, calculated our own weights using Shionoya's production estimates.
9. See Akasaka (1964).

9

The Course of Capital Accumulation

When the course of capital accumulation derived from the theoretical model shown in Part I for the period since the Meiji Restoration in 1868 is tested empirically, a number of difficulties are encountered. First, while the model assumes that all industries can be classified into three sectors, there are a number of sectors which cannot be so classified, partly due to statistical constraints. In what follows we adopt the classification developed in the preceding chapter: the *S*-sector consists of agriculture, forestry, and fishery;[1] the *X*-sector (light industry) of food-processing, textiles, lumber and related products, printing and publishing, and miscellaneous manufacturing industries; and the *Z*-sector (heavy industry) of machinery, chemical, ceramics, and metal industry. This is the basic way of tracing capital accumulation in this chapter, but occasionally we include gas, electricity, and transportation in the *Z*-sector. Where this is done it will be noted.

Major sectors not covered are mining, communication, finance, distribution, and services, due mostly to statistical constraints. We hold that mining belongs to the *X*-sector, while the other sectors may be considered factors affecting the efficiency of the manufacturing industries. We have already argued that social overhead capital influences production functions in our model, and these sectors are considered to play a similar role.

A second reason why some sectors had to be excluded was the unavailability of suitable data on capital stock data for the pre–World War II period. We therefore use data on generator capacity and factory buildings, as available, although these are by no means ideal data series for capital stock.

These defects notwithstanding, we try in this chapter to achieve several tasks. First, by observing the course of capital accumulation in both the *X*- and *Z*-sectors, we try to explain the situation from a historical perspective. Secondly, we test whether the dynamic path of capital accumulation which is derived from our model can be supported by the actual path. As we denoted capital stock as K_x and K_z, hereafter we will refer to the diagram as the K_x–K_z graph. Thirdly, we try to identify the critical period when industrialization took off, in the sense that the *Z*-sector went into a phase of self-sustaining growth. Another way of putting it is that we try to identify until what year the *Z*-sector might have shrunk in the absence of subsidies and favourable exogenous factors such as wars in other regions and countries. Fourthly, we test our hypothesis with regard to investment allocation: considering only private investment financed by profit income (a simplifying assumption), we hypothesize that if the profit rate is higher in the *X*-sector, all *X*-sector profits will be reinvested in the

X-sector, and a proportion of *Z*-sector profits will also be invested in the *X*-sector. This proportion will be an increasing function of the profit rate differential, namely using the notation in Part I $R_x - R_z$.

As some part of labour income is also saved, the assumption that profit is the sole source of investment is obviously a simplification for the theoretical analysis, and some way is needed of bridging the gap between this assumption and reality. Since in the Japanese economy a large amount of income from the *S*-sector was invested in the *X*- and *Z*-sectors, this must also be kept in mind when we test the investment allocation.

An interesting point in the empirical study will be to measure how quickly investment responds to the profit rate differentials. Also, if investors react to the expected differential in the profit rate rather than the actual differential, then the observed investment allocation will differ from that assumed in the model.

9.1. Light Industry (*X*-Sector)

As already explained, the *X*-sector in our empirical test consists of food-processing, textiles, lumber and related products, printing and publishing, and miscellaneous manufacturing industries. Generator capacity and data on factory buildings are taken as proxies of capital stock data; in general we use the Minami estimates[2] and Ishiwata estimates[3], respectively, although in some instances we used our own estimates. The figures obtained are presented in Tables 9.1–9.4 for five-year intervals over the entire period, giving a long-run perspective of capital accumulation.

Towards the end of the Tokugawa era (1603–1867), factories were constructed both by *han* (local *daimyo* or lords) and by the government, but in both cases the numbers and the scale of operation were small; data on generating capacity indicate that even the *X*-sector was still in its infancy. As well known by textbooks on economic history of Japan, most factories had been owned and operated by the government until 1880 when the government published 'General Principles of Factory Selling to Private Sectors' (*kojo haraisage gaisoku*). It was only after some troublesome transition period on the result of this policy change that private manufacturing industries became a common place. Therefore, the period before 1885 was that of cradle days even for *X*-sector. The growth rates for the five-year periods 1885–90 and 1890–5 appear very high (as shown in Table 9.1), but this is merely a reflection of the low base figures.

Let us consider the composition of capital accumulation in the *X*-sector. Table 9.5 shows the generator capacity by industry for 1890 and 1910, and factory buildings for 1900 and 1920. In terms of generator capacity, textile sectors accounted for 83.5% and 75.3% of the *X*-sector in 1890 and 1910, respectively; in terms of factory buildings, the ratio was somewhat lower, being

TABLE 9.1. Capital accumulation (1885–1920): Horse-power of prime-movers in manufacturing industries (1,000 HP)

	Total	*X*-sector	*Z*-sector
1885	1.4	1.0	0.4
1890	21.9	16.4	5.5
1895	63.4	46.0	17.4
1900	69.0	44.5	24.5
1905	135.8	85.6	50.2
1910	320.2	174.3	145.9
1915	635.8	334.5	301.3
1920	1,447.5	574.6	872.9

Note: *X*-sector: food, textiles, wood and wood products, printing and binding, and miscellaneous; *Z*-sector: metal and metal products, machinery, ceramics, and chemicals.

TABLE 9.2. Capital accumulation (1895–1920): value of factory buildings (1930 prices ¥m.)

	Total	*X*-sector	*Z*-sector
1895	81.2	51.2	30.0
1900	88.1	53.4	34.7
1905	127.3	72.1	55.2
1910	190.9	98.9	92.0
1915	239.9	115.6	124.3
1920	332.6	162.4	170.2

Note: *Z*-sector includes gas and electricity utilities in addition to the items in the *Z*-sector listed in note to Table 9.1.

Source: Ohkawa *et al.* (1966).

TABLE 9.3. Capital accumulation (1920–1940): horsepower of prime-movers in manufacturing industries (1,000 HP)

	Total		*X*-sector		*Z*-sector	
	Original series	Modified series	Original series	Modified series	Original series	Modified series
1920	1447	1868	574	590	873	1278
1925	2594	2594	917	917	1677	1677
1930	3723	4370	1418	1453	2305	2918
1935	5222	5693	1669	1694	3553	3999
1940	10,632	11,198	2231	2268	8401	8930

Notes: *X*- and *Z*-sectors as Table 9.1. Modified series includes prime-movers of government factories.

Source: Minami (1965) for original series.

TABLE 9.4. Capital accumulation (1920–1940): value of factory buildings (1930 prices, ¥m.)

	X-sector		*Z*-sector		Total	
	Isiwata estimate	Government factories with more than 5 persons	Isiwata estimate	Government factories with more than 5 persons	Isiwata estimate	Government factories with more than 5 persons
1920	162	90	171	49	333	139
1925	183	95	203	58	386	153
1930	257	117	335	81	592	198
1935	338	149	454	117	792	266
1940	473	205	737	221	1,210	426

Notes: *X*- and *Z*-sectors as Table 9.1, with gas and electricity utilities included in the *Z*-sector. Isiwata estimate covers only private factories. Factories with more than 5 persons exclude gas and electricity utilities.

Source: Ohkawa *et al.* (1966).

TABLE 9.5. Composition of capital stock in *X*-sector

	Horsepower of prime-movers (1,000HP)		Capital stock of factory buildings (¥m.)	
	1890	1910	1900	1920
Textiles	13.7	131.3	43.4	116.3
	(83.5)	(75.3)	(81.3)	(71.6)
Food	2.0	19.2	7.7	32.3
	(12.2)	(11.0)	(14.4)	(19.9)
Wood and wood products	0.1	16.8	0.8	0.8
	(0.1)	(9.6)	(1.5)	(0.5)
Printing and binding	0.1	2.7	0.6	2.4
	(0.1)	(1.5)	(1.1)	(1.5)
Miscellaneous	0.5	4.3	0.9	10.6
	(0.3)	(0.2)	(1.7)	(6.5)
Total	16.4	174.3	53.4	162.4

Note: Percentages are given in parantheses.

Sources: Ohkawa *et al.* (1966); Minami (1966).

51.3% (1920). On the basis of these figures, however, it would seem reasonable to take the textile industry as representative of the *X*-sector. Fig. 9.1 is a graphic presentation of the data in tables 9.1–9.4, and it confirms that the rapid increase of capital stock in the 1880s was due to the construction of spinning factories, especially in the first half of the decade.

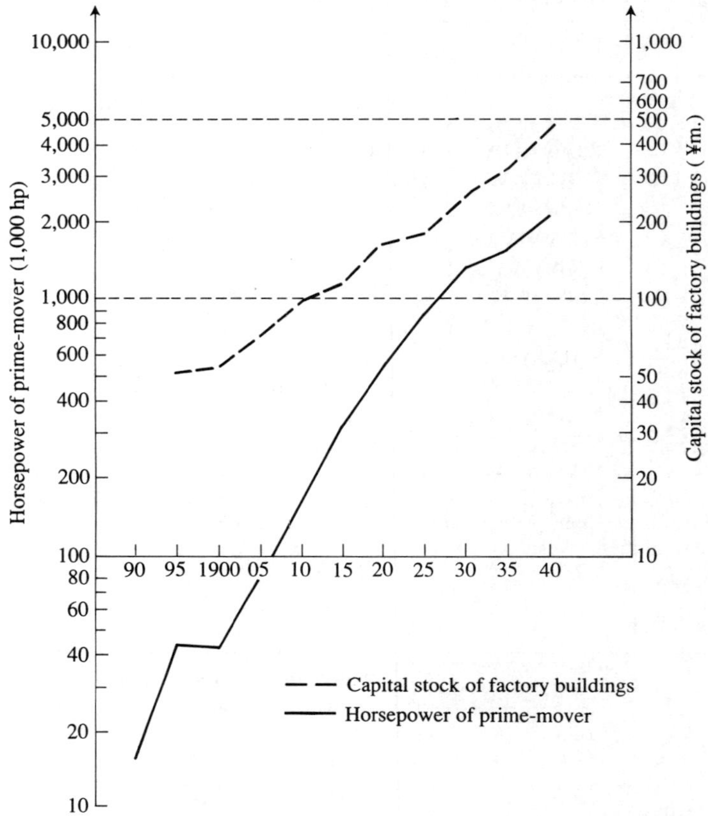

Note: Based on the data of Table 9.1.

FIG. 9.1 Historical developments: horsepower of prime-mover and capital stock of factory buildings in *X*-sector.

Table 9.6 presents the development of spinning factories between 1885 and 1890. Many of those opened in the first half of the decade were small-scale factories with some 2,000 spindles, but in the latter part of the decade the scale drastically expanded e.g. 15,000 spindles at Tenma Bouseki in 1886 and 11,520 at Hirano Bouseki in 1886. Correspondingly, generator capacity in the *X*-sector jumped from 1,000 hp in 1885 to 46,000 hp in 1895, and by 1890 the volume of cotton yarn produced in Japan equalled the volume imported. Iwanami Shoten (1968) has shown that import substitution progressed to such a degree that by 1897 exports of cotton yarn exceeded imports. As Table 9.7 shows, domestic production of cotton yarn in 1880 was only 1.3 million pounds, whereas imports totalled 38.1 million pounds. But in less than twenty years, the same industry became a most important export industry.

While their importance declined as new *X*-sector industries developed, the

TABLE 9.6. Spinning factories, 1882–1885

	Factory (or company name) and place	No. of spindles
1882	TAMASHIMA, OKAYAMA	4,000
	KUWAHARA, OSAKA	2,000
	ICHIKAWA, YAMANASHI	15,000
	OSAKA, OSAKA	2,000
	SHIMOMURA, OKAYAMA	2,000
	OKAYAMA, OKAYAMA	2,000
1883	TOYOI, NARA	2,000
	MIYAGI, MIYAGI	2,000
1884	SHIMADA, SHIZUOKA	2,000
	ENSHU HUTAMATA, SHIZUOKA	2,000
	NAGASAKI, NAGASAKI	2,000
1885	SHIMOTSUKE, TOCHIGI	2,000
	NAGOYA, AICHI	4,000

Source: Iwanami Shoten (1968).

TABLE 9.7. Production, imports, and exports of cotton yarns (million pounds)

	Production	Imports	Exports
1880	1.3	38.1	0
1885	6.4	28.6	0
1890	41.9	42.5	—
1895	146.7	19.5	4.7
1900	258.2	12.1	83.5
1905	362.2	3.0	107.0
1910	453.9	0.4	139.1
1915	688.1	0.2	230.4
1920	726.8	2.0	122.0
1925	974.7	1.5	124.3
1930	1,009.9	3.6	23.8
1935	1,424.3	7.2	38.9

Source: Kajinishi (1964), Appendix Table III.6.

tea and silk industries were other representative of the *X*-sector in our model. As the availability of data for the former is more limited and as the latter in any case played a more important role in industrial development, we briefly review the development of silk industry in this period.

In 1868, the year of the Meiji restoration, silk products and tea were the only exports, and silk was by far the more important (Asahishimbun 1930); even in 1880, silk accounted for more than 30% of the total value of Japanese exports

(Table 9.8). While the importance of silk declined as the cotton industry grew, silk nevertheless remained the single most important export throughout the period before the World War II.

The construction of factories for silk spinning was also ahead of that for cotton spinning. As early as 1870, the Ministry of Private Affairs (Minbu) employed a French engineer to help construct the Tomioka Seisijo (Tomioka Silk Factory). The factory started operation in 1872 and was privatized in 1893. In the same year, a private group called Ono-gumi opened a factory in the Nagano prefecture using the Italian technique, with ninety-six employees.

TABLE 9.8(*a*). Composition of Japan's exports by selected commodities (¥m.)

	Tea	Raw silk	Cotton yarns	Cotton Fabrics	Total exports
1880	7.32	8.61	—	0.03	28.40
	(25.8)	(30.3)	(—)	(0.1)	(100.0)
1890	6.07	13.86	—	0.17	56.60
	(10.7)	(24.5)	(—)	(0.3)	(100.0)
1900	8.10	44.66	—	5.72	204.43
	(4.0)	(21.9)	(—)	(2.8)	(100.0)
1910	13.41	130.18	0.10	20.46	458.43
	(2.9)	(28.4)	(0.0)	(4.5)	(100.0)
1920	16.35	382.22	5.57	335.27	1,948.39
	(0.8)	(19.6)	(0.3)	(17.2)	(100.0)
1930	8.24	416.65	0.40	272.12	1,469.85
	(0.6)	(28.4)	(0.0)	(18.5)	(100.0)

Note: Percentages are given in parentheses.
Source: Toyo Keizai Shinposha (1935).

TABLE 9.8(*b*). Volume and value of raw silk exports

	Export volume (1,000 kin)	Export value (¥'000)	Unit price (¥/kin)
1880	1,462	8,607	5.89
1885	2,457	13,034	5.30
1890	2,110	13,859	6.57
1895	5,882	47,862	8.14
1900	4,631	44,657	9.64
1905	7,279	71,999	9.89
1910	14,846	130,833	8.81
1915	17,814	152,030	8.53
1920	17,469	382,717	21.91
1925	43,800	879,657	20.08
1930	47,700	419,107	8.79

Note: 1 kin = 0.6 kg.
Source: Kajinishi (1964), Appendix Table IV.5.

The same group had already started a spinning mill using the same technique in Tsukiji, Tokyo with sixty employees. The construction of further modern, Western-style silk factories continued throughout the 1870s.

By 1900, then, the *X*-sector, represented by the silk and cotton industries, had thus firmly established its position as an export sector. As it developed however, a pattern of business cycles emerged, typified by price declines in the time of recession (see Fujino 1965: chs. 16–17). Thus in the 1900s there were several corporate mergers, and the Japan Cotton Spinners' Association (Dainippon Menshi Boseki Dogyo Rengokai) decided to make the fourth joint reduction of production since 1890.

As we saw in Chapter 8, the Japanese *X*-sector had increased its share in the world market as it expanded—from 2.9% in 1899 to 4.6% in 1913, and 10.8% in 1929 (Maizels 1963: 342)[4]—and therefore the terms of trade it faced were negatively related to production. Furthermore, the frequent business cycles fluctuations grew to influence the profits of the *X*-sector. Reflecting these facts, Japan's *X*-sector stagnated in the latter part of the pre–World War II period except for the period of military control of the late 1930s.

9.2. Heavy Industry (*Z*-Sector)

In comparing capital stock in the *Z*-sector with production, for the sake of statistical consistencies we used aggregated capital stock data for the chemical, metal, machinery, and ceramic industries. In tracing the long-run capital accumulation, however, we included electricity and transportation sectors in the *Z*-sector because the criteria of economies of scale and independence of terms of trade seemed to be applicable to them.

Japan's *Z*-sector started to develop fairly late relative to the *X*-sector (see Table 9.1). Even though using generator capacity as a proxy for capital stock tends to distort capital development upwards, our figures show that the *Z*-sector lagged behind the *X*-sector as late as 1915. Only in 1920 did *Z*-sector generator capacity exceed that of the *X*-sector, by which time the *X*-sector frequently suffered excess capacity. From 1920 on, the growth in power in the *Z*-sector was largely higher than that in the *X*-sector.

Table 9.9 shows the generator capacity of the *Z*-sector in more detail. Metal, machinery, and chemical industries, respectively, accounted for some 30% of the generator capacity in the early years, although the share of the machinery industry surpassed that of the metal industry in later years, largely due to the adoption of foreign technology. It was only after 1920, however that the *Z*-sector, led by the machinery industry, took off for sustained growth.

Readers are reminded that our definition of take-off is not that of Rostow, which defines it as the stage when the savings rate exceeds 12%–13%. Accordingly, he considered that the Japanese economy had taken off much earlier than we do on the basis of our definition, i.e. that take-off is achieved when the

TABLE 9.9. Composition of capital stock in Z-sector

	Horsepower of prime-movers (1,000 HP)		Capital stock of factory buildings (¥m.)	
	1900	1930	1910	1940
Metal and metal products	8.8	534.6	2.4	38.5
	(35.9)	(23.2)	(2.6)	(5.4)
Machinery	5.0	700.2	7.4	186.1
	(20.4)	(30.4)	(8.0)	(26.1)
Ceramics	1.7	293.2	33.6	211.0
	(6.9)	(12.7)	(36.6)	(29.5)
Chemicals	9.0	777.1	5.5	62.2
	(36.7)	(33.7)	(6.0)	(8.7)
Gas and electricity utilities	—	—	43.1	239.2
	(—)	(—)	(47.0)	(33.5)
Z-sector (total)	24.5	2,305.1	92.0	737.0

Note: Percentages are given in parentheses.

Sources: Ohkawa *et al.* (1966); Minami (1966).

Z-sector starts to attain self-sustaining growth. As we shall see later, Arthur Lewis's (1955) 'turning point' concept is also different from our take-off, despite its apparent similarity, as it refers to a shift from an unlimited supply of labour to a limited supply of labour. The notion is meaningful in itself, but it is a different concept. It is possible according to our model that the *Z*-sector expands in a self-sustaining manner even though there is a labour surplus in the *S*-sector. Thus, take-off in our terminology differs both from that of Rostow and Lewis's 'turning point' concept.

Prior to World War II, Japan's *Z*-sector was heavily dependent on government enterprises. When the new Meiji government took office, most of the metal and machinery factories were owned either by *daimyo* (local feudalistic lords) or the Tokugawa government. As for foreign trade, almost all transactions were under the control of foreign traders (*gaisho*). Thus in the Meiji era almost all prime ministers saw it as their mission to regain control of foreign trade, including sovereignty to decide tariff rates and to develop national maritime carriers. It was for this reason that ship-building and marine transportation enjoyed preferential government treatment for many years. In 1875 the government transferred thirteen imported carrier ships to Mitsubishi Marine Transportation free of charge, with annual subsidies of ¥250,000 for operation and ¥15,000 for crew for more than fifteen years. This strategy was further strengthened in 1896 by the enactment of the Law Encouraging Marine Transportation and the Law Encouraging Shipbuilding. In consequence, Japanese-owned ships accounted for 6% of the world total by 1924 (see Tables 9.10 and 9.11).

Let us briefly review the development of Japan's *Z*-sector in the early Meiji period chronologically.

TABLE 9.10. Growth of Japanese shipping, 1870–1940

	No. of ships	Gross tonnage	Index
1870	25	25	100
1875	149	68	272
1880	210	66	264
1885	228	89	356
1890	325	143	572
1895	528	331	1,324
1900	859	524	2,096
1905	1,390	932	3,728
1910	846[a]	1,146[a]	4,584[a]
1915	1,155[a]	1,826,068[a]	136[a]
1920	1,940[a]	2,995,878[a]	223[a]
1925	2,087[a]	3,919,807[a]	291[a]
1930	2,060[a]	4,316,804[a]	321[a]
1935	2,146[a]	4,085,650[a]	304[a]
1940	1,819	5,682,705	422

Note: For 1870 – 1910, 1870 = 100; for 1915 – 1940, 1911 = 100.

Sources: [a] Lloyd's statistics; Kaneko (1964): 164.

TABLE 9.11. Number of ships in the world and Japan

	(1) World	(2) Japan	(2)/(1) %
1914	49,092	1,708	3.5
1919	50,928	2,325	4.6
1924	64,023	3,843	6.0

Sources: Lloyd's statistics; Kaneko (1964).

In 1880, the government opened the Kamaishi Steel Mill with its two 2.5-ton blast furnaces. The establishment of today's Onoda Cement and Nihon Railways, Japan's first private railway, were approved by the government in 1881. In 1886 the Hyogo Ship Yard was leased (and later sold) to a private company, and in the same year Tokyo Electric Light, which later became Tokyo Electric, started operation. All these were private in the initial phase.

It was only after 1896, however, that machine manufacturing flourished; many new companies were established in that year, including Osaka Gas, Nihon Kisha Seizo (train manufacturer), Uraga Dockyard, and Kawasaki Shipbuilding. The private companies created in the first five years of the twentieth century are listed in Table 9.12. The symbol of the emergence of the *Z*-sector in Japan, however, was the National Yawata Steel Mill, which started to operate in 1901, but it was only late in World War I that the *Z*-sector really expanded rapidly.

TABLE 9.12. Date of establishment of businesses in Z-sector

1900 (Meiji 33)	International Oil Company established Tokyo Electric Railway Co (Tokyo Denki Tetsudou) established
1901 (Meiji 34)	The Sumitomo took over Nihon Chukou-zyo and began Sumitomo Chukou-zyo
1902 (Meiji 35)	Tanaka factory (at Kouriyama) opened carbide factory Tokyo Street Railway (Tokyo Sigai Tetsudou) Co Ltd established
1903 (Meiji 36)	Mitsui Trading Company established the shipbuilding section at Moji (ancestor of present Mitsui Shipbuilding & Engineering Co Ltd) Shinagawa White Brick (Shinagawa Sirorenga) Co Ltd established Tokyo Electric Railway (Tokyo Densya Tetsudou) Co Ltd began to operate between Sinbashi and Shinagawa Osaka municipal Electric Railway (Osaka Shiei Denkikidou) began to operate between Hanazonobashi and Chikkou-umetatechi
1904 (Meiji 37)	Asano Cement opened rotary kiln made in USA Yokohama Railway (Yokohama Tetsudou) Co Ltd established Shibaura Seisakujyo Co Ltd established Takarada Sekiyu Co Ltd acquired Nippou Sekiyu Co Ltd Nihon Sekiyu and Takarada Sekiyu made equal joint contributions to the foundation of Kokuyukyoudou-hanbaizyo
1905 (Meiji 38)	Osaka Lines Ltd began to operate line between Osaka and Dalian. Furukawa Mining Co Ltd established Mitsubishi-Goushi established Kobe Shipbuilding Co Ltd Kobe Steel Ltd established Nippon Yusen Kaisha opened the line between Hakodate and Otomari Tokyo Electric Co Ltd began technical co-operation with USGE

From Fig. 9.2, which plots log values of factory generator capacity and the value of factory buildings as proxies of *Z*-sector capital stock, it is apparent that the *Z*-sector grew very fast between 1910 and 1920, a period of expansion stimulated by World War I. None the less, it did not reach the stage of take-off for self-sustaining growth; this can be seen from the fact that with the recovery of the European economies after World War I there was no further growth (see Fig. 9.3). The figure indicates that until 1934 government-owned factories occupied a large share in the total *Z*-sector. Thus, the development of the *Z*-sector still depended on government support. At this stage it is guessed that it was in the latter half of 1920s, at the earliest, that *Z*-sector might have reached the take-off stage.

Table 9.13, which provides data on the external trade of individual *Z*-sector industries, indicates the situation in international competitiveness and the development of import substitution. For example, iron and steel exports were nil or negligible until 1910, and even in 1930 exports were only 10% of imports.

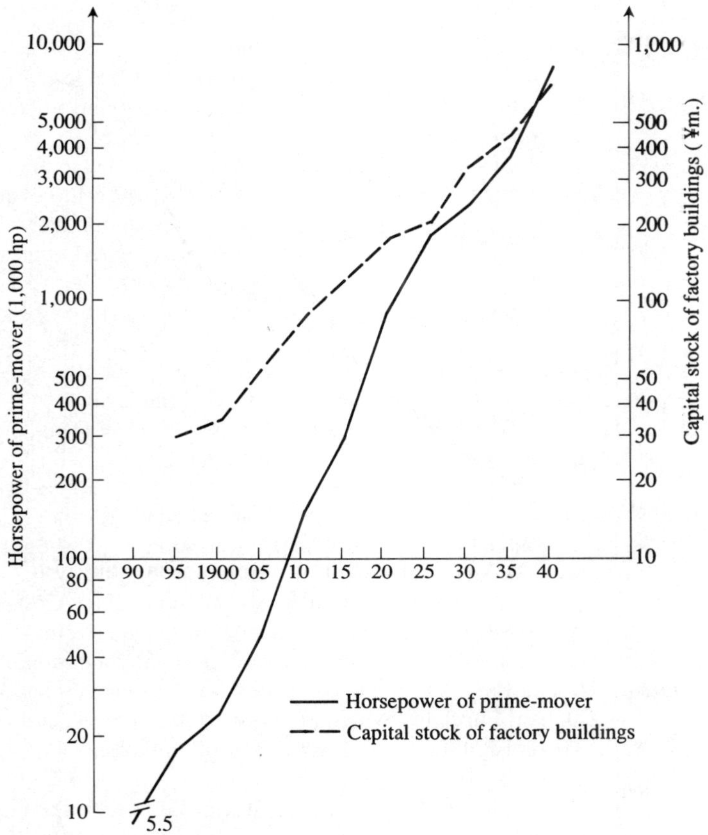

Note: Based on the data of Table 9.1.

FIG. 9.2 Historical developments: horsepower of prime-mover and capital stock of factory buildings in *Z*-sector.

We can therefore infer that domestic production was far short of domestic demand. While the export/import ratio for metal products was rising, that of machines was extremely low even in 1930; and although the figure for the chemical industry looks fairly high, in reality the chief chemical product exported was matches.

We can summarize this brief review of *Z*-sector development before World War II as follows:

1. Prior to World War I the *Z*-sector was still at an embryonic stage. Despite direct government operation and/or large subsidies, the success of the *Z*-sector was by no means assured.
2. From 1914 to 1920, Europe's industries stagnated because of war damage. The subsequent strong demand for *Z*-sector goods stimulated Japan's *Z*-sector, which created a firm base for later development. Many

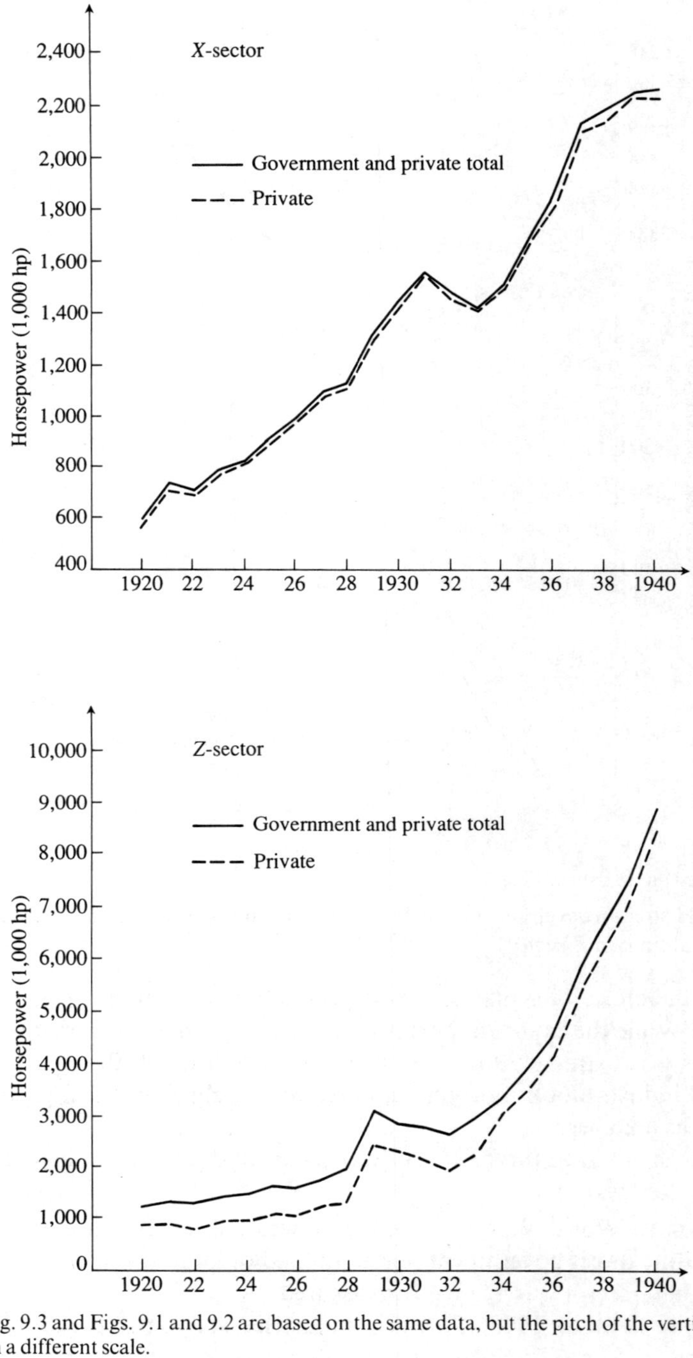

Note: Fig. 9.3 and Figs. 9.1 and 9.2 are based on the same data, but the pitch of the vertical line (hp) is on a different scale.

FIG. 9.3 Capital accumulation.

TABLE 9.13. Imports and exports of manufactured products for Z-sector (¥m.)

	Iron		Metal Manufactures		Vessels		Machinery		Chemicals[a]	
	I	E	I	E	I	E	I	E	I	E
1890	2.2	—	3.4	0.4	0.7	—	7.3	0.1	2.0	4.2
				(11.8)		(—)		(1.4)		(210.0)
1895	6.0	—	4.7	0.9	4.7	0	14.4	0.3	5.2	7.6
				(19.2)		(—)		(2.1)		(146.2)
1900	21.7	—	12.7	1.0	2.6	0.1	17.8	0.6	7.9	11.0
				(7.9)		(3.9)		(3.4)		(139.2)
1905	35.7	—	11.3	2.3	7.7	0.8	38.1	3.1	17.1	17.2
				(20.4)		(10.4)		(8.1)		(100.6)
1910	32.5	0.1	11.1	3.5	0.9	0.3	23.6	3.4	26.0	18.4
		(0.3)		(31.5)		(33.3)		(14.4)		(70.8)
1915	34.4	0.5	4.1	7.8	2.6	0.4	14.7	10.0	30.6	30.8
		(1.5)		(190.2)		(15.4)		(68.0)		(100.7)
1920	264.3	13.5	47.0	38.4	0.3	15.6	149.5	50.0	141.0	69.0
		(5.1)		(81.7)		(5,200.0)		(33.4)		(48.9)
1925	102.8	6.8	25.5	22.2	4.6	1.9	156.6	29.7	131.2	51.7
		(6.6)		(87.1)		(41.3)		(19.0)		(39.4)
1930	93.6	9.1	16.1	22.4	0.2	2.9	125.1	35.3	106.2	29.3
		(9.7)		(139.1)		(1,450.0)		(28.2)		(27.6)

Note: I − Imports, E − Exports. Figures in parantheses = E/I × 100.

[a] Matches made up the bulk of chemical exports.

Source: Toyo Keizai Shinposh (1935).

machine industries, steel mills, and chemical factories were established in this period, including Hitachi, Mitsubishi Engines (today's Mitsubishi Heavy Industry), and Nippon Soda, all of which started operation in 1920.

3. In the 1920s, the boom of war-time demand receded and Japan's Z-sector stagnated as Europe's industries recovered. Z-sector profit rates were still below those of the X-sector, so the Z-sector could not attract investment of X-sector profits as the profit rate of the X-sector was higher.
4. The Depression continued into the early 1930s but, as will be seen later, the Z-sector profit rate far exceeded that of the X-sector at this time, thus accelerating the growth of the Z-sector. A further contributory factor to this situation was the Manchuria Incident (1931) and the outbreak of the Japan–China War in 1937, which put the Japanese economy on to a war-time footing.
   Our tentative conclusion is that the take-off of the Z-sector had started in the early 1930s despite these difficulties, because the Z-sector profit rate had almost caught up with that of the X-sector at this period.
5. As will be shown later, R_z, the Z-sector profit rate, which had surpassed R_x in the early 1930s, bottomed in 1932 due to the Depression but then rose above R_x again. Evidence of profit rates suggest that development of Z-sector after 1933 took off as is suggested by Fig. 9.4 (ii). As a matter of fact the data on profit rates are scattered, some indicating an earlier take-off. So we conclude that the Z-sector took off in the early 1930s, since R_z was higher than R_x in 1928. Thus the drop of R_z during 1930 through 1932 is considered as the influence of Depression.

9.3. The $K_X - K_Z$ Graph

Having investigated the development of the X- and Z-sectors separately, we now test the dynamic path of capital accumulation in both sectors by plotting actual capital stock–the 'K_x–K_z' graph.

We first provide a bird's-eye view of long-term capital accumulation over the entire seventy-year period (1868–1940) and then focus in more detail on the critical periods when capital accumulation approached a trajectory of take-off. In providing an overview, we must first justify our selection of proxy variables for capital stock, i.e. generator capacity and the value of factory buildings. If we are to establish a proportional relationship between capital stock and real output, review of the path of real output in the two sectors is a necessary preliminary. Even if there is no fixed coefficient, such a review can provide a general indication of the trend so long as the path that the coefficients follow in each sector is parallel. Our graph of $X - Z$ will thus be an approximation of the path of $K_x - K_z$. For similar reasons, a graph of $L_x - L_z$, that is, the number of employees in the X- and Z-sectors, will be another approximation if the variation of capital–labour ratios behaves in a similar way in both sectors.

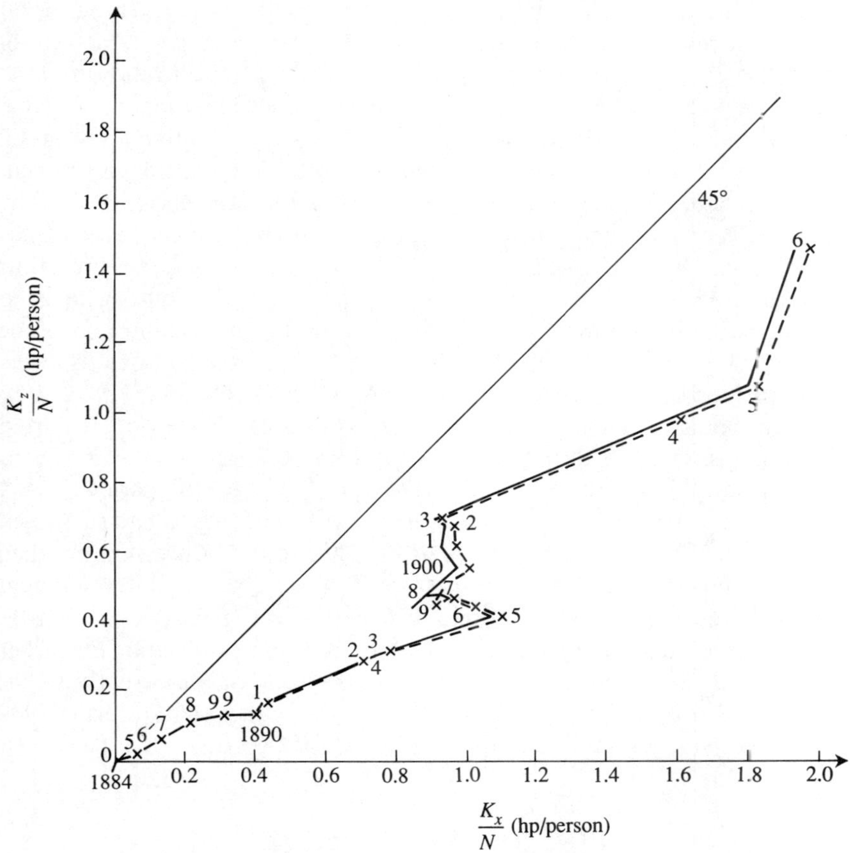

Note: Line shows *X*-sector excluding 'miscellaneous manufacturing'; broken line shows *X*-sector including this.

FIG. 9.4(i) Growth path 1884–1906: horsepower of prime-mover per population.

Figs. 9.4 (i) and (ii) plot the generator capacity of the *X*- and *Z*-sectors for 1884–1906 and 1906–36, respectively. As they are based on the Minami estimates, they show the path of expansion in private factories (note that the solid line in the graph does not include 'miscellaneous manufacturing'). Generator capacity is expressed on a per capita basis since this gives a more accurate reflection of capital accumulation in comparison with population growth. Can we explain the take-off of the *Z*-sector by contrasting these graphs with Fig. 9.5, which is a reproduction of the phase diagram of Part I?

As we have already seen, the size of the *Z*-sector was negligible compared with the *X*-sector in the early Meiji era. In other words, the point (K_x, K_z) must have been located far below the minimum critical effort curve. It is thus inferred that if industrial production had been subject to only market forces, the point (K_x, K_z) would have fallen to $(K_x^*, 0)$, which corresponds to a poverty

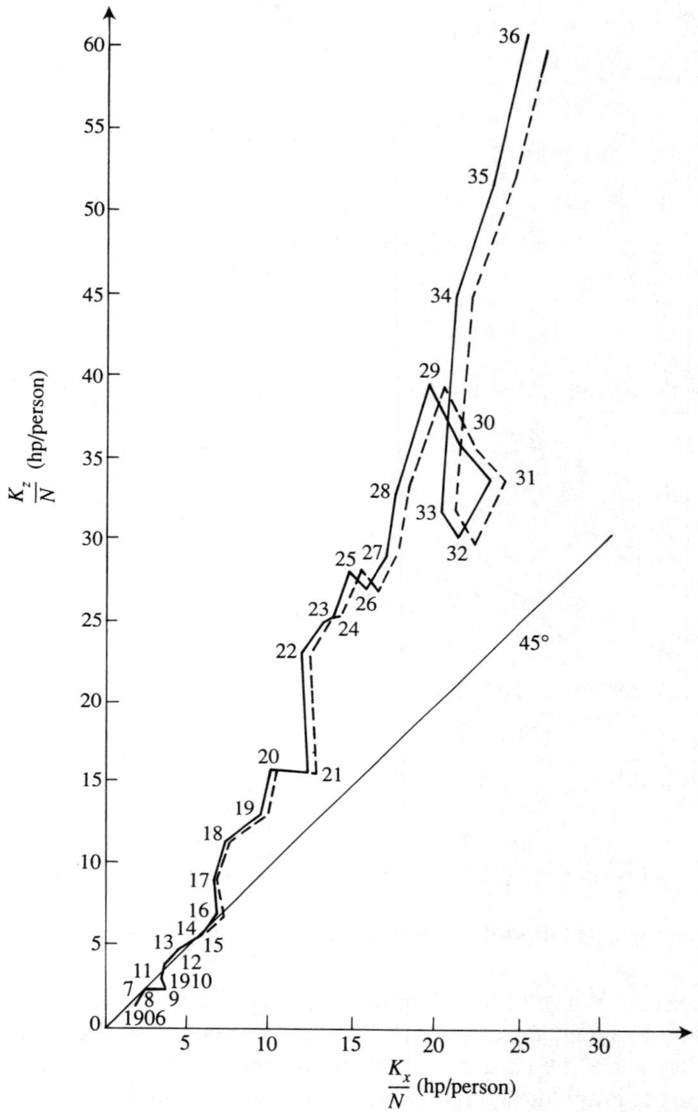

Note: Line shows X-sector excluding 'miscellaneous manufacturing'; broken line shows X-sector including this.

FIG. 9.4(ii) Growth path 1906–36: horsepower of prime-mover per population.

trap. However, as the government intervened heavily in the market through tax, subsidies, and financial measures, such downward movement took place only occasionally.

From 1884 through 1890, (K_x, K_z) moved horizontally east, indicating that X-sector expansion was predominant. Then from 1890 through 1895, the

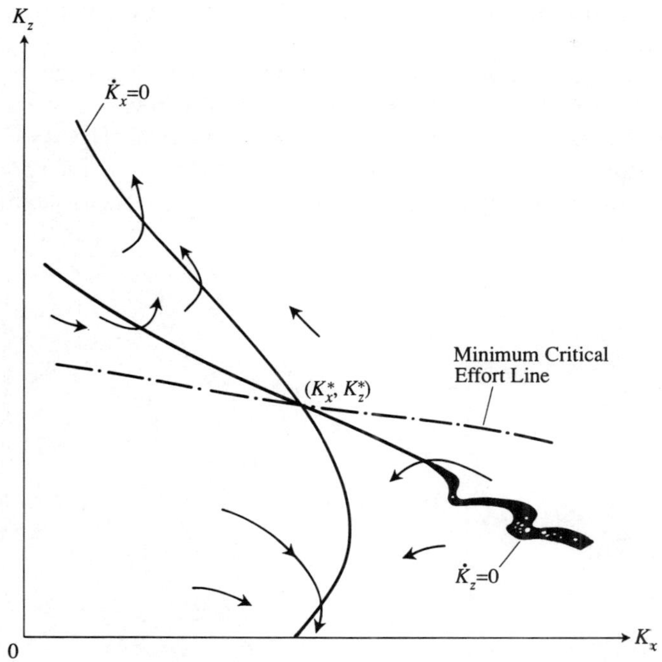

FIG. 9.5 Minimum Critical Effort Line.

X-sector continued to expand but the Z-sector also started to grow. This was the period after the Sino-Japanese War. After 1895, however, there was an absolute decline in per capita generator capacity in the X-sector, reflecting a periodic recession in the cotton-spinning industry. At the same time, (K_x, K_z) moved north because of the creation of new enterprises in the Z-sector.

During several years in the early 1890s both the X- and the Z-sectors expanded substantially, reflecting the Russo-Japanese War. The point K_x, K_z expressed in per capita terms was now located towards the north rather than the east, whereas previously it had been located far below the 45°-line.

In the period between the end of the Russo-Japanese War (1905) and World War I, Z-sector growth was slow, and K_x, K_z often moved horizontally east. It is obvious that World War I favoured the growth of Japan's Z-sector in Japan: Fig. 9.4 (ii) indicated that (K_x, K_z) moved rapidly north from 1915 through 1920. Although there was some period of stagnation in Japan while the European industries recovered from war damage, the Z-sector expanded more rapidly than the X-sector until the Great Depression. This is one of the reasons we conclude that the Z-sector was about to take-off before the depression.

The Great Depression seriously affected both the X- and the Z-sectors; our diagram reflects this in that K_x, K_z moved south or west and sometimes south-west during the period 1929–33. Although it was impossible to plot all points in Fig. 9.4 (ii), it should be emphasized that the Z-sector expanded very rapidly from 1934 to 1940.

Let us next trace the path of capital accumulation by using factory building statistics. Fig. 9.6 plots (K_x, K_z) points based on the Ishiwata's estimates (see n. 3), in Ohkawa *et al.* (1966) which cover only private capital. The solid line relates to manufacturing sectors only, while the broken line plots (K_x, K_z) points where K_z includes the electricity and gas sectors. The value of factory buildings per

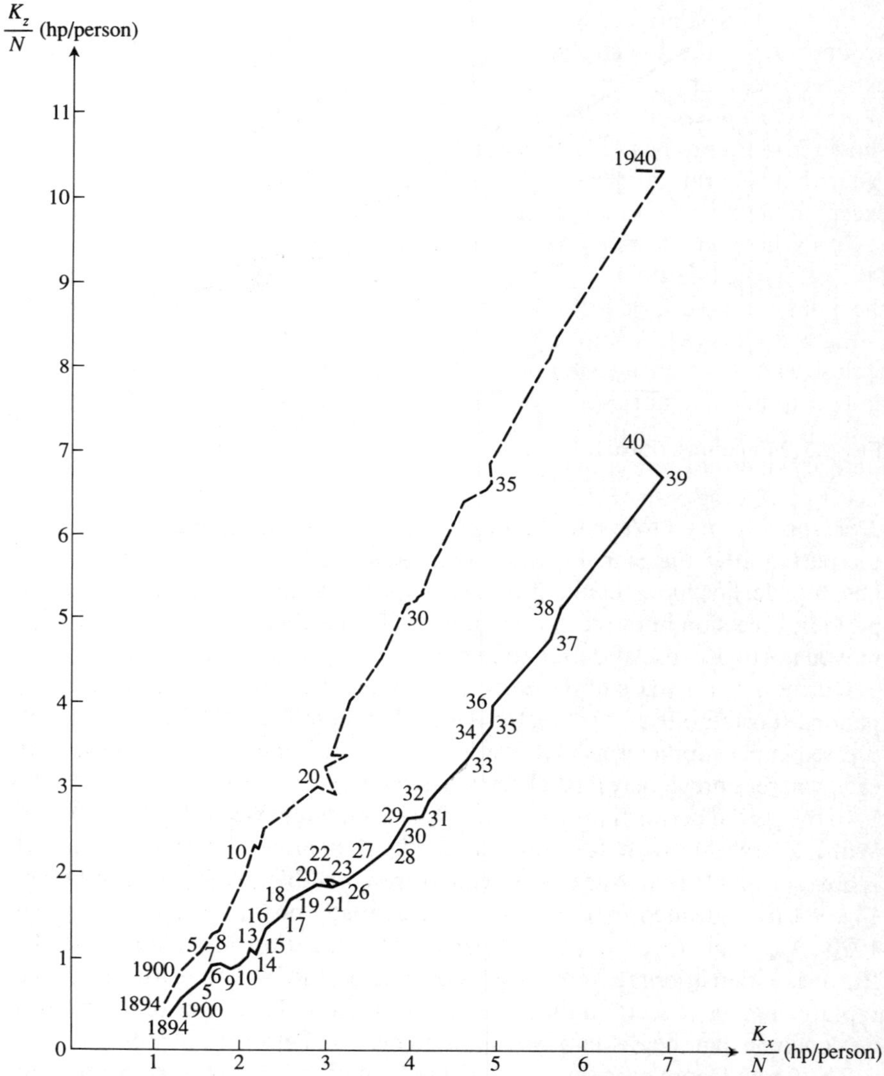

Note: Line shows Z-sector of manufacturing only; broken line shows Z-sector including electricity and gas industries.

FIG. 9.6 Growth path 1894–1940: capital stock of factory building.

capita of the population was only about ¥1.00 from 1894 to 1900, and accumulation was slow. Factory building data also show that K_z expanded rapidly during World War I, stagnated in the early 1920s, and grew rapidly after the mid-1930s as the economy recovered from the Great Depression.

If we trace production points (X, Z) using Shionoya estimates at constant prices for 1934–6, the growth path appears more dramatic since production fluctuates more wildly than capital stock. Fig. 9.7 plots such production points from 1874 to 1939. It is clear that until 1890 growth was mostly in the X-sector, as the point moves east. Later, business cycles emerged in the X-sector, and around 1900–7 the X-sector production shrank, the Z-sector growth was also extremely slow at this time. The importance of the Russo-Japanese War and World War I in stimulating the expansion of the Japanese Z-sector is clearly shown in our graph as the point moved north in the corresponding periods. After that, the rate of growth of the Z-sector exceeded that of the X-sector except in the early 1920s and during the Great Depression.

From these observations, we can infer that Japan's Z-sector was poised for take-off in the late 1920s. Since it is our main task in this chapter to identify the point of take-off, we focus our analysis on the period from 1920 to 1940.

Fig. 9.8 plots (K_x, K_z) in terms of generator capacity, the solid line tracing exclusively that of private enterprises while the broken line covers both private and government factories. It shows that government factories accounted for a large share of the Z-sector until 1925. As capital accumulation in government factories does not reflect market mechanisms, it is uncertain whether the Z-sector was on the path of self-sustaining growth.

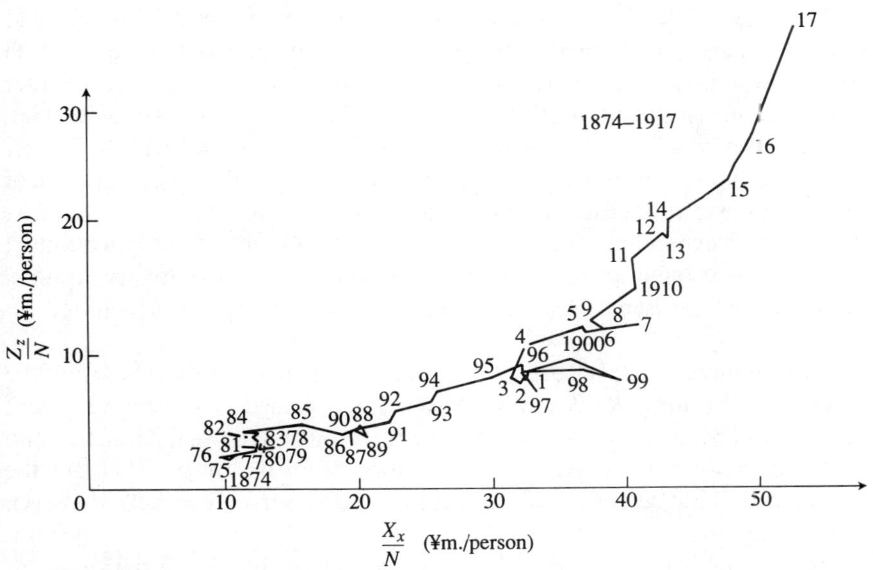

FIG. 9.7(i) Growth path 1874–1917: real per capita production.

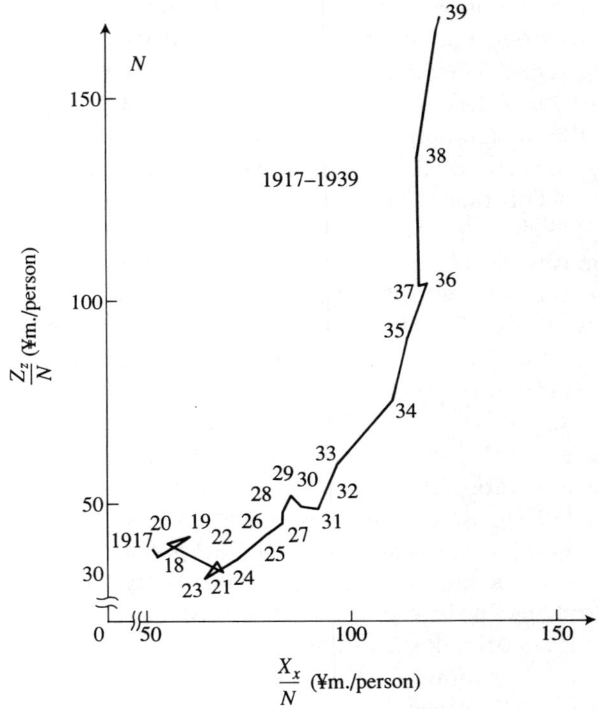

FIG. 9.7(ii) Growth path 1917–39: real per capita production.

To be more concrete, in 1920 the government presence in the X-sector was negligible: generator capacity per capita in the X-sector was 10 6 hp of which 10.4 hp were in private enterprises. In the Z-sector the government presence was far more significant: of the total generator capacity per capita of 23.0 hp, only 15.7 hp were in private enterprises. In other words, K_z was 46% larger than it would have been with private enterprises only. This was in fact one of the minimum critical efforts of the government that pushed (K_x, K_z) north. As the Z-sector expanded after 1925, however, the weight of the government enterprises was reduced significantly: by 1933 generator capacity per capita in the Z-sector had risen to 41.3 hp of which private enterprises accounted for 34.0 hp.

The minimum critical efforts of the government aimed at the take-off of the Z-sector, ie. pushing (K_x, K_z) upward, were not confined to direct government operation of businesses, however: other efforts—through fiscal, financial, and trade policies—will be investigated in more detail in Chapter 10. But the reasons that the Z-sector took off in the late 1920s seem to have been that even considering private enterprises alone, K_z had not only climbed above the minimum critical effort curve, but also above $\dot{K}z = 0$ or the $dKz/dt = 0$ curve; R_z (the Z-sector profit rate) became comparable to R_x, or the X-sector profit rate,

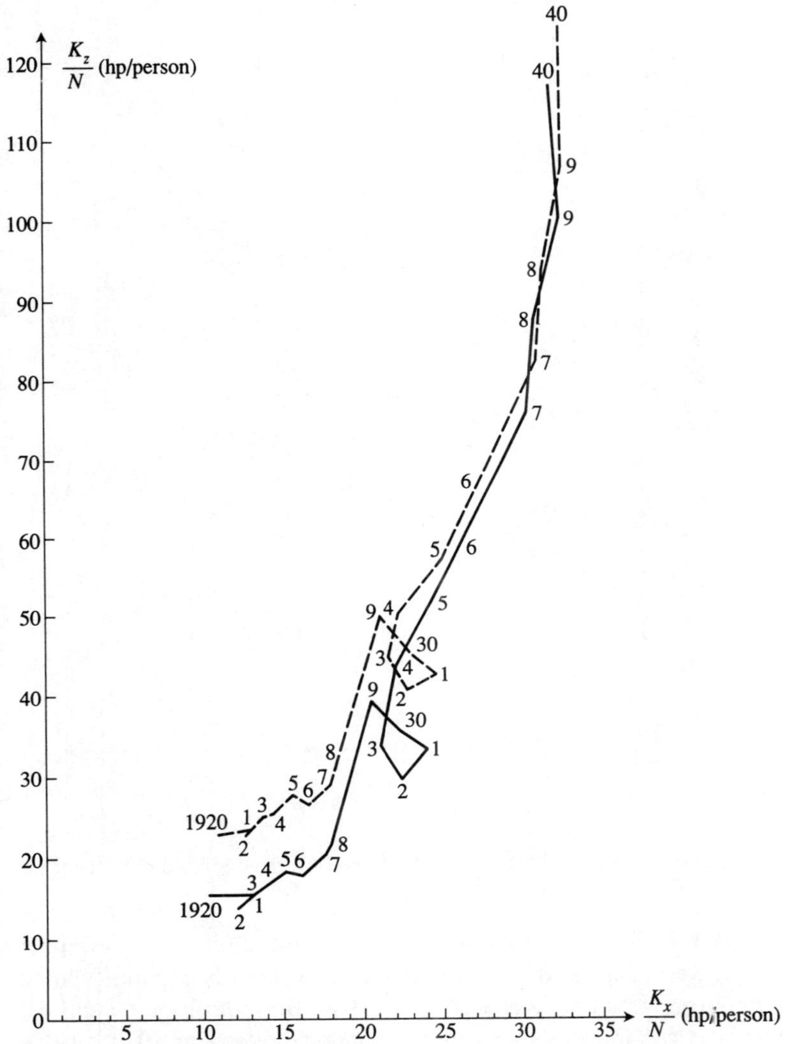

Note: Line shows private factories; broken line includes government factories as well.

FIG. 9.8 Growth path 1920–40: horsepower of prime-mover.

which meant that some portion of the X-sector profits were invested in the Z-sector.[4]

Figs. 9.9, 9.10 (i), and 9.10 (ii) plot (K_x, R_z) in terms of the value of factory buildings and the number of employees once, for private enterprises only, and once for the whole sector. These diagrams also support our argument that the Z-sector took off in the late 1920s. The per capita value of Z-sector factory buildings shrank in the early 1920s and stagnated at the time of the Great Depression, but in the 1930s (K_x, K_z) moved rapidly north rather than east.

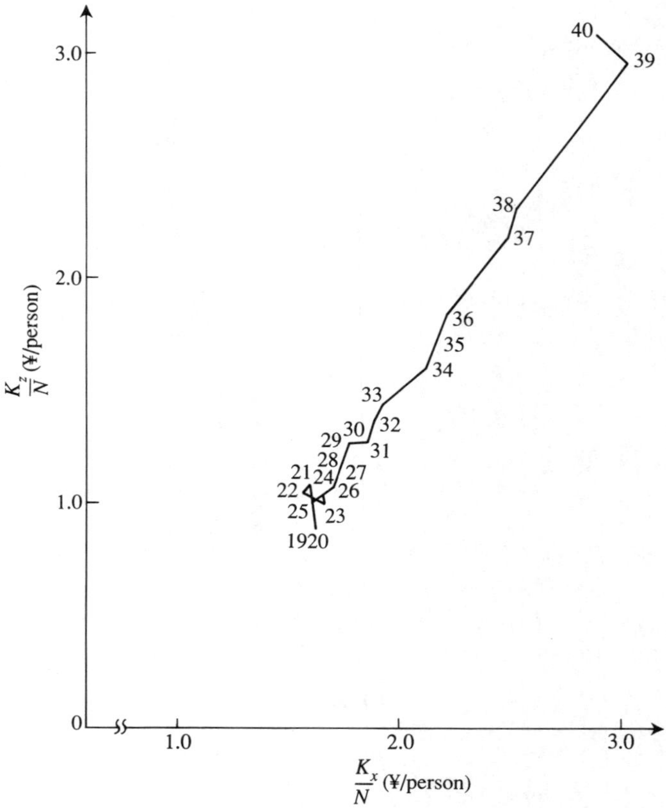

FIG. 9.9 Growth path 1920–40: factory building (government and private factories).

Unfortunately, we cannot see whether the point is ultimately established in the north-west, that is whether the Z-sector eventually expands while the X-sector contracts, as the data cease from 1940. But both Fig. 9.9 and Fig. 9.10 show that at least there was some movement north-west from 1939 to 1940.

A more dramatic change in the structure is seen in Fig. 9.10: the plot of (L_x, L_z), that is, the number of employees of the two sectors indicates a shift north-west from 1937 on.

Despite these interesting facts, however, we should be careful not to draw hasty conclusions because the Japanese economy went on to a war-time footing at this time, and market mechanisms were largely distorted by government controls. Although our model suggests that capital accumulation would have been orientated more towards the Z-sector in the 1930s so long as the assumptions about production technology and terms of trade were satisfied, we do not know for sure because the story of heavy industrialization in Japan was interrupted by World War II.

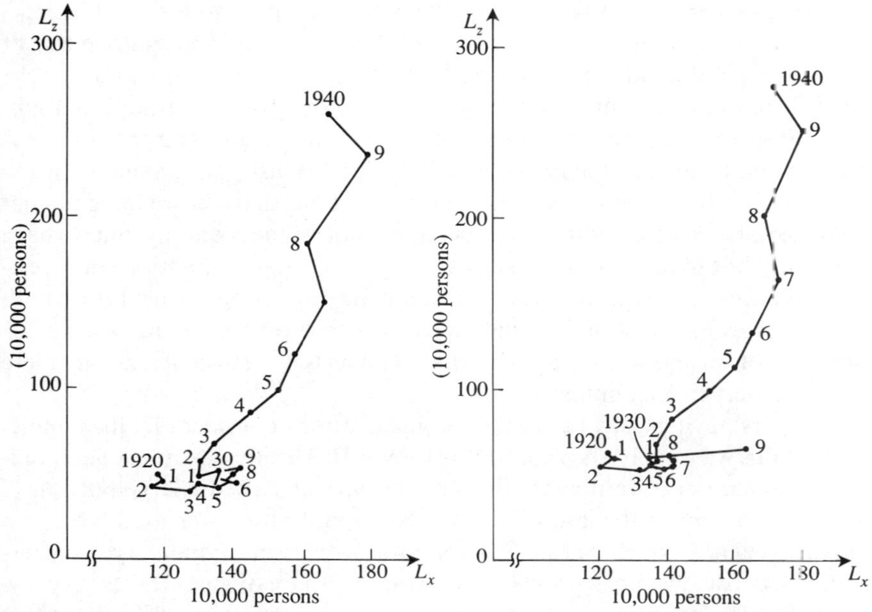

(i) 1920–40: private factories
(ii) 1920–40: government and private factories.

FIG. 9.10 Growth path: number of employees.

9.4. Profit Rates and Investment Allocation

The investment allocation mechanism established in Part I is general enough to be accepted as a theoretical hypothesis. The only possibility of this hypothesis not being substantiated is if investors choose to adopt longer-term perspectives rather than react to current inter-sectoral differentials in profit rates. To incorporate such an expectation, we must explicitly offer an alternative hypothesis regarding the formation of expectations. As stated in Part I, however, it is difficult to propose a definite expectation pattern without losing universality, and it is much more difficult to derive a dynamic capital accumulation path under such an assumption.

Several empirical difficulties have already been mentioned in relation to our assumptions, namely: (*a*) the model assumes that the only source of investment is the profit made in the *X*- and *Z*-sectors; (*b*) analysis based on differential equations in the theory must be converted to an analysis of discrete time making a proper time-period as the unit of time measure; (*c*) if the economy has significant tertiary sectors, such as distribution, services, and so on, we shall be unable to test our theory because we would have to ignore several major destinations as well as sources of investment.

As regards the first problem, it is to be noted that the *S*-sector surplus, especially that of landowners, was large, and that it was an important source of investment in the industrial sectors. Moreover, the tax on farm land was the most important tax source for transferring surpluses from the traditional subsistence sector to the industrial sectors. The role of the government, however, will be investigated in the next chapter. As for the last point, namely the exclusion of distribution and service sectors from the analysis, we imagine that these sectors did account for a significant portion of the economy, but we have no choice but to neglect them, both as sources for and destinations of investment, because the requisite data are unavailable. By dropping them from both sides, however, the issue becomes solely whether or not the net surplus or shortage of savings was significant. In other words, the effect of ignoring these sectors is very much diminished.

Workers' savings may be another problem, but we consider that the importance of this was much less prior to World War II. The introduction of compulsory purchase of government bonds might have absorbed a large portion of workers' savings in the late 1930s, but the capital raised was used largely to finance defence expenditure rather than for investment in industry. Furthermore, according to Lockwood (1954), workers' share of income was very low in 1930: 43%, as compared with the share of entrepreneurs (28%) or income from assets (29%).[5] In fact, it is natural that workers' savings are of little importance in the early stage of economic development because at that time the supply of labour at the subsistence wage level is unlimited, and wages, and thus savings, are correspondingly low.

In today's LDCs, great importance is attached to the role of foreign capital in development. Our theory also suggests that the introduction of foreign capital would push (K_x, K_z) towards the north, east, or north-east. As capital import is important as one of the means of minimum critical effort, we shall discuss this further in Chapter 10. Our conclusion on this point, however, is that the role of foreign capital in Japanese economic development prior to World War II was not large. This conclusion coincides with those of Lockwood (1954) and Baba and Tatemoto (1967).

For the above reasons, therefore, our analysis is limited to how the profits realized in the *X*- and *Z*-sectors only were reinvested in the two sectors in response to the profit rate differential.

As regards problem (*b*), using profit data from Mitsubishi Keizai Kenkyusho (1930–), which have the advantage of consistency of profit and investment data, permits biannual examinations. But industry coverage is limited. Profit data estimated from macro-economic statistics are available for annual intervals. Although using this data expands the coverage of the industrial sectors, the lack of equivalent capital stock data is a severe constraint. Thus we face a dilemma.

These approaches have their limits but they permit a first approximation for the testing of our theory. Our conclusion is that both the relationship between profit rates and capital stock in the two sectors and the hypothesis with regards to the allocation of investment between them which was shown in Part I are supported by the facts whichever data set are used.

Constraint (*c*) means we must confine ourselves to testing our hypothesis on a weak criterion. The hypothesis says that as long as R_x is not smaller than R_z, all *X*-sector profits will be reinvested in that sector, as will a portion of the *Z*-sector profits which will not be decreasing with respect to $R_x - R_z$. Since we cannot cover all industrial sectors in the empirical test, it is obviously impossible to test whether or not all *X*-sector profits are re-invested; and nor is it possible to confirm that total investment is equal to the total profit from the *X*- and *Z*-sectors. This is not an insuperable defect in the theory, but an inevitable abstraction to permit a clear result to be derived in the theoretical analysis. In reality, it will be sufficient for us to test whether the actual process of investment allocation is largely consistent with our hypothesis.

9.4.1. Outline of the empirical testing

Before going into the details of testing, which inevitably involve cumbersome matters of data, it would seem useful to explain what parts of the theory we are going to test and summarize our assumptions and interim results:

1. Our model assumes that *X*-sector production technology is subject to constant returns to scale and that the terms of trade *vis-à-vis* the *S*-sector decline as the production of *X*-goods increases. On the other hand, it is assumed that *Z*-sector production enjoys increasing return to scale, and that the similarly defined terms of trade of the *Z*-sector is independent of the level of production of *Z*-sector goods. While there are some additional assumptions on the nature of production functions, they are the normal assumptions consistent with rigorous analysis. The interim results indicate that profit rates (in terms of *S*-sector products) and real capital stock are mutually related in the following manner:

$$\frac{\mathrm{d}R_X}{\mathrm{d}K_X} < 0, \quad \frac{\mathrm{d}R_Z}{\mathrm{d}K_Z} \geq 0.$$

   Equality, however, does not hold for all K_z, but only for a specific value. The first step of our test is to confirm these results.[6]

2. The second step is to examine the relevance of our investment allocation equations. The model states that reinvestment of profit to the two sectors is subject to the following equations:

$$\dot{K}_X + \mu K_X = \xi_X(R_X - R_Z) \cdot R_X \cdot K_X + \zeta_X(R_X - R_Z) \cdot R_Z \cdot K_Z \quad (9.1)$$

$$\dot{K}_Z + \mu K_Z = \xi_Z(R_X - R_Z) \cdot R_X \cdot K_X + \zeta_Z(R_X - R_Z) \cdot R_Z \cdot K_Z \quad (9.2)$$

$$\xi_X + \xi_Z = 1, \ \xi_X, \ \xi_Z \geqq 0$$

$$\zeta_X + \zeta_Z = 1,\ \zeta_X \geqq 0,\ \zeta_Z \geqq 0$$

where $\xi_X = 1$ for $R_X - R_Z \geq 0$
function ξ_X is non-decreasing with respect to $R_X - R_Z$, and
$\zeta_X = 0$ for $R_X - R_Z \leqq 0$
function ζ_X is non-decreasing with respect to $R_X - R_Z$
where R, μ, K, ξ, and ζ stand, respectively, for profit rate, depreciation and stock of capital, and investment allocation ratios which are defined as the functions of profit rate differentials.

To paraphrase this, profits realized in the two sectors are re-invested in the following ways. If the X-sector profit rate is higher than that of the Z-sector, all X-sector profit, $R_x K_x$, will be re-invested in the X-sector, as will part of the profit of the Z-sector, $\xi_x \cdot R_z \cdot K_z$. This proportion, ξ_x, will increase as $R_x - R_z$ expands, or more precisely, will not decrease as $R_x - R_z$ grows, and *vice versa*. In the special case when $R_x = R_z$, the profits of each sector will be re-invested in the original sector.

Because of very severe data constraints, we make the test on a weak criterion, ie. whether or not the relative size of investment in the two sectors corresponds positively to the profit rate differential, that is, whether $\dot{K}_x - \dot{K}_z$ is positively related to $R_x - R_z$. Of course, when the capital stock in one sector is very large relative to that in the other at a given point in time, and the absolute level of profit is much larger even if the profit rate is lower, the scale of investment in that sector may exceed that in the other despite an outflow of profits to the second sector. Equations (9.1) and (9.2) above do not exclude such a case, but the results of our empirical tests seem to support our hypothesis despite this possibility.

3. While we argued in the previous section that the Z-sector had taken off by the late 1920s, in the sense that the Z-sector started self-sustaining expansion, we can strengthen our argument by comparing the profit rates of the two sectors. In fact, we discovered that $R_x - R_z$ became negative around 1933 and $\dot{K}_z$ became larger than $\dot{K}_x$, that is, investment in the Z-sector surpassed that in the X-sector. This led us to conclude that the Z-sector had approached the stage of take-off several years before this.

9.4.2. Testing with data for large enterprises

Using data from the balance sheets and profit and loss statements of large enterprises, for estimating profit rates and investment in the X- and Z-sectors has the clear advantage that the data are mutually consistent even if the coverage of the sector is not complete. The data by Mitsubishi Keizai Kenkyusho (1930–) are available only from the 1930s on, however, so for the early period we have to rely on data that are less complete. Using the Overview of Japanese Economic Statistics, (Asahishimbun 1930), we compiled data on profit rates and paid-up capital for both the X- and Z-sectors. These statistics were collected from a sample of companies surveyed by the Nihon Kangyo Bank—170 companies in the early 1920s, and more thereafter. The only problem is that data are limited to registered capital and paid-up capital, which is not enough to trace capital accumulation as a whole. Even so, the data from this source do afford us a bird's-eye view over the 1920s, as will be described below.

Table 9.14, which compares paid-up capital at the end of business years and the ratio of profit to paid-up capital, indicates that the *Z*-sector profit rate exceeded that of the *X*-sector from 1925 on, whereas in 1922 R_x was 1.80% and R_z was 1.46%. As throughout the period the absolute capital stock was much larger in the *X*-sector, however, the *amount* of investment in the *X*-sector exceeded that in the *Z*-sector (compare cols. 2 and 4). If we compare the net increase in paid-up capital of the *X*- and *Z*-sectors with profits, (cols. 8 and 9) as estimated by applying the profit rate to the total paid-up capital, we find that the net increase in paid-up capital in the *Z*-sector became much larger than *Z*-sector profits. This implies that some *X*-sector profits were re-invested in the *Z*-sector.

While the profit rate data we used were for the period September–February, we estimated profits by applying this rate to the year-end paid-up capital, thereby underestimating profit by some 30% since March–August profits tend to be correspondingly higher. Had the profit estimates been modified accordingly, the ratio of total profit in the *X*- and *Z*-sectors as a source of investment in the *X*- and *Z*-sectors will increase. Since there is no direct short-run correspondence between net investment and an increase in paid-up capital, however, we shall not go into details at this moment.

If we apply a weak criterion for our test, namely that the ratio of net *Z*-sector investment to that in the *X*-sector is positively related to $R_z - R_x$, then the data support our hypothesis, in that $R_x - R_z$ became negative and $\mathrm{d}K_x/\mathrm{d}K_z$ declined. Or, more strictly, the net flow of reinvestment (estimated by the sign of $[R_i \cdot K_i - \mathrm{d}K_i]$ (where $i = x, z$), as indicated by the arrows in Table 9.14, supports our hypothesis that the Z-sector took off in the mid-1920s.

For the 1930s, we used the Mitsubishi data on 190 large companies which permitted comparison of net increases of total assets and fixed assets, not paid-up capital, with realized profits. From Table 9.15, which shows the profit rate differential and investment allocation from 1930 to 1939 for half-yearly periods, the following findings may be summarized:

1. Whether total assets or fixed assets were used to measure profit rates, *X*-sector profit rates fell below those of the *Z*-sector after around 1933, whereas before that year the *X*-sector profit rates were definitely higher.
2. Whether total assets or fixed assets were used to estimate net capital increases, $\dot{K}_x$ was larger than $\dot{K}_z$ until 1932 but smaller thereafter. One has to be cautious, however, because the period 1929–33 was greatly influenced by the Depression, and there might have been extreme fluctuations.

TABLE 9.14. Profit rates and distribution of investment (¥m.)

	Paid-up capital[a]				Rate of profit[b] over paid-up capital (%)		(7) (5)–(6)	Profit		Direction of profit and investment	
	(1) *X*-sector	(2) change	(3) *Z*-sector	(4) change	(5) *X*-sector	(6) *Z*-sector		(8) *X*-sector	(9) *Z*-sector	*X*-sector	*Z*-sector
1922	4,169		1,326		1.80	1.46	0.34	75.2	19.4		←
1923	4,235	66	1,321	–5	1.69	1.51	0.18	71.5	20.0		←
1924	4,337	102	1,331	10	2.02	1.48	0.54	87.4	19.7		←
1925	4,528	191	1,345	14	1.23	1.42	–0.19	55.7	19.1		→
1926	4,691	163	1,438	93	1.13	1.34	–0.21	53.0	19.3		→
1927	4,894	203	1,542	104	1.30	1.17	0.13	63.8	18.1		←
1928	5,005	111	1,591	49	0.97	1.11	–0.14	48.7	17.6		

Notes: *X*-sector: dyeing and weaving, foods and beverages, and miscellaneous manufactures; *Z*-sector: chemicals and machinery.

[a] Paid-up capital at year end.

[b] Rate of profit is second half of the year (Sept.–Feb.).

Source: Asahishimbun (1930).

Nevertheless, we can firmly conclude that the Z-sector profit rate definitely exceeded that of the X-sector after 1934.

3. The difference between net investment in the two sectors as measured by fixed assets is largely positively related to the profit rate differential; that is, $(\dot{K}_x - \dot{K}_z)$ is a positive function of $(R_x - R_z)$. This is easily confirmed by Fig. 9.11. However, when total asset data are used, the relation becomes ambiguous (see Fig. 9.12). This is probably because the short-term assets included in total assets depend more on financial sources other than the X- and Z-sectors. In particular, they may reflect large firms' short-term borrowing from small-scale firms and absorbing the latters' savings surplus, a distortion due to the fact that our data cover only large companies (see Fig. 9.12).

Although the difference in net increases in capital in the two sectors can thus be related to the profit rate differential, this does not directly prove our hypothesis on investment allocation. This is because the hypothesis is related to the *proportion* of profit re-invested in each sector, not to the *absolute size* of re-investment. A direct testing will face a difficulty as the coverage of data on profits and investment is limited to that of large firms. Additional sources of investment and leakage of savings both make it meaningless to test the hypothesis if they are significantly large. A major factor in making the residual larger is borrowing and lending between large and small enterprises. Thus, we cannot make a strict test of our hypothesis, but on the other hand we obtain information about how much large firms depend on small ones to help finance their investment, or, conversely, the extent to which large firms finance the investment of small firms.

Table 9.16 tries to answer this question. The ratio of the net increase in the fixed assets of large firms in a given year to the net profit in the previous year was below 100% in the Z-sector until around 1933, rose above 100% for a few years thereafter, but declined again in 1936–7. The equivalent ratio in the X-sector had already been lower than 100% for many years. This suggests that large firms had financed the investment of small enterprises in the years when the ratio was below 100% in the Z-sector. Large enterprises may have established many subsidiaries in this period.

If we compare net increases in total assets with net profit in the Z-sector, it turns out that the ratio dKz/dt over $R_{z-1} \cdot K_{z-1}$ rose sharply after 1933, reaching 300% and 400%. In the X-sector, the ratio stayed below 100% for some years, but occasionally exceeded 100%, probably because large enterprises depended for short-term borrowing on the savings of individuals and small firms which were deposited with commercial banks. Looking at the last column, we find that the relationship between the ratio of reinvestment to absolute profit and the profit rate differential largely accords with our hypothesis. Most importantly, the Z-sector had already established its advantage in attracting investment.

TABLE 9.15. Investment allocation and profit differential (¥m.)

	Net profit / fixed assets		Differential of profit rate (%)	Change of fixed assets		(6)	Net profit / fixed assets		Differential of profit (%) (9)	Increase of total assets		(12)
	(1) *X*-sector	(2) *Z*-sector	(3) (1)–(2)	(4) dK_X	(5) dK_Z	(4)–(5)	(7) *X*-sector	(8) *Z*-sector	(7)–(8)	(10) $dK_{X'}$	(11) $dK_{Z'}$	(10)–(11)
1930 2nd half	1.44	1.19	0.25				0.84	0.65	0.19			
1931 1st half	3.77	–0.09	3.86	–3.0	–14.0	11.0	2.18	–0.05	2.23	5.0	–29.0	34.0
2nd half	3.42	1.01	2.41	–1.0	–5.0	4.0	2.02	0.56	1.46	–39.0	–30.0	–9.0
1932 1st half	4.35	1.49	2.86	2.0	5.0	–3.0	2.49	0.82	1.67	57.0	6.0	51.0
2nd half	4.78	2.99	1.79	–4.0	–14.0	10.0	2.76	1.59	1.17	–19.0	30.0	–49.0
1933 1st half	5.52	5.42	0.10	–11.0	–17.0	6.0	3.09	2.82	0.27	30.0	14.0	16.0
2nd half	5.62	6.45	–0.83	–8.0	120.0	–128.0	3.14	3.32	–0.18	–10.0	257.0	–267.0
1934 1st half	6.11	6.97	–0.86	–3.0	–87.0	84.0	3.30	3.22	0.08	52.0	70.0	–18.0
2nd half	6.21	7.86	–1.65	6.0	344.0	–338.0	3.33	4.00	–0.67	26.0	468.0	–442.0
1935 1st half	6.15	8.45	–2.30	37.0	25.0	12.0	3.33	4.16	–0.83	51.0	140.0	–89.0
2nd half	6.59	8.08	–1.49	29.0	131.0	–102.0	3.55	3.87	–0.32	61.0	352.0	–291.0
1936 1st half	6.58	7.89	–1.31	42.0	93.0	–51.0	3.48	3.74	–0.26	111.0	229.0	–118.0
2nd half	6.75	8.12	–1.37	38.0	67.0	–29.0	3.57	3.75	–0.18	76.0	234.0	–158.0
1937 1st half	7.83	9.00	–1.17	14.0	123.0	–109.0	3.78	3.94	–0.16	221.0	500.0	–279.0
2nd half	7.57	9.64	–2.07	40.0	89.0	–49.0	3.68	3.73	–0.05	69.0	783.0	–714.0
1938 1st half	7.29	9.97	–2.68	–4.0	243.0	–247.0	3.50	3.71	–0.21	25.0	861.0	–836.0
2nd half	7.30	9.89	–2.59	40.0	269.0	–229.0	3.46	3.60	–0.14	111.0	867.0	–756.0
1939 1st half	7.76	9.67	–1.91	4.0	229.0	–233.0	3.63	3.46	0.17	25.0	752.0	–727.0
2nd half	8.39	9.29	–0.90	43.0	291.0	–248.0	3.61	3.24	0.37	318.0	1,034.0	–716.0

Note: The profit rate here is the ratio of half-yearly profits to assets at the end of the business period, and differs from the normal definition. More detailed definition is given in the appendix to this chapter.

Source: Mitsubishi Keizai Kenyusho (1930–).

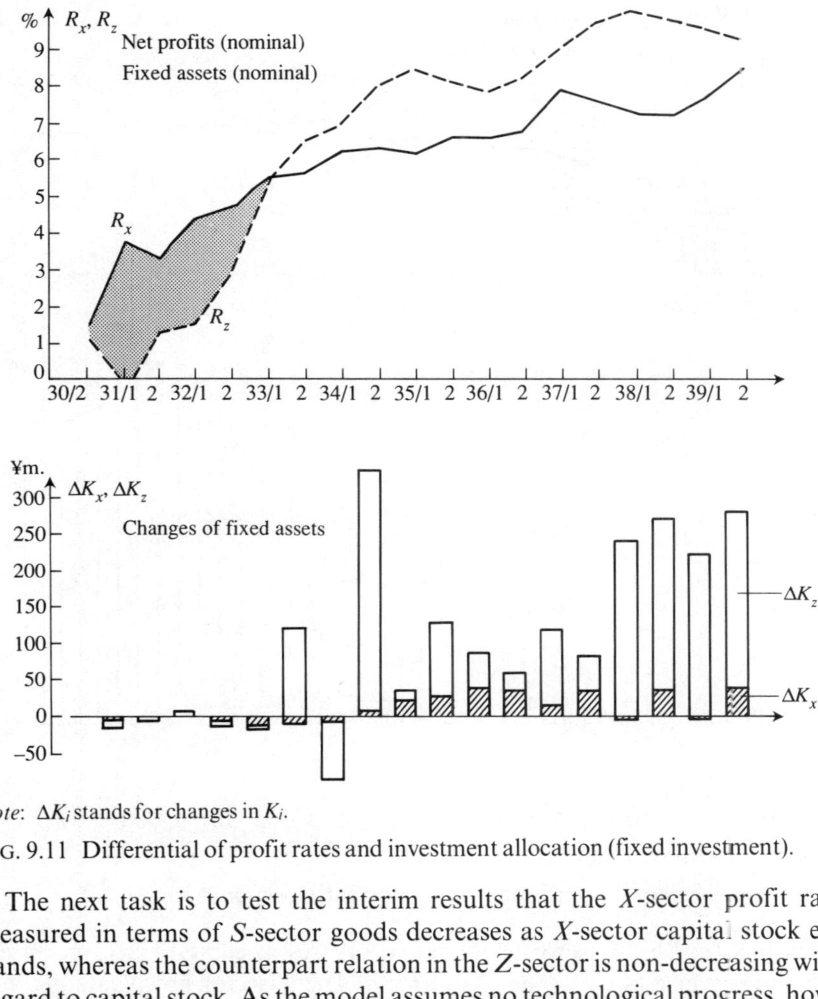

Note: ΔK_i stands for changes in K_i.

FIG. 9.11 Differential of profit rates and investment allocation (fixed investment).

The next task is to test the interim results that the X-sector profit rate measured in terms of S-sector goods decreases as X-sector capital stock expands, whereas the counterpart relation in the Z-sector is non-decreasing with regard to capital stock. As the model assumes no technological progress, however, an empirical test may not support our hypothesis unless we subtract the effects of technical progress. Needless to say, if technological progress is significant, the profit rate may rise as capital stock increases even though terms of trade decline and production is subject to constant returns to scale. In actuality, however, technological progress in the 1930s does not seem to have been so dominant as to overwhelm these two negative factors.

Table 9.17 and the corresponding Fig. 9.13, show that dR_x/dK_x is negative and that dR_z/dK_z is non-negative except for the period of the Great Depression. The plotted points (K_x, R_x) clearly follow a negative course; the distribution of points (K_z, R_z) is slightly more irregular, but the general trend is none the less positive.[7]

To sum up, using the data of large corporations permits us to conclude the

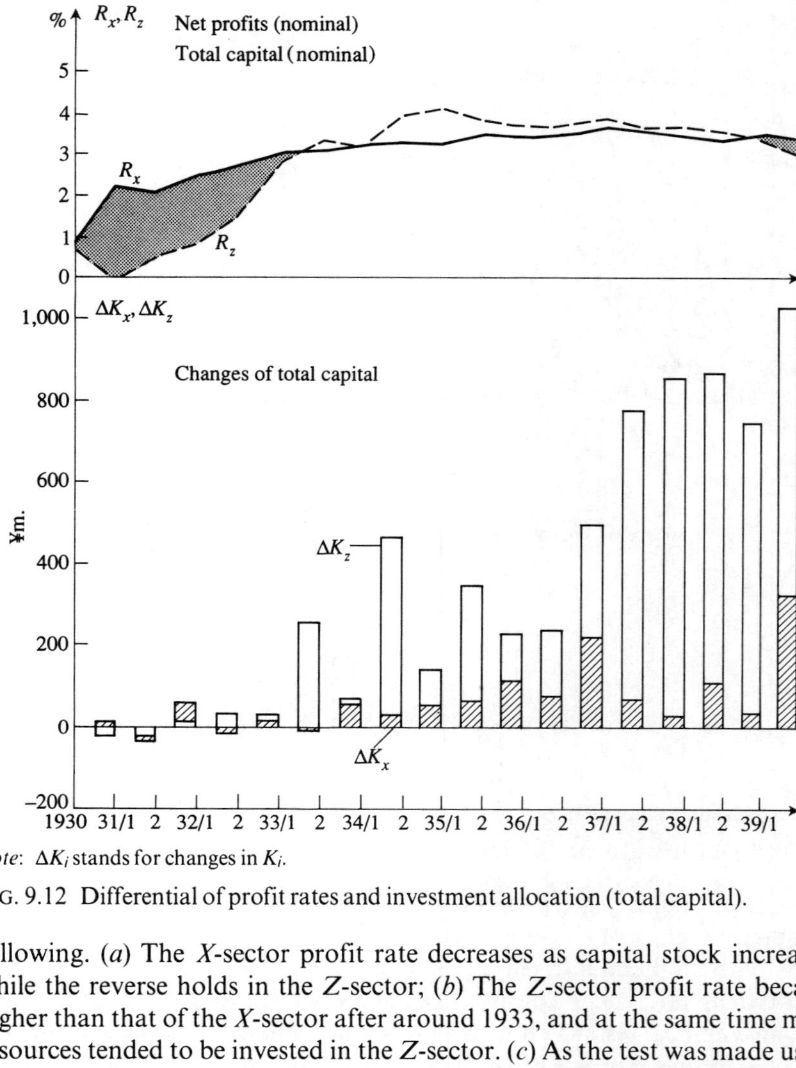

Note: ΔK_i stands for changes in K_i.

FIG. 9.12 Differential of profit rates and investment allocation (total capital).

following. (*a*) The *X*-sector profit rate decreases as capital stock increases, while the reverse holds in the *Z*-sector; (*b*) The *Z*-sector profit rate became higher than that of the *X*-sector after around 1933, and at the same time more resources tended to be invested in the *Z*-sector. (*c*) As the test was made using biannual data, the adjustment of the investment allocation to the changing profit rate differential implied that the adjustment takes place rather quickly; using the annual data might be more appropriate if the response of investment allocation to the profit rate differential took more time. (*d*) For financing total assets, large firms depended more on the savings of others, but for fixed assets they were largely self-financing, and often financed other firms' investment too.

9.4.3. Testing with global factory data

We now describe the results achieved in applying the same test using global data on all factories. We used the data for production, employees, and capital

TABLE 9.16. Re-investment ratios from profits (¥m.)

	Net profit		Change in fixed assets		Change in total assets		Differential of profit rates to		Change of fixed assets/Net profit %		Change of total assets/Net profit %	
							Fixed assets	Total assets				
	X-sector	*Z*-sector	*X*-sector	*Z*-sector	*X*-sector	*Z*-sector	R_X–R_Z %	R_X–R_Z %	*X*-sector	*Z*-sector	*X*-sector	*Z*-sector
1930 2nd half	13.9	13.0					0.25	0.19				
1931 1st half	36.2	–0.9	–3.0	–14.0	5.0	–29.0	3.86	2.23	—	—	13.8	—
2nd half	32.8	10.9	–1.0	–5.0	–39.0	–30.0	2.41	1.46	—	—	—	—
1932 1st half	41.7	16.1	2.0	5.0	57.0	6.0	2.86	1.67	4.8	31.1	136.7	37.3
2nd half	45.7	31.8	–4.0	–14.0	–19.0	30.0	1.79	1.17	—	—	—	94.3
1933 1st half	52.2	56.6	–11.0	–17.0	30.0	14.0	0.10	0.27	—	53.0	57.5	24.7
2nd half	52.7	75.2	–8.0	120.0	–10.0	257.0	–0.83	–0.18	—	159.6	—	341.7
1934 1st half	57.1	75.3	–3.0	–87.0	52.0	70.0	–0.86	0.08	—	—	91.1	93.0
2nd half	58.5	111.9	6.0	344.0	26.0	468.0	–1.65	–0.67	10.3	307.4	44.4	418.2
1935 1st half	60.1	122.4	37.0	25.0	51.0	140.0	–2.30	–0.83	61.6	20.4	84.9	114.4
2nd half	66.4	127.6	29.0	131.0	61.0	352.0	–1.49	–0.32	43.7	102.7	91.9	275.9
1936 1st half	69.0	132.0	42.0	93.0	111.0	229.0	–1.31	–0.26	60.9	70.5	160.9	173.5
2nd half	73.3	141.1	38.0	67.0	76.0	234.0	–1.37	–0.18	51.8	47.5	103.7	165.8
1937 1st half	86.1	167.6	14.0	123.0	221.0	500.0	–1.17	–0.16	16.3	73.4	256.7	298.3
2nd half	86.3	188.0	40.0	89.0	69.0	783.0	–2.07	–0.05	46.3	47.3	80.0	416.5
1938 1st half	82.9	218.7	–4.0	243.0	25.0	861.0	–2.68	–0.21	—	111.1	30.2	393.7
2nd half	85.8	243.6	40.0	269.0	111.0	867.0	–2.59	–0.14	46.6	110.4	129.4	355.9
1939 1st half	90.9	260.4	–4.0	229.0	25.0	752.0	–1.91	0.17	—	87.9	27.5	288.8
2nd half	102.0	277.3	43.0	291.0	318.0	1,034.0	–0.90	0.37	42.2	104.9	311.8	372.9

Source: Tables A.9.2. and A.9.3.

TABLE 9.17. Statistics for testing $dR_x/dK_x < 0$, $dR_z/dK_z \geqq 0$

Business year		X-sector				Z-sector				Deflator (price)		Profit in terms of agricultural prices	
		Nominal net profit rate, R'_X	Net profit rate in terms of agricultural prices, R'_x	Nominal fixed capital, K'_X	Fixed capital at constant price, K_X	Nominal net profit rate, R'_Z	Net profit rate in terms of agricultural prices, R_Z	Nominal fixed capital, K'_Z	Fixed capital at constant prices, K_Z	Agriculture, P_s	Capital stock, P_k	Fixed assets at constant prices % $\hat{R}_X$	$\hat{R}_Z$
1930	2nd half	1.44	2.01	962	1,062	1.19	1.66	1,091	1,204	71.8	90.6	1.82	1.50
1931	1st half	3.77	5.68	959	1,341	—	—	1,077	1,506	66.4	71.5	4.06	—
	2nd half	3.42	5.15	958	1,340	1.01	1.52	1,072	1,499	66.4	71.5	3.68	1.09
1932	1st half	4.35	5.69	960	1,116	1.49	1.95	1,077	1,252	76.4	86.0	4.89	1.68
	2nd half	4.78	6.26	956	1,111	2.99	3.91	1,063	1,236	76.4	86.0	5.38	3.36
1933	1st half	5.52	6.92	945	957	5.42	6.79	1,046	1,060	79.8	98.7	6.83	6.70
	2nd half	5.62	7.04	937	949	6.45	8.08	1,166	1,181	79.8	98.7	6.95	7.97
1934	1st half	6.11	6.40	934	925	6.97	7.31	1,079	1,069	95.4	100.9	6.46	7.38
	2nd half	6.21	6.51	940	932	7.86	8.24	1,423	1,410	95.4	100.9	6.57	8.31
1935	1st half	6.15	6.17	977	991	8.45	8.47	1,448	1,469	99.7	98.6	6.08	8.35
	2nd half	6.59	6.61	1,006	1,020	8.08	8.10	1,579	1,601	99.7	98.6	6.52	7.99
1936	1st half	6.58	6.27	1,048	1,043	7.89	7.52	1,672	1,663	104.9	100.5	6.30	7.56
	2nd half	6.75	6.43	1,086	1,081	8.12	7.74	1,739	1,730	104.9	100.5	6.46	7.78
1937	1st half	7.83	6.69	1,100	779	9.00	7.69	1,862	1,320	117.0	141.1	9.44	10.85
	2nd half	7.57	6.47	1,140	808	9.64	8.24	1,951	1,383	117.0	141.1	9.13	11.63
1938	1st half	7.29	5.76	1,136	766	9.97	7.88	2,194	1,479	126.5	148.3	8.54	11.69
	2nd half	7.30	5.77	1,176	793	9.89	7.82	2,463	1,661	126.5	148.3	8.56	11.60
1939	1st half	7.76	4.79	1,172	788	9.67	5.97	2,692	1,810	162.1	148.7	7.12	8.88
	2nd half	8.39	5.18	1,215	817	9.29	5.73	2,983	2,006	162.1	148.7	7.70	8.52

Note: Deflators are common to first and second half of business year; 1934–6 = 100.

R'_X stands for current profit of X-sector over K'_X.

$$R_X = \frac{R'_X}{P_S} \times 100.$$

$$\hat{R}_X = R_X \times \frac{P_K}{100}.$$

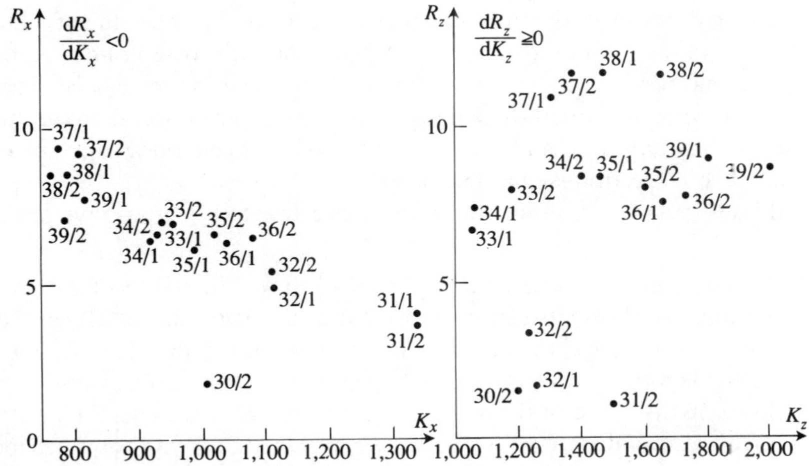

Notes: R_z is omitted from figure because of negative value at 1st half of 1931.
R_i: profit in terms of agriculture price/fixed assets of sector i at constant prices.
K_i: fixed assets of sector i at constant prices.

FIG. 9.13 Tests for $dR_x/dK_x < 0$ and $dR_z/dK_z \geq 0$.

stock (in the final years, proxy variables for capital stock) which were used to estimate production functions in Chapter 8. This is for the sake of consistency among the data. The deflator used for estimating profit in terms of *S*-sector goods was the same as that used in the previous section. The total wage bill was calculated from the daily wage rate (Ohkawa *et al.*, 1967) and the assumed 310 work days per year. Nominal profit was then estimated by subtracting the total wage bill from the value added, which was estimated by multiplying the value added ratio to the value of sales of each sector. Nominal profit was then converted to real profit measured in terms of *S*-sector goods.

Estimating the value added presented some difficulties. To obtain a reasonable ratio of value added to sales value, we assumed certain ratios of raw and intermediate material input to production, as shown in Table 9.18. These ratios were estimated for the years 1930, 1935, and 1940. As we ignored the cost of energy and power, however, the figures obtained were overestimates,[8] particularly in the *Z*-sector which is more energy intensive. Thus, we estimated nominal profit as nominal production minus material costs minus wage costs. While the model assumed that the price of *Z*-sector goods was unity in terms of *S*-sector goods, the above procedure took into consideration changes in the prices of *Z*-sector goods.

The most difficult step in such an empirical analysis of aggregated data is making a reasonable capital stock data. As explained earlier, there are no sector-by-sector data on capital stock. Neither generator capacity nor the value of factory buildings are really appropriate indices for this purpose, but nothing else was available. The results of our investigation of the relation between these two indices is given in the statistical appendix to this chapter (Section

9.6). Clearly, however, the ratio of generator capacity to the value of factory buildings is lower in the X-sector. Over time, the ratio rose in both sectors, implying that production was increasingly mechanized as time passed. Therefore, a true series of capital stock data would have grown faster than the index of factory building value and slower than the index of generator capacity. Our test was, therefore, done separately on the two data sets.

Tables 9.19 and 9.20 show the profit rates estimated by the above procedures.

1. *Test on generator capacity data.* Fig. 9.14 shows the results of the test on the relationship between profit per unit of power and total generator capacity. It suggests that the data support our hypothesis. As the Z-sector is more power-intensive, however, profit per unit of power was lower in the Z-sector throughout the period. None the less, the Z-sector profit rate tended to rise over time, whereas that of the X-sector declined.

Comparing the rate of profit per unit of generator capacity in the X- and Z-sectors in nominal terms, the difference $R_x - R_z$ shrank over time but always remained positive (Fig. 9.15). As regards total power, the Z-sector overtook the X-sector from the mid-1920s, and remained overwhelmingly ahead from then on except during the Depression. This strengthens our argument that the Z-sector was at the stage of take-off in the late 1920s.

2. *Test on factory building data.* Fig. 9.16, which shows real profit over the value of factory buildings at 1934–6 prices, shows clearly that R_z increases as K_z expands, but that the relation between R_x and K_x is ambiguous: although apparently positive from 1920 to 1932, from 1933 to 1939 it appears to have been negative.

Fig. 9.17 compares the profit rate differential with $\dot{K}_x$ and $\dot{K}_z$. According to this, $R_x - R_z$ became negative around 1925 and remained so until 1933. After that, R_z exceeded R_x, and by far more. Correspondingly, $\dot{K}_z$ became much larger than $\dot{K}_x$ after 1933. Since measuring the profit rate in terms of factory buildings may overestimate the Z-sector profit rate, with the effect that R_z exceeded R_x in 1925, $\dot{K}_z$ stayed lower than $\dot{K}_x$ after this turning point. Despite this contradictory evidence, we can argue that the Z-sector took off around the mid-1920s since take-off is possible even if R_x is larger than R_z, so long as the margin is small.

To sum up, when power data are used as a proxy for capital stock, $dR_x/dK_x < 0$ appears definite, but $dR_z/dK_z \geqq 0$ is slightly unclear. With factory buildings data, in contrast, the former relation looks ambiguous, but the latter becomes very clear. If the real variable of capital stock follows a path between the two proxy variables, which as already explained is quite probable, the above results suggest that $dR_x/dK_x < 0$ and $dR_z/dK_z \geq 0$ hold. Following a similar reasoning, if the true profit rate could be estimated using real capital stock data, it would turn out that R_z overtook R_x in the late 1920s or early

TABLE 9.18. Ratios of material input (factories with more than 5 employees)

		X-sector						
		Textiles	Food	Wood products	Printing	Miscellaneous	Total	Average ratio (%)
1930	Production (¥m.)	2,257	954	165	192	88	3,656	
	Material consumed (¥m.)	1,606	514	122	63	53	2,358	
	Ratio of material input (%)	71.2	53.9	73.9	32.8	60.2	64.5	
1935	Production (¥m.)	3,498	1,168	252	223	197	5,338	
	Material consumed (¥m.)	2,650	699	178	111	123	3,761	67.1
	Ratio of material input (%)	75.8	59.8	70.6	49.8	62.4	70.5	
1940	Production (¥m.)	4,976	2,465	1,022	342	542	9,018	
	Material consumed (¥m.)	3,331	1,448	693	151	350	5,973	
	Ratio of material input (%)	66.9	58.7	67.8	44.2	64.6	66.2	

		Z-sector						
		Chemicals	Ceramics	Metal products	Machinery	Miscellaneous	Total	Average ratio (%)
1930	Production (¥m.)	909	174	568	629	—	2,280	
	Material consumed (¥m.)	568	54	346	257	—	1,225	
	Ratio of material input (%)	62.5	31.0	60.9	40.9	—	53.7	
1935	Production (¥m.)	1,815	313	1,991	1,359	—	5,478	
	Material consumed (¥m.)	1,039	99	1,291	513	—	2,942	50.5
	Ratio of material input (%)	57.2	31.6	64.8	37.7	—	53.7	
1940	Production (¥m.)	4,623	776	5,900	8,443	—	19,742	
	Material consumed (¥m.)	2,494	293	3,630	2,301	—	8,718	
	Ratio of material input (%)	53.9	37.8	61.5	27.3	—	44.2	

Source: MITI (1962).

TABLE 9.19. Profit rates of *X*-sector (government and private factories with more than 5 employees)

	Production at current prices (¥m.)	Nominal wages (¥m.)	Profit, net of estimated material costs (67% of production)	Horse-power (1,000)	(1) Profit/HP ¥m/1000hp	Factory building (¥m.)	(2) Profit/Factory building	Profit in terms of agricultural prices (1)	Profit in terms of agricultural prices (2)	No. of employees in private and government factories ('000s)	Average wage (¥m.)
1920	5,164	392	1,312.1	590	2,224	90.5	14.50	1,583	10.32	1,232,389	318
1921	5,342	467	1,295.9	735	1,763	89.9	14.41	1,293	10.57	1,245,652	375
1922	5,278	434	1,307.7	707	1,850	89.3	14.64	1,694	13.41	1,216,970	357
1923	5,196	457	1,257.7	792	1,588	95.3	13.20	1,297	10.78	1,338,748	341
1924	5,523	474	1,348.6	830	1,625	94.0	14.35	1,154	10.19	1,343,914	353
1925	5,856	488	1,444.5	917	1,575	95.1	15.19	1,173	11.31	1,393,666	350
1926	5,842	509	1,418.9	989	1,435	99.8	14.22	1,164	11.53	1,430,529	356
1927	5,705	504	1,378.6	1,088	1,267	103.4	13.33	1,148	12.07	1,426,715	353
1928	5,806	501	1,415.0	1,131	1,251	106.8	13.25	1,159	12.28	1,440,602	348
1929	5,904	540	1,408.3	1,322	1,065	112.5	12.52	1,008	11.84	1,616,396	334
1930	4,192	414	969.4	1,453	667	116.7	8.31	929	11.57	1,349,760	307
1931	3,924	390	904.9	1,576	574	121.2	7.47	864	11.25	1,367,012	285
1932	4,302	381	1,038.7	1,493	696	123.6	8.40	911	10.99	1,374,176	277
1933	5,208	389	1,329.6	1,437	925	128.9	10.31	1,159	12.92	1,424,285	273
1934	5,962	415	1,552.5	1,508	1,030	143.6	10.81	1,080	11.33	1,531,373	271
1935	6,229	434	1,621.6	1,694	957	148.9	10.89	960	10.92	1,608,394	270
1936	7,035	455	1,866.5	1,856	1,006	154.7	12.07	959	11.51	1,659,325	274
1937	8,206	504	2,204.0	2,133	1,033	174.1	12.66	883	10.82	1,739,072	290
1938	8,947	516	2,436.5	2,194	1,111	178.5	13.65	878	10.79	1,697,089	304
1939	11,688	612	3,245.0	2,257	1,438	212.9	15.24	887	9.40	1,805,069	339
1940	12,577			2,268		205.3				1,730,300	

TABLE 9.20. Profit rates of Z-sector (government and private factories with more than 5 employees)

	Production at current prices (¥m.)	Nominal wages (¥m.)	Profit, net of estimated material costs (51% of production) (¥m.)	Horse-power (1,000)	(1) Profit/HP ¥m/1000HP	Factory building (¥m.)	(2) Profit/Factory building	Profit in terms of agricultural prices (1) ¥/*HP*	(2) ¥/¥	No.of employees in government and private factories('000s)	Average wage (¥)
1920	3,468	340.3	1,359.0	1,278	1,063.4	49.2	27.6	757	19.6	619,752	549
1921	2,461	376.2	829.7	1,338	620.1	60.1	13.8	450	10.1	599,096	628
1922	2,570	350.8	908.5	1,315	690.9	59.4	15.3	633	14.0	548,961	639
1923	2,238	339.9	756.7	1,451	521.5	57.5	13.2	426	10.9	534,456	636
1924	2,464	367.8	839.6	1,486	565.0	58.6	14.3	401	10.1	557,331	660
1925	2,519	354.1	880.2	1,677	524.9	58.3	15.1	391	11.2	533,335	664
1926	2,764	387.9	966.5	1,639	589.7	61.8	15.6	478	12.7	563,765	688
1927	2,875	408.7	1,000.0	1,802	554.9	63.8	15.7	503	14.2	590,616	692
1928	3,106	505.0	1,016.9	2,034	450.0	66.8	15.2	417	14.1	621,180	813
1929	3,358	505.4	1,140.0	3,149	362.0	79.4	14.4	343	13.6	655,462	771
1930	2,759	385.1	966.8	2,918	331.3	81.2	11.9	461	16.6	600,782	641
1931	2,291	345.9	776.7	2,814	276.0	82.1	9.5	416	14.3	585,301	591
1932	2,762	405.2	948.2	2,724	348.1	88.4	10.7	456	14.0	676,445	599
1933	3,872	535.6	1,361.7	3,029	449.6	96.0	14.2	563	17.8	818,933	654
1934	4,888	608.5	1,786.6	3,427	521.3	108.5	16.5	546	17.3	989,422	615
1935	5,908	697.0	2,197.9	3,999	549.6	116.7	18.8	551	18.9	1,144,514	609
1936	6,925	801.4	2,591.8	4,633	559.4	127.8	20.3	533	19.4	1,335,616	600
1937	9,733	1,020.1	3,749.1	5,815	644.7	153.1	24.5	551	20.9	1,634,761	624
1938	14,635	1,305.1	5,866.0	6,700	875.5	163.4	35.9	692	28.4	2,001,741	652
1939	16,639	1,710.2	6,442.9	7,621	845.1	209.6	30.7	521	18.9	2,511,291	681
1940	18,779			8,930		220.7				2,745,945	

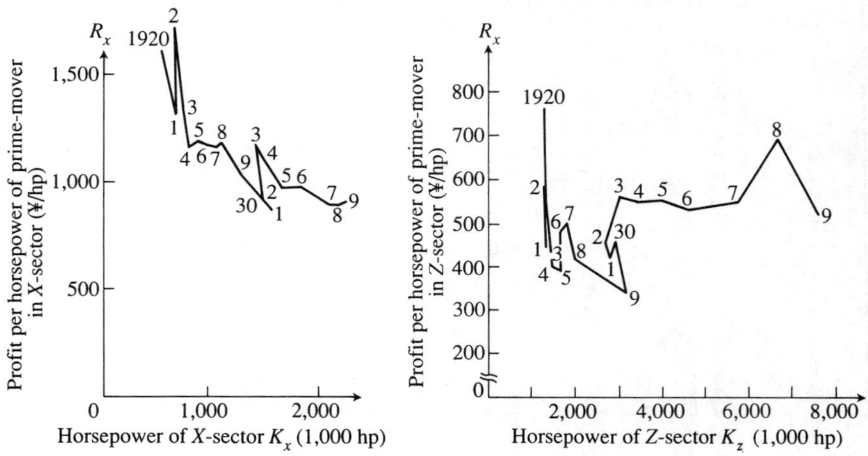

FIG. 9.14 Relation of real profit and capital (horsepower of prime-mover).

1930s. Since the take-off of the *Z*-sector is theoretically possible if R_x exceeded R_z only in a modest scale, it would seem that the *Z*-sector took off in the mid- or late 1920s.

9.5. Some Reservations

1. *Limitations due to unclassified industrial sectors.* If further statistical data were to become available such that all industries could be classified as either *S*-, *X*-, *or Z*-sector, our analysis of reinvestment of profits would be much more complete. Even if it were to prove difficult to incorporate the distribution sector into our *X*- and *Z*-sectors, it would be possible to quantify how much profit came from and went to 'the other sector'.

2. *Incompleteness of investment data on private factories.* The Mitsubishi statistics are excellent as regards consistency but their coverage is limited to large companies. The exclusion of small- and medium-scale firms is a limitation.

A further constraint was that the available data on factories employing five persons or more made no distinction between government factories and private ones. If the investment by government factories was not responsive to profit rates, this would have weakened our results. We hope to retest our hypotheses using data on private factories only when they become available.

3. *Influence of subsidies to industries.* As will be discussed in Chapter 10, the government gave heavy industries tax reliefs and subsidies. If our calculations had been made after tax and subsidies, the *Z*-sector profit rate might have been higher. More precisely, while the effects of subsidies are reflected in our profit at current prices tax effects are not, and after-tax profit rates must have been higher in the *Z*-sector.

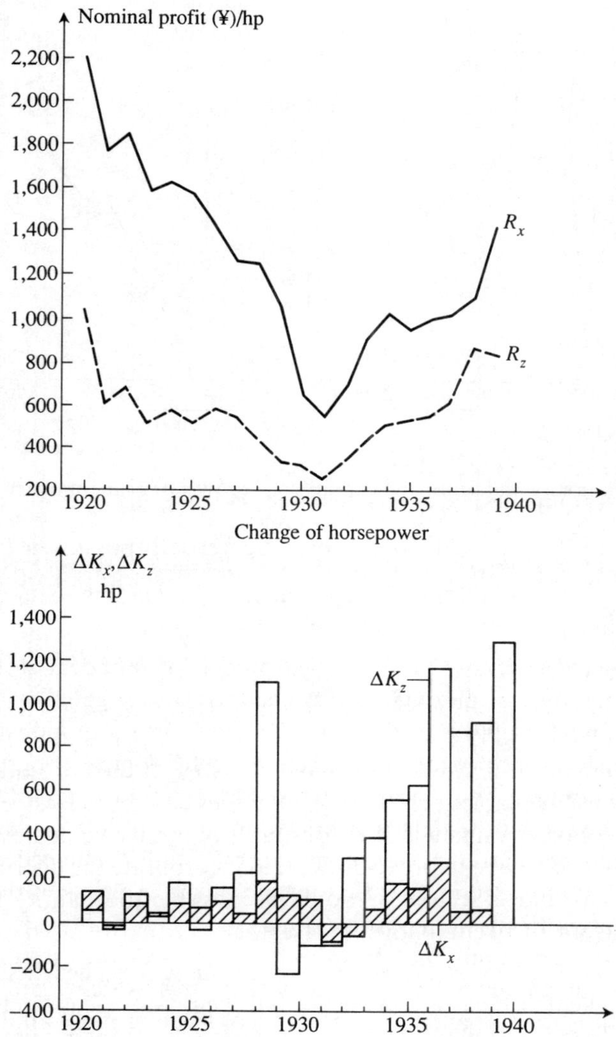

Notes: ΔK_i stands for changes in K_i.

Sources: Tables 9.19 and 9.20.

FIG. 9.15 Differential of nominal profit rates and direction of investment (horsepower).

4. *Lack of comprehensive data on capital stock*. This is the most serious limitation of our test. Due to the necessity of using proxy data such as generator capacity and the value of factory buildings, our analyses of investment allocation and profit rate differentials are weak. Although we advanced tentative conclusions regarding the increase in the Z-sector profit rate, it was extremely difficult to relate profit rate differentials to the allocation of profit between the two sectors, principally because the value of factory buildings accounts for

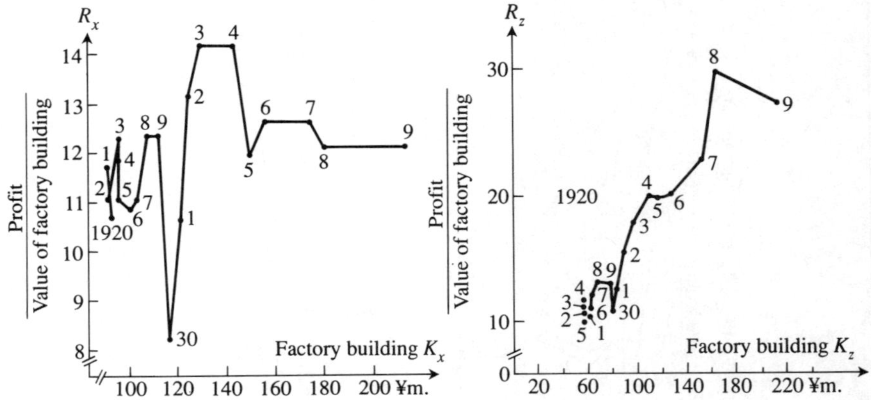

Sources: Tables 9.19 and 9.20.

Note: 'real' means in terms of agricultural price.

FIG. 9.16 Real profit rates and capital stock (factory building).

only a small proportion of all fixed assets. If the data base were expanded to include the value of all production facilities, including equipment and machines, our empirical test would be very much improved.

5. *Technological progress*. Our theoretical model does not deal explicitly with technological progress, although its effect has been investigated as a modification of the model. Analysis of the long-run development of industries, however, obviously cannot ignore the effects of technological progress. As we have not separated technological progress, our empirical test is interpreted as having been made on the data which have included technological progress. Therefore, the fact that our curve connecting R_x and K_x showed a declining slope from 1920 to 1940 implies that technological progress in the X-sector was not sufficient to give R_x a positive slope.

In contrast, our finding that R_z was non-decreasing with regards to K_z is interpreted to mean that it was technological progress which together with increasing returns to scale and constant terms of trade brought about the final result.

9.6. Statistical Appendix

1. *Per capita capital stock*. Data on the generator capacity of private and government factories and the value of factory buildings per capita are shown in Table 9.A.1 and Fig. 9.18. Details are given in Chapter 8. Population statistics are taken from Nippon Ginko Tokeikyoku (1966).

2. *Wage rates*. Wage rates are the average of male and female daily wage rates by industry, as given in Ohkawa *et al*. (1967). Employment figures were

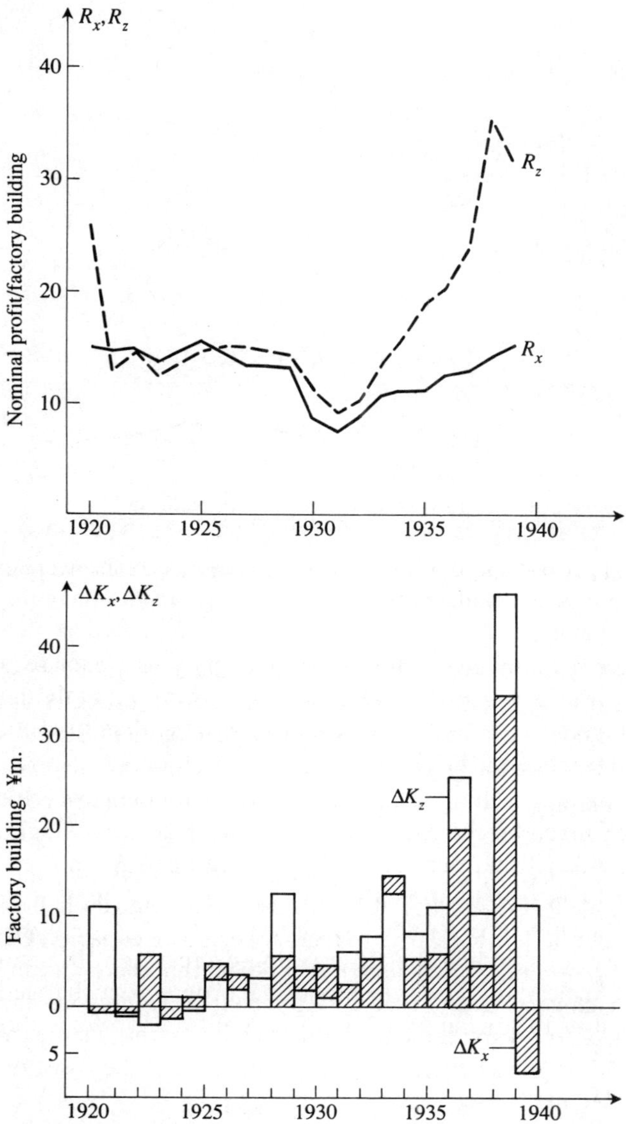

Note: ΔK_i stands for changes in K_i.

FIG. 9.17 Differentials of profit rates and direction of investment (factory building).

obtained from MITI (1963). This series was not always consistent, in that government factories were sometimes excluded, but for estimating average wage rates this defect is not serious. Weighted averages were then used to estimate separate wage rates for the *X*- and *Z*-sectors. The annual wage rate was then estimated by multiplying the daily wage rate by 310 working days. This

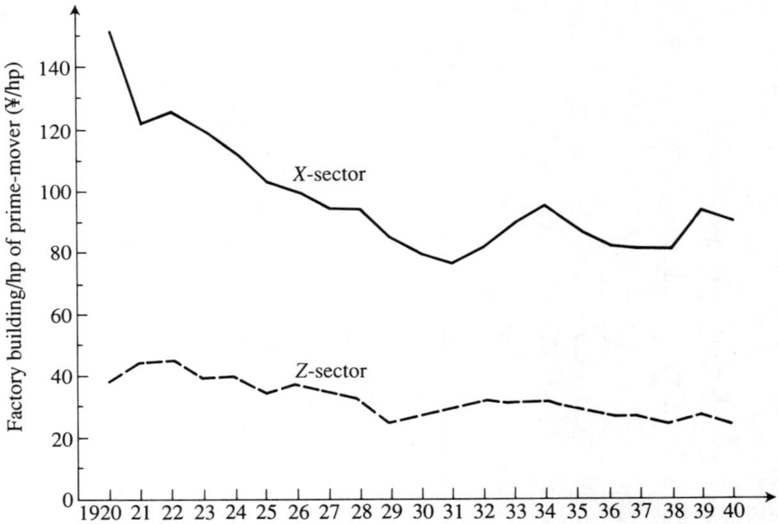

FIG. 9.18 Levels and changes of factory building/hp (factories of government and private with more than 5 employees).

figure was then multiplied by the number of employees in each sector in order to estimate the total wage bill by sector. Note, however, that though the number of employees covers both private and government factories, family firms are excluded (see Section 8.2.3).

3. *The Mitsubishi statistics on Japanese business.* A summary of these data is given in Tables 9.A.2 and 9.A.3.

4. *Relations between data on generator capacity and on the value of factory buildings.* Table 9.A.4 and Fig. 9.19 illustrate the long-run behaviour of the value of factory buildings relative to factory generator capacity. Over the long term this ratio declined rather steadily, implying that there was an increase in the use of machinery; except that during the Great Depression the ratio rose slightly, reflecting a displacement of some machines (though the value of factory buildings increased modestly). Overall, the ratio in the *Z*-sector was about three times that in the *X*-sector.

TABLE 9.A.1. Data for Tables 9.19 and 9.20

	Horsepower of private factories				Horsepower of government and private factories				Stock value of buildings of government and private factories				Population ('000s)
	K_X	R_X/N HP/person	K_Z	K_Z/N HP/person	K_X	R_X/N	K_Z	K_Z/N	K_X (¥'000)	K_X/N ¥/person	K_Z	K_Z/N ¥/person	
1920	574,545	10.36	872,940	15.74	589,616	10.63	1,278,304	23.04	90,456	1.63	49,238	0.89	55,473
1921	718,286	12.80	879,084	15.66	734,911	13.10	1,338,406	23.85	89,903	1.60	60,053	1.07	56,120
1922	689,558	12.13	801,996	14.11	707,179	12.44	1,315,227	23.14	89,352	1.57	59,447	1.05	56,840
1923	771,950	13.41	941,355	16.35	791,650	13.75	1,451,177	25.20	95,334	1.66	57,487	1.00	57,580
1924	811,186	13.90	983,279	16.85	830,425	14.23	1,485,535	25.46	94,066	1.61	58,637	1.00	58,350
1925	896,570	15.15	1,091,887	18.45	917,386	15.50	1,676,862	28.34	95,130	1.61	58,272	0.98	59,179
1926	967,440	16.07	1,084,589	18.01	989,165	16.43	1,638,634	27.22	99,834	1.66	61,818	1.03	60,210
1927	1,066,689	17.45	1,268,285	20.74	1,087,516	17.79	1,801,842	29.47	103,422	1.69	63,835	1.04	61,140
1928	1,107,560	17.84	1,337,228	21.54	1,131,043	18.22	2,034,235	32.77	106,811	1.72	66,825	1.08	62,070
1929	1,286,684	20.45	2,487,354	39.53	1,321,582	21.00	3,149,378	50.05	112,478	1.79	79,393	1.26	62,930
1930	1,417,761	22.20	2,305,201	36.09	1,452,671	22.74	2,917,604	45.68	116,737	1.83	81,180	1.27	63,872
1931	1,552,259	23.93	2,199,960	33.91	1,576,401	24.30	2,813,785	43.38	121,254	1.87	82,138	1.27	64,870
1932	1,473,531	22.36	1,990,488	30.21	1,493,222	22.66	2,723,664	41.34	123,554	1.88	88,420	1.34	65,890
1933	1,410,797	21.09	2,276,107	34.03	1,437,458	21.49	3,028,567	45.28	128,944	1.93	96,021	1.44	66,880
1934	1,487,576	21.98	3,011,913	44.50	1,507,735	22.27	3,427,486	50.66	143,578	2.12	108,480	1.60	67,690
1935	1,668,814	24.30	3,553,102	51.75	1,694,100	24.67	3,998,809	58.24	148,898	2.17	116,734	1.70	68,662
1936	1,821,411	26.17	4,182,134	60.10	1,856,452	26.68	4,632,927	66.57	154,657	2.22	127,794	1.84	69,590
1937	2,100,030	29.98	5,358,881	76.51	2,133,417	30.46	5,814,870	83.02	174,143	2.49	153,099	2.19	70,040
1938	2,142,583	30.38	6,214,314	88.11	2,193,504	31.10	6,700,354	95.00	178,463	2.53	163,397	2.32	70,530
1939	2,237,512	31.58	7,160,328	101.06	2,257,028	31.86	7,620,735	107.56	212,954	3.01	209,622	2.96	70,850
1940	2,230,879	31.24	8,400,672	117.66	2,268,451	31.77	8,929,604	125.06	205,250	2.87	220,716	3.09	71,400

TABLE 9.A.2. *X*-sector (¥m.)

		Net profit					Fixed assets					Total assets				
		(1) Dyeing and weaving	(2) Food	(3) Wood and wood products	(4) Printing, binding	(5) *X*-sector total	(1)	(2)	(3)	(4)	(5)	(1)	(2)	(3)	(4)	(5)
1930	2nd half	1,237	11,852	119	684	13,892	538	401	1	22	962	923	698	4	29	1,654
1931	1st half	22,340	13,037	100	694	36,171	534	402	1	22	959	906	721	3	29	1,659
	2nd half	21,805	10,519	101	350	32,775	535	401	1	21	958	894	694	4	28	1,620
1932	1st half	25,647	15,495	13	567	41,722	535	403	1	21	960	913	733	3	28	1,677
	2nd half	27,426	17,711	115	436	45,688	535	398	1	22	956	909	718	3	28	1,658
1933	1st half	30,879	20,637	40	612	52,168	535	387	1	22	945	953	704	3	28	1,688
	2nd half	32,453	19,578	49	618	52,698	540	374	1	22	937	962	684	3	29	1,678
1934	1st half	33,984	22,304	56	734	57,078	549	363	1	21	934	1,018	680	3	29	1,730
	2nd half	32,653	24,967	54	784	58,458	564	354	1	21	940	1,036	688	3	29	1,756
1935	1st half	34,254	25,797	63	3	60,117	608	348	1	20	977	1,076	699	3	29	1,807
	2nd half	38,306	27,258	824		66,388	657	330	19		1,006	1,163	674	31		1,868
1936	1st half	36,995	31,093	876		68,964	700	329	19		1,048	1,269	679	31		1,979
	2nd half	39,926	32,469	930		73,325	743	324	19		1,086	1,296	728	31		2,055
1937	1st half	50,222	34,699	1,181		86,102	759	322	19		1,100	1,501	743	32		2,276
	2nd half	49,457	35,510	1,304		86,271	795	326	19		1,140	1,521	788	36		2,345
1938	1st half	48,626	32,949	1,294		82,869	781	335	20		1,136	1,493	838	39		2,370
	2nd half	50,315	34,031	1,473		85,819	817	338	21		1,176	1,558	883	40		2,481
1939	1st half	52,021	37,334	1,535		90,890	815	336	21		1,172	1,584	880	42		2,506
	2nd half	59,993	40,101	1,900		101,994	856	337	22		1,215	1,830	947	47		2,824

Note: Separation of columns (3) and (4) is not possible after 1935.

Source: Mitsubishi Keizai Kenkyusho (1930—).

TABLE 9.A.3. *Z*-sector (¥m.)

		Net profit					Fixed assets					Total assets				
		(1) Ceramics	(2) Chemicals	(3) Machinery	(4) Metals	(5) *Z*-sector total	(1)	(2)	(3)	(4)	(5)	(1)	(2)	(3)	(4)	(5)
1930	2nd half	967	13,753	194	–1,894	13,020	196	513	204	178	1,091	276	1,000	448	294	2,018
1931	1st half	1,633	5,346	–7,049	–866	–936	195	506	199	177	1,077	275	1,003	423	288	1,989
	2nd half	2,702	6,577	491	1,105	10,875	194	508	198	172	1,072	271	996	411	281	1,959
1932	1st half	3,283	12,246	–1,798	2,352	16,083	191	485	197	204	1,077	268	961	406	330	1,965
	2nd half	6,117	18,982	1,645	5,049	31,793	189	478	197	199	1,063	264	986	407	338	1,995
1933	1st half	9,185	26,219	8,284	12,955	56,643	186	465	200	195	1,046	268	952	438	351	2,009
	2nd half	9,927	33,949	13,452	17,875	75,203	188	491	293	194	1,166	276	979	649	362	2,266
1934	1st half	10,868	41,767	16,958	5,664	75,257	191	476	294	118	1,079	287	1,019	691	339	2,336
	2nd half	11,927	39,047	19,945	40,976	111,895	201	501	309	412	1,423	300	1,062	738	704	2,804
1935	1st half	12,773	38,992	22,470	48,161	122,396	206	516	300	426	1,448	316	1,103	773	752	2,944
	2nd half	12,938	44,387	27,964	42,359	127,648	206	590	305	478	1,579	326	1,246	852	872	3,296
1936	1st half	12,129	44,492	31,247	44,106	131,974	206	631	318	517	1,672	339	1,335	937	914	3,525
	2nd half	11,069	50,917	34,060	45,091	141,137	209	656	324	550	1,739	356	1,421	1,022	960	3,759
1937	1st half	12,862	58,520	37,935	58,280	167,597	205	742	344	571	1,862	374	1,607	1,186	1,092	4,259
	2nd half	12,779	66,477	43,493	65,270	188,019	206	778	377	590	1,951	406	1,873	1,483	1,280	5,042
1938	1st half	12,879	73,679	54,601	77,579	218,738	205	854	473	662	2,194	419	2,063	1,931	1,490	5,903
	2nd half	12,934	81,274	66,619	82,821	243,648	206	930	573	754	2,463	432	2,266	2,395	1,677	6,770
1939	1st half	14,136	86,673	75,764	83,844	260,417	205	983	647	857	2,692	452	2,429	2,736	1,905	7,522
	2nd half	14,970	86,889	87,290	88,101	277,250	211	1,017	745	1,010	2,983	480	2,598	3,225	2,253	8,556

Source: Mitsubishi Keizai Kenkyusho (1930—).

TABLE 9.A.4. Factory building and horsepower

	X-sector (1) Horsepower of prime-mover (1,000 HP)	(2) Factory building (¥m. at 1934 prices)	(2)/(1)	*Y*-sector (3) Horsepower of prime-mover (1,000 HP)	(4) Factory building (¥m. at 1934 prices)	(4)/(3)
1920	589	90.5	153.7	1,278	49.2	38.5
1921	735	89.9	122.3	1,338	60.1	44.9
1922	707	89.3	126.3	1,315	59.4	45.2
1923	792	95.3	120.3	1,451	57.5	39.6
1924	830	94.1	113.4	1,486	58.6	39.4
1925	917	95.1	103.7	1,677	58.3	34.8
1926	989	99.8	100.9	1,639	61.8	37.7
1927	1,088	103.4	95.0	1,802	63.8	35.4
1928	1,131	106.8	94.4	2,034	66.8	32.8
1929	1,322	112.5	85.1	3,149	79.4	25.2
1930	1,453	116.7	80.3	2,918	81.2	27.8
1931	1,576	121.3	77.0	2,814	82.1	29.2
1932	1,493	123.5	82.7	2,724	88.4	32.5
1933	1,437	128.9	89.7	3,029	96.0	31.7
1934	1,508	143.6	95.2	3,427	108.5	31.7
1935	1,694	148.9	87.9	3,999	116.7	29.2
1936	1,856	154.7	83.4	4,633	127.8	27.6
1937	2,133	174.1	81.6	5,815	153.1	26.3
1938	2,194	178.5	81.4	6,700	163.4	24.4
1939	2,257	212.9	94.3	7,621	209.6	27.5
1940	2,268	205.3	90.5	8,930	220.7	24.7

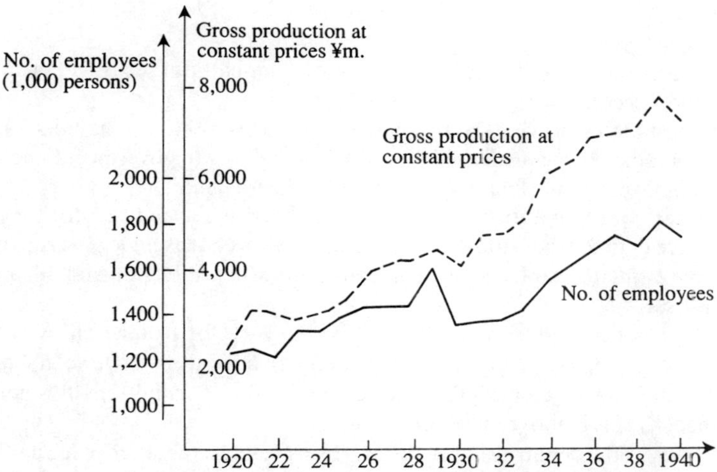

FIG. 9.19(i) *X*-sector: real gross production and number of employees (1920–40).

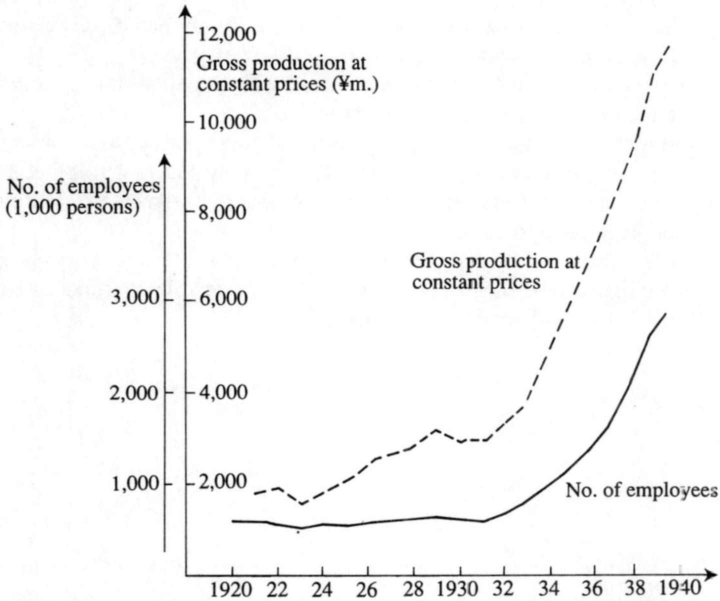

FIG. 9.19(ii) *Z*-sector: real gross production and number of employees (1920–40).

Notes

1. As data for forestry and fishery are scarce, the agricultural sector is taken to represent the S-sector in our empirical test.
2. Minami estimates provide data by industry for 1884–1940 on the horsepower of generators (the 'Prime mover', in his terminology) in both government and private factories, but excluding family firms. See sec. 9.3.3 for details.
3. Ishiwata estimates give gross values for factory buildings by industry for 1894–1940. The source data for the estimates include family firms but exclude government factories. We estimated capital stock data consistent with production and other related data, see sec. 9.6.
4. In our theoretical model, it is possible for the Z-sector to take-off even though $R_z < R_X$. If (K_X, K_z) is located above the minimum critical effort curve, K_z can grow in self-sustaining way. This is the necessary and sufficient condition. It is not necessary that (K_X, K_z) is above the $K_z/dt = 0$ curve.
5. Clark (1951) offers contrasting evidence. He argues that the share of labour income in the major nations was as follows (figures in the parentheses are the relevant years): UK 70.3% (1924) and 72.2% (1938); US 71.9% (1929), West Germany 65.9% (1928), France 59% (1929) and 72.5% (1938), in Japan 67.9% (1934–5). Thus, although the share in Japan was low relative to the UK and the US, it was not so significantly low compared with other industrial countries.
6. Rigorous statistical tests can be applied to these two inequalities but we confine ourselves to plotting R_X, K_X, and R_z, K_z on the R–K space.
7. The deflators used are the cultivated agricultural goods' price index for expressing profit in terms of S-sector goods, and the deflator for producers' durable equipment. On the former, see ch. 8. The latter is from Ohkawa *et al.* (1967: 158). The base year for these deflators is the average for 1934–6.
8. MITI (1962) provides data for energy and power inputs in physical terms. Had we multiplied these data by the prices, we would have been able to estimate the cost of these inputs, but we did not enter into such details.

10

Critical Efforts and Roles of the Government

10.1. Roles of Foreign Capital

The contemporary debate on development strategies for LDCs has assigned a key role to foreign capital. Although Prebisch (1964) considered trade more important than aid, economists, such as Chenery and Strout (1965) and several institutions, including the UNCTAD (1968), argued that foreign capital would be a critical factor in alleviating the resource constraints that an LDC would face in trying to accelerate economic growth.

Clearly, our model gives a much more specific, and therefore much clearer, picture of the part that foreign capital can play in economic development than that given by so-called 'gap model' (Chenery and Strout 1965) which consists of a single sector.

To restate, the minimum critical effort by government in our model can be summarized as follows:

The point which represents the combination of capital stock (K_x, K_z) that is located below the minimum critical effort curve when a traditional economy is in transition to capitalistic development. If the government leaves industrial development to market forces, this point will eventually drop to (K_x^*, 0) as heavy industry phases out and only a certain size of light industry survives. If foreign capital were to be invested in the *X*-sector, the initial point of capital stock would move east and rise above the curve; if it were to be invested in the *Z*-sector it would move north, again crossing it from below. There are many other ways to push the economy above the minimum critical effort curve, but the importation of foreign capital is one of the most important development strategies.

Although capital import generally plays a key role in economic take-off, we conclude that this factor was not so important in the development of the Japanese economy prior to World War II. A more important factor was the acquisition of know-how, whether by inviting foreign engineers, sending Japanese students and engineers abroad, or technological partnerships with foreign enterprises.

10.1.1. The early period (1868–1901)

The balance of payments for the period from 1868 to 1901 have been estimated by Baba and Tatemoto (1967: 272–9). The simplified summary presented in Table 10.1 indicates that capital transactions were limited to government

TABLE 10.1. Japanese balance of payments estimated by Tatemoto
(¥m.)

	Current balance						Capital balance			
	Balance of goods trade	Balance of freight	Balance of investment income	Balance of government transaction	Balance of other items	(1) Current balance	(2) Balance of private capital	(3) Balance of government capital	Monetary gold	(1)+(2)+(3) Total balance
1868	15.0	–1.5	–0.4	0.5	–10.1	3.5	—	0.4	—	3.8
1869	–7.8	–3.0	–0.9	0.2	–1.2	–12.7	—	–0.4	—	–13.1
1870	–19.2	–5.0	–1.6	0.2	–2.9	–28.5	—	4.0	—	–24.6
1871	–3.9	–3.2	–1.3	–0.1	–1.2	–9.7	—	–0.2	—	–9.9
1872	–9.1	–3.9	–1.5	–0.8	–1.9	–17.2	—	–3.1	(0.8)	–20.2
1873	–6.5	–4.7	–2.0	–1.8	–0.9	–15.9	—	11.1	(2.0)	–4.8
1874	–5.2	–3.8	–2.2	–2.0	–7	–20.2	—	–0.5	(12.9)	–20.7
1875	–13.3	–4.4	–2.4	–1.0	–3.4	–24.5	—	–0.7	(15.2)	–25.2
1876	–1.2	–3.6	–2.2	–0.9	0.4	–7.5	—	–0.7	(3.3)	–8.3
1877	–6.6	–4.1	–2.3	–0.7	–1.9	–15.6	—	–0.8	(8.1)	–16.4
1878	–11.7	–4.6	–2.4	–0.5	–2.0	–21.2	—	–0.8	(7.0)	–22.0
1879	–9.9	–4.3	–2.8	–0.3	–1.6	–18.9	—	–0.9	(10.4)	–19.9
1880	–13.8	–4.8	–3.0	–0.5	–1.1	–23.2	—	–10.0	(10.7)	–24.2
1881	–4.7	–3.9	–3.6	–0.0	–0.1	–12.3	—	–1.3	(6.1)	–13.5
1882	–4.4	–3.2	–3.0	0.4	10.1	–0.1	—	–1.0	(–1.5)	–1.1
1883	2.4	–2.8	–2.7	0.5	3.8	1.2	—	–0.5	(–2.1)	–1.6
1884	0.3	–2.3	–2.0	0.1	10.8	6.9	—	–0.5	(0.4)	–7.4
1885	3.6	–1.9	–1.6	0.3	1.1	1.5	—	–0.5	(–3.3)	–2.0

Table 10.1 (*contd.*)

1886	11.5	–1.4	–1.7	0.5	–0.4	8.5	—	–0.6	(0.3)	7.9
1887	–0.7	–2.9	–1.7	0.5	9.8	5.0	—	–0.7	(1.8)	–5.6
1888	0.3	–5.0	–1.9	0.9	3.8	–1.9	—	–0.7	(–0.9)	–2.7
1889	4.0	–3.6	–2.3	0.6	4.5	3.2	—	–0.8	(–9.0)	2.4
1890	–25.1	–5.3	–2.9	2.1	4.0	–27.2	—	–0.8	(12.6)	28.0
1891	–16.6	–3.2	–2.4	1.3	40.4	19.5	—	–0.8	(–12.4)	18.7
1892	19.8	3.2	–2.4	4.1	–3.5	21.2	—	–1.0	(–13.2)	20.2
1893	1.4	6.3	–2.3	4.1	–5.6	3.9	—	–1.2	(1.1)	2.7
1894	4.2	10.2	–1.3	1.5	–21.6	–7.0	—	–1.6	(7.6)	8.3
1895	6.9	9.2	–2.0	1.3	106.4	121.8	—	–120.9	(21.4)	0.9
1896	–53.8	–10.3	–2.0	0.6	92.7	27.2	—	–1.8	(–27.5)	25.4
1897	–56.2	–10.8	–1.4	1.2	9.6	–57.6	—	112.5	(–62.2)	54.9
1898	–111.7	–15.0	0.3	0.6	87.0	–38.8	—	—	(44.4)	38.8
1899	21.7	–5.9	–2.0	–0.9	16.0	28.9	—	—	(–9.0)	28.9
1900	44.6	–12.3	–6.4	–2.3	–73.4	–49.8	—	—	(45.2)	–49.8
1901	22.1	2.9	–7.4	–2.3	–27.8	–18.3	—	—	(3.1)	18.3

Note: Current account in 1895 includes ¥119m, which is the transfer of war reparations for Sino-Japanese war.

Source: Baba and Tatemoto (1967): 272–9.

borrowing,[1] and also were restricted in their extent. Thus, in the early Meiji era the new government borrowed $1 million worth of silver from the Boldwin Company of Holland, $ 400,000 of which were used to repay a debt to a British company. In 1870 the government issued a sterling-denominated bond of £1 million (¥488 million) at 9% interest, and there was a further sterling-denominated bond issue in 1873, of £240 million at 7% interest. Thereafter the government refrained from foreign borrowing until 1897 when funds had to be raised to finance the war with Russia.

The capital account of the Japanese balance of payment was thus marked by occasional loans and repayments, but the amounts involved never became significant; a more important source of foreign capital was the reparations paid by China for the Sino-Japanese War, which was not a capital transaction but a transfer of gold. It should be noted, however, that the occasionally foreign borrowing was not directed to investment in the *Z*-sector but mostly to the construction of infrastructure: it was not foreign capital that prevented the *Z*-sector from falling into the stagnation trap, but the government budget in a more general sense.

Table 10.2 shows the amount of foreign debt outstanding, as calculated from *The Detailed Statistics of Japanese Trade*. According to this, Japan's foreign debt was some ¥43 million in 1897, and ¥141 million in 1900, respectively 26.4% and 69.1% of exports. As there are no data available on national wealth before 1900, let us compare these foreign debt figures with national income—according to Yamada Yuzo's estimates (Nippon Ginko Tokeikyoku 1966) ¥1,933 million and ¥2,519 million respectively. The ratio of debt to national income was thus 2.2% in 1897 and 5.6% in 1900, meaning that the burden of interest payment was negligible in comparison with national income.

10.1.2. The later period (after 1901)

In the absence of statistical data for the period 1901–4, let us again look at the chronology of borrowing.

In 1902, the Yokohama city government concluded a contract with the British Samuel company on a bond issue ¥900,000 to finance the construction of a water supply system. The same company also underwrote a Japanese government bond that year of ¥3.5 million for the construction of Osaka harbour. The Japan Industrial Bank, which was then a government bank, took the first step in foreign borrowing by issuing the bond in London at a 5% interest rate.

The Russo-Japanese War drastically increased Japan's foreign borrowing. In May and November of 1904, the government announced its intention to issue sterling-denominated bonds of £10 million and £12 million respectively at 6% interest in the London and New York markets. Two further government bonds were issued at 4.5% interest to raise £30 million, and a £50-million bond was issued at 4% interest.

The balance of payments after 1904 is shown in Table 10.3. In the early

TABLE 10.2. Outstanding foreign investments in Japan (¥m.)

	National loans			Local loans (issued abroad)	Debentures (issued abroad)	Foreign loans on banks and companies in Japan (estimated)	Total
	Issued abroad	Home issued (under-written)	Sold to foreigners (estimated)				
1897	0	43 (100)	n.a.	0	0	n.a.	43 (100)
1900	98 (69.5)	43 (30.5)	n.a.	*	0	n.a.	141 (100)
1905	1,142 (80.8)	93 (6.6)	160 (11.3)	4 (0.3)	10 (0.7)	5 (0.4)	1,414 (100)
1910	1,447 (81.4)	0	108 (6.1)	85 (5.9)	109 (6.1)	28 (1.6)	1,777 (100)
1915	1,493 (77.1)	0	72 (3.7)	177 (9.1)	167 (8.6)	28 (1.4)	1,937 (100)
1920	1,428 (84.9)	0	34 (2.0)	140 (8.3)	48 (2.9)	31 (1.8)	1,681 (100)
1925	1,500 (75.5)	0	9 (0.5)	125 (6.3)	332 (16.7)	22 (1.1)	1,987 (100)
1930	1,567 (63.5)	0	84 (3.4)	245 (9.9)	456 (18.5)	114 (4.6)	2,466 (100)
1933	1,421 (68.2)	0	50 (2.4)	231 (111)	383 (18.4)		2,085 (100)

Note: Percentage of total in parentheses.

* = less than 0.5.

Source: Toyo Keizai Shinposha (1935): 696.

years, the influx of capital was used mostly to finance the war; this is reflected in a large increase in borrowing in 1905. Later, around 1910, a few municipal governments issued bonds to secure capital for the construction of infrastructure. Thus in 1909, the city of Osaka issued a £3-million bond in London, and the city of Kyoto issued a FF 45-million bond in Paris. To repay the debt borrowed earlier, the Ministry of Finance issued two bonds in 1910, one of FF 45 million and the other of £11 million, both at 4% interest.

After this time there was no significant borrowing except in the period following the Kanto earthquake that devasted Tokyo in 1923. By the 1930s, Japan had changed from a net borrower to a net lender, although loans at this time were limited to its overseas colonies.

Prior to 1945, therefore, Japan borrowed abroad not to finance investment in the *Z*-sector but to finance wars and the construction of infrastructure. As explained in Part I, however, the construction of infrastructure also facilitates the take-off of the *Z*-sector. In this sense, foreign capital was useful in development.

TABLE 10.3. Capital account and investment income (¥m.)

	(1) Net foreign capital inflow	(2) Of which foreign investment in Japanese private firms	(3) Net Japanese investment abroad	(4) (1)—(3) Net capital balance	(5) Investment income outflow	(6) Investment income inflow	(7) (5)—(6) Net
1902	15	n.a.	0		6	15	
1903	33	n.a.	0		6	12	
1904	227	0	2	224	21	15	6
1905	993	15	*	993	41	19	22
1906	Δ77	13	7	Δ84	81	19	62
1907	63	34	6	57	69	35	34
1908	58	60	2	56	74	38	36
1909	103	5	11	92	78	41	37
1910	216	9	10	206	83	43	40
1911	Δ11	39	17	Δ28	87	43	44
1912	88	1	30	58	91	48	43
1913	116	19	9	107	101	55	46
1914	9	3	13	Δ4	103	48	55
1915	Δ42	0	8	Δ50	93	49	44
1916	Δ127	0	7	Δ134	83	88	5
1917	Δ55	0	47	Δ102	86	127	Δ41
1918	Δ50	1	n.a.	Δ50	87	160	Δ73
1919	18	8	n.a.	18	92	192	Δ100
1920	Δ41	0	n.a.	Δ41	90	150	Δ60
1921	Δ104	0	n.a.	Δ104	89	157	Δ68
1922	Δ27	0	n.a.	Δ27	86	127	Δ41
1923	40	90	Δ174	214	72	140	Δ68
1924	230	102	Δ8	238	89	143	Δ54
1925	87	155	Δ24	111	113	152	Δ39
1926	23	44	Δ14	37	113	135	Δ22
1927	Δ27	64	59	Δ86	106	144	Δ38
1928	24	186	Δ1	23	111	134	Δ23
1929	Δ77	92	16	93	126	155	Δ29
1930	Δ28	69	120	Δ148	119	128	Δ9
1931	Δ217	78	16	Δ233	102	108	Δ6
1932	Δ106	77	Δ5	Δ101	130	178	Δ48
1933	21	35	42	Δ21	148	191	Δ43
1934	Δ4		180	Δ184	23	125	Δ102
1935	Δ18		354	Δ372	27	135	Δ108
1936	Δ102		168	Δ270	34	121	Δ87
1937	Δ51		514	Δ565	134	132	2
1938	660		701	Δ41	211	171	40
1939	Δ72		1,077	Δ1,149	74	57	17
1940	4		1,305	Δ1,301	70	103	Δ33

Note: Data from 1934 to 1940 are not necessarily comparable with previous years as they come from different sources. Δ indicates outflow.

Source: Nippon Ginko (1966) and Toyo Keizai Shinposha (1935).

In 1905, Japan's outstanding foreign debt was roughly ¥1.4 billion, or 9.9% of the total wealth of the nation which in that year amounted to about ¥14.2 billion excluding land (Nippon Ginko 1966). By 1910 the ratio had declined to 9.1% as net debt increased to ¥1.78 billion while wealth was ¥19.63 billion, and by 1930 it had declined still further, to 3.6%, as debt had risen to ¥2.47 billion while wealth had risen to ¥69.1 billion (see Table 10.2).

In 1930 Japan became a net creditor. The census of wealth, albeit a rough estimation, confirmed that Japan became a net lending country during World War I (Nippon Ginko 1966); and Lockwood (1954: 34) shared this view. Thus, the Japanese government allowed Russia, the United Kingdom, and France to issue bonds on the Tokyo market during the war. Thus Japan exported capital during this period, and, after the recovery of Europe from war damage, the Japanese overseas colonies became the destination of capital outflow.

Now let us look at the forms in which Japan borrowed from abroad. Tables 10.2 and 10.3, substantiate the fact that it was almost always either the central government or major municipal governments that borrowed from abroad: in 1920, for example, the central government accounted for roughly 87% of the outstanding foreign debt, and municipal governments for a further 8%. Only 5% was borrowed by the private sector, including short-term bank borrowing. By 1925 this figure had risen to 16.7%, but in absolute terms this hardly represented conspicuous expansion as the net outstanding value of capital borrowed by the private sector was still only about ¥400–500 million.

The annual flows of capital importation, in the second column of Table 10.3, show that it was only a short period between 1923 and 1925, ie. the immediate post-earthquake period, that Japan actively raised funds from abroad. In order to recover from the damage by the earthquake, the government had to import capital in a large scale. Without this earthquake, the economy could have taken off much earlier and more easily.

After the earthquake, foreign borrowing by private firms remained at around ¥60–70 million per annum until 1933. The net increase in the fixed assets of the 190 major manufacturing enterprises surveyed by Mitsubishi Keizai Kenkyusho (1930–) was ¥84 million, and the net increase in their total assets was ¥290 million. Thus, the importance of foreign capital was by no means negligible.

If we compare the net increases in the fixed and total assets of the 299 companies (Mitsubishi Keizai Kenkyusho 1930–), including transportation, warehousing, and department stores, they expanded by ¥224 million and ¥541 million respectively in 1933. As capital imports in that year by the private sector totalled ¥35 million (Table 10.3), foreign capital accounted for 15.6% and 6.5% respectively. However, Emi and Rosovsky (Nippon Ginko Tokeikyoku 1966: 34) estimated that total capital formation by the private sector in 1933 was ¥2,191 million. The ratio of foreign capital to total private capital formation was therefore less than 2%.[2]

10.1.3. Conclusions

Foreign capital was less significant in the economic development of Japan prior to World War II than in the development of contemporary LDCs. In fact, the majority of foreign borrowing was used to finance military expenditures, and the remainder was invested in infrastructure construction rather than directly in the Z-sector. From being a net debtor country in the early stage of economic development, however, by the end of World War I Japan became a net creditor. Then capital export grew to concentrate on that to the Japanese colonies.

This rapid change in the financial position of Japan, which started industrialization later than other countries but within half a century had become a net creditor, had much to do with the international environment. World War I and colonization were historical factors that affected the above development. Without the militarization of the economy and competition for colonies, the rate of capital accumulation would have been much faster. Without capital outflows, our point (K_x, K_z) would have been located further north or north-east. On the other hand, as these circumstances created a demand for Z-sector goods and thereby raised R_z, the direction of the final effects is unclear.

Table 10.4 shows the long-run evolution of the inflow of foreign capital and domestic investment. In 1870, gross domestic investment was roughly equal to the inflow of foreign capital, but this was a temporary phenomenon due to a large inflow of foreign capital at a time when domestic investment was still generally

TABLE 10.4. Gross fixed domestic capital formation (¥m., current prices)

	Total	Government	(Non-military)	Private	Foreign capital
1870	3.7	3.7	3.6	—	4.0
1875	14.5	11.9	7.7	2.7	–0.7
1880	20.5	11.8	9.9	8.7	–10.0
1885	30.7	23.0	15.6	7.7	–0.5
1890	83.5	24.9	17.8	58.6	–0.8
1895	154.3	49.0	25.7	105.4	–120.9
1900	273.3	142.2	83.0	131.1	0.0
1905	345.1	152.1	74.7	193.1	993.0
1910	491.5	243.6	180.0	247.9	206.0
1915	610.5	254.1	179.3	356.3	–50.0
1920	3,258.6	1,034.9	658.1	2,223.7	–41.0
1925	2,381.5	1,118.5	875.8	1,263.0	111.0
1930	2,132.8	1.010.1	836.7	1,122.7	–148.0
1935	3,479.1	1,353.8	887.0	2,125.3	–372.0
1940	14,664.9	5,368.1	1,173.3	9,296.8	–1,301.0

Note: – indicates outflow.

Sources: Emi (1971); foreign capital inflow based on Data estimated by Baba and Tatemoto (1967) and on Nippon Ginko (1966).

low. From then until the Russo-Japanese War, however, the inflow of foreign capital remained negligible but domestic investment expanded very rapidly.

The only exception was the inflow of gold, which was not capital inflow but a transfer due to war reparation by Jing Dynasty. On two occasions foreign capital flew into Japan in a large scale: war financing at the time of Russo-Japanese War, and foreign borrowing to finance the recovery from the damage by the Great Earthquake. In normal years, however, the capital imports totalled some ¥60–70 million, and the size relative to domestic investment declined rapidly. In the 1910s, while domestic investment reached some ¥600–700 million, capital movements, both outflows and inflows, remained negligible.

In the 1930s, Japan changed from being a net debtor to a net lender country. Even in the account of return to investment, incomes from foreign investment exceeded return to foreign capital. All in all, we conclude that foreign capital did not play a great role in Japanese industrialization.

10.2. Operation of Government Enterprises and Subsequent Privatization

We have emphasized that the government played an important role in economic development prior to World War II. We shall now investigate in greater detail the ways in which the government promoted the development of the Z-sector, and specifically the direct operation of Z-sector enterprises and the subsequent privatization by selling off such enterprises at reasonable prices.

We start by clarifying government's role in the context of our model. First, we follow the institutional evolution, tracing the changes in shares of government enterprises in both *X*- and *Z*-sectors and then investigating the process by which privatization occurred. Accordingly, Section 10.2.1 discusses institutional frameworks of government enterprises together with their areas of activity. Section 10.2.2 focuses on the performance of government enterprises. Lastly, Section 10.2.3 analyses the effects of privatization on capital accumulation.

To summarize a central conclusion, in the early period of economic development the government placed importance on the direct operation of enterprises in the *Z*-sector. Proportionally, therefore, government involvement in both production and capital stock was higher in the *Z*-sector than in the *X*-sector. The subsequent privatization of these enterprises on favourable terms further stimulated investment in the *Z*-sector at the expense of the *X*-sector.

10.2.1. Institutional changes in government businesses

In preparation for what follows, let us begin with a brief review of the history of institutional changes related to government enterprises.[3]

The Meiji government took over the factories and other productive facilities of the old *daimyo* (feudal lords) and also established new factories and mines (Koyama 1943: 55). Those taken over were mostly weapon factories, though some produced other goods. Many of the latter had originally been established for military purposes but had been converted to the production of other goods in the latter part of the Tokugawa era (see Koyama 1943: 53).

In 1868, the new Meiji government created ministries of Private Affairs, Finance and Defence, each of which controlled those of the factories that had been taken over that related to its own sphere of interest. Two years later, however, in 1870, the Ministry of Industry was established by separating its activ ities from the Ministry of Private Affairs, and in 1872 the Ministry of Defence was split to create a Ministry of the Army and a Ministry of the Navy. In 1873 a Ministry of Internal Affairs was created, and in 1881 parts of this and of the Ministry of Industry were merged to create a Ministry of Agriculture and Commerce. (The other responsibilities of the Ministry of Industry were taken over by the Ministry of Communication which was created in 1885.) In 1924, the Ministry of Agriculture and Commerce was abolished, and separate ministries were created for Industry, Commerce, and Agriculture and Forestries (Emi and Shionoya 1965: 48–9).

Fig. 10.1 summarizes the division of labour among these ministries and the institutional changes prior to World War II. Classifying the activities of the factories under the various ministries, mining, manufacturing of silk, cotton and wool, and printing belong to our *X*-sector. Steel mills, civil engineering, machine, and tool sectors,[4] railways and ship-building are in the *Z*-sector, and agriculture and livestock are in the *S*-sector. This can be only a rough classification, however, because manufacturing and defence ministries alike engaged in industrial activities that had both *X*- and *Z*-sectors components, and some of the mines belonged to the Army and Navy ministries.

The correspondence between ministries, factories, and our economic classification is thus not always clear-cut; for example, a factory under the Ministry of Finance produced chemical products, and sometimes a factory passed from the control of one ministry to another. In general terms, however, the allocation of activities between ministries roughly corresponded to the structure shown in this figure.

The accounting of these government enterprises was sometimes included in the general budget and sometimes under special accounts; just before World War II there were 117 such special accounts (Toyo Keizai Shinposha 1950: 433). These accounts, while changing over time through new establishment, merger, and abolishment, were made to manage the activities of government enterprises. Roughly speaking, there were four categories of special accounts: (*a*) *Sagyo* special accounts (largely, production activities); (*b*) *Jigyo* special accounts (project management); (*c*) finance special accounts, and (*d*) facility management accounts (Emi and Shionoya 1965: 72).

For most government enterprises, production and marketing were controlled

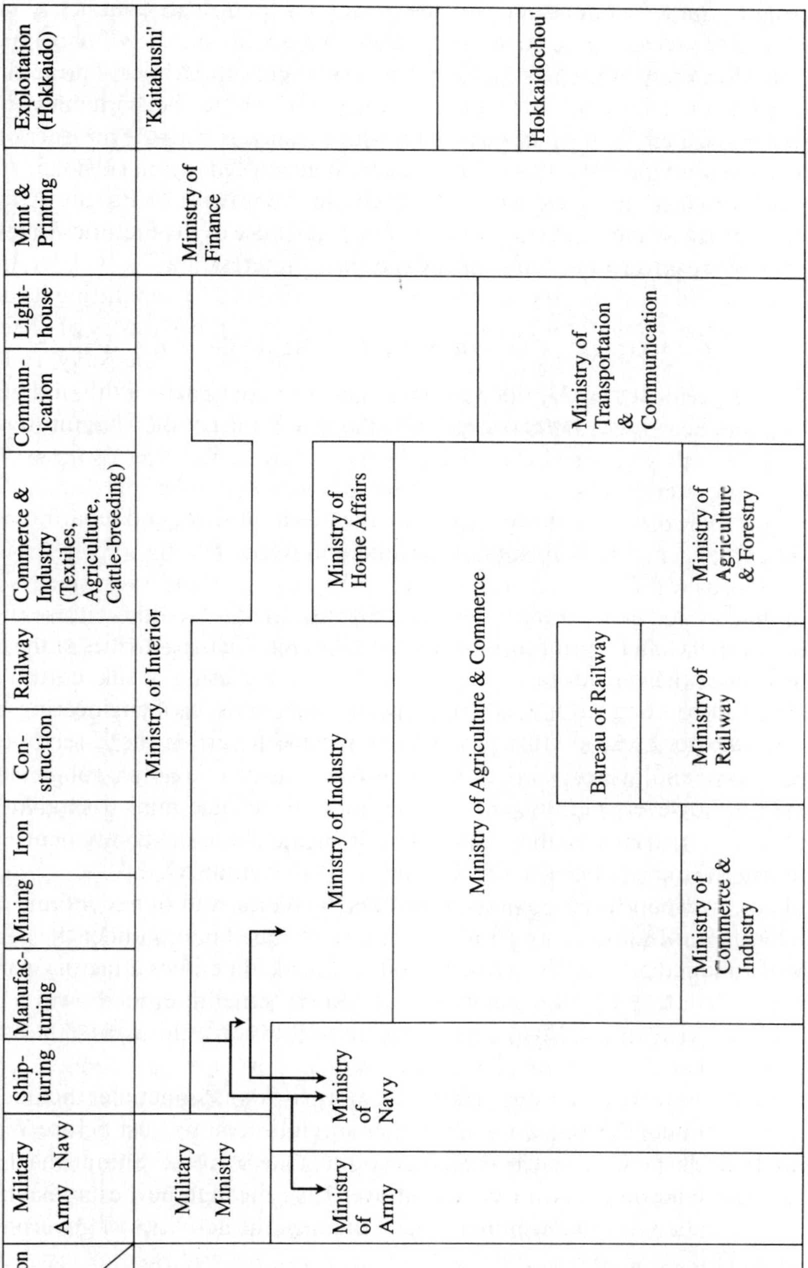

FIG. 10.1 Relations of division of government factories among the ministries.

through a *sagyo* account and finance through a finance account. Only after 1890 did the government adopt the system of general and special accounting which characterizes today's budget system. Roughly speaking the general account covers taxes and current expenditures and special accounts deal with government activities which involve business services by government units. While the government had a similar system before 1890, special accounts had not been divided into such categories (Hayashi 1965: 12).

Keeping these facts in mind, we now turn to a review of the business performance of these government enterprises and their privatization.

10.2.2. Operation of government enterprises in the early Meiji era

Over the period 1868–71, the new Meiji government took over the factories which had been owned and operated by the feudal lords. These factories produced not only weapons but also machines, metals, and chemicals for explosive products, and the government regarded it as a matter of some urgency to upgrade their production capacity.

The total generator capacity of the enterprises that had been owned by the feudal lords was 274 hp (Koyama 1943: 48–9). By 1883, the government factories had a total generator capacity of 2,509 hp. In other words, within fifteen years of the Meiji restoration,[5] the government had increased generator capacity more than eight-fold.

The largest of the industrial concerns taken over by the Meiji government was the Yokosuka Steel Mill which had eleven branch factories and a total generator capacity of 180 hp. It produced steam ships, weapons, agricultural machines and tools, leather-processing tools, equipment for the processing of seed-oils, and so on. Most of the other non-weapon factories were small, but a few were of fairly large scale. The Lord of Satsuma (a *daimyo* who ruled the Kagoshima area) had owned four sugar refineries and a Western-style cotton-spinning factory, the Kagoshima Spinning Factory.[6] The latter had 3,648 steam-operated spindles and 200 workers and produced 48 *kan* of cotton yarn per day (1 *kan* = 3.75 kg) (Koyama 1943: 51). Since the total number of spindles in the first year of the Meiji period was only 5,456 (Kajinishi 1965: 44), the Kagoshima factory accounted for more than half the national production.

Of the factories taken over from the *daimyos*, the *Z*-sector factories were mostly put under the ministries of Defence and Industry, and the *X*-sector factories, typically textile factories, were managed by the Ministry of Private Affairs, which was later divided into the ministries of Internal Affairs and Agriculture and Commerce. In addition, many new factories were set up by the government in Hokkaido, the Northern island, to encourage its development.

Table 10.5 compares the generator capacity of government and private factories. The raw data for the early period are unsatisfactory because institutional responsibilities often shifted and because data for factories under the ministries of Army and Navy were often unavailable. In addition, from around

TABLE 10.5(*a*). Horsepower of prime-movers of government factories at early Meiji period

	Total	Government factories							Private factories			Government percentage of total
		Manufacture in Ministry of Industries	Mining in Ministry of Industries	Ministry of Finance	Ministry of Agriculture and Commerce	Ministry of Army	Ministry of Navy	'Kaitakushi'	Total	*X*-sector	*Z*-sector	
1881	3316	506	977	n.a.	0	(1,334)		(499)				
1882	3486	465	(977)	n.a.	135	(1,410)		(499)				
1883	4112	472	(977)	489	190	(1,485)		(499)				
1884	4047	386	(906)	487	208	860	701	499	1,069	648	421	74.6
1885	3093	0	(894)	482	208	567	781	161	1,352	926	426	61.9
1886	2550	0	(894)	285	178	489	528	176	3,212	2,322	882	34.0
1887	2642	0	(894)	292	265	550	641	n.a.	8,109	5,211	2,723	17.7
1888	2028	0	(387)	291	5	582	763	n.a.	12,833	7,924	4,587	11.3
1889	2303	0	(387)	555	5	587	769	n.a.	17,631	11,989	5,209	9.8
1890	2901	0	(387)	548	5	978	982	n.a.	21,944	15,956	5,549	10.3
1891	2877	0	(387)	548	5	875	1,061	n.a.	23,085	16,514	6,098	9.7
1892	2807	0	(387)	548	5	920	946	n.a.	40,052	27,793	11,543	5.7
1893	3505	0	(387)	908	5	1,125	1,080	n.a.	45,173	31,338	13,029	6.5

Note: n.a. is treated as nil in total. Private factories exclude mining and public utilities. Values in parentheses are estimates obtained from various statistics.

Source: Minami (1965).

TABLE 10.5(*b*). Horsepower of prime-movers of main privatized mines

Name of mine	Horsepower	Date of privatization
Nakakosaka Iron Mine	n.a.	1884.7
Kosaka Silver Mine	65	1884.9
Innai Silver Mine	6	1884.12
Aji Copper Mine	12	1885.3
Okatsu Gold Mine	n.a.	1885.6
Kamaishi Iron Mine	472	1887.12
Miike Coal Mine	35	1888.8
Horonai Coal Mine	n.a.	1889.12
Soda Gold Mine	56	1896.9
Ikuno Silver Mine	311	1896.9
Temiya	20	
TOTAL	977	

Source: Minami (1970).

1882, the government embarked on privatization. For all these reasons, the figures for government enterprises fluctuated wildly. After making suitable adjustments,[7] we arrive at the following conclusions. In 1881, the generator capacity of government enterprises was approximately 3,300 hp, most of which was accounted for by enterprises under the ministries of Army and Navy and mines. By 1884 it had reached about 4,000 hp, of which 3,000 hp was in manufacturing sectors. As private manufacturing in the *X*- and *Z*-sectors combined used about 1,000 hp in that year, we can conclude that the government sector accounted for three-quarters of total manufacturing activities. Although no data are available on the private manufacturing sector prior to 1884, it is conjectured that there was less private industrial activity at that time, and that the government sector would therefore have occupied a much larger share from 1868 to 1883. If we include the Kamaishi Mine, a government-owned mine with steel production capacity, in the manufacturing sector, the above share would be still higher. Thus, in the first fifteen years of the Meiji period, most industries were operated by the government, and particularly in the *Z*-sector.

We now turn to government expenditure on industry. Generally speaking, the government invested far more in mining and railway construction, than in manufacturing, but within the manufacturing sector a far greater proportion of government investment was directed to the *Z*-sector. As we have already seen (Fig. 10.1), responsibility for the government's industrial activities were split between several ministries: Defence, Industry, Private Businesses, Internal Affairs, and Agriculture and Commerce. Since a large part of the accountings of the ministries of Defence and Private Business have not been made public, we focus our examination on the other three ministries.

Table 10.6 shows the cumulative expenses of the Ministry of Industry on the industries under its control from 1870 to 1885. The industries under its

TABLE 10.6. Cumulative expenditure by Ministry of Industries, 1870–1885 (¥'000)

	Cumulative	%
Current expenditure	12,934	
Operating expenditure		
Mines		
Sado Mine	1,286	
Ikuno Mine	1,656	
Kosaka Mine	547	
Miike Mine	755	
Okatsu Mine	149	
Kamaishi Mine	2,376	
Aji Mine	535	
Innai Mine	226	
Nakakosaka Mine	85	
Well-drilling	45	
Cost for new operation	1,663	
Subtotal	9,327	31.4
Railways		
Construction cost for Tokyo–Yokohama	2,979	
Kobe–Osaka	4,446	
Seikyo–Osaka	2,767	
Kyoto–Kobe	76	
Cost for new operation	3,634	
Repayment for public debt of Nakasendo line	299	
Subtotal	14,293	48.1
Manufacturing		
Iron and steel	62	
Hyogo Shipbuilding	816	
Akabane manufacturing	550	
Construction cost for Nagasaki Shipbuilding	628	
Hukagawa manufacturing	100	
Shinagawa Glass Manufacture	294	
Subtotal	2,452	8.3
Telecommunications	3,638	12.2
TOTAL	29,712	100.0

Source: Kajinishi (1965).

supervision were mining (including petroleum), railways, manufacturing, and telecommunication. Railways received 48% of the total allocation, mining 31%, and telecommunication 12%, and manufacturing only 8%. Since we shall be considering railway construction in the section on infrastructure, we focus first on mining.

The Meiji government had requisitioned the Sado Island gold mines and the Ikuno silver mines, both of which had been controlled by the Tokugawa government, and in 1873, two years after the *Haihan Chiken* (abolition of the old han ruled by Daimyo and creation of prefectures), it took over all the major mines of the *daimyos*.[8]

The reasons why the new government owned and operated major mines were threefold. First, it wanted to present a model for the private sector to follow as regards the modernization of equipment and management. Secondly, it wanted to promote the export of metals as a source of foreign exchange (Japan was at this time a major exporter of copper). Thirdly, it wanted to develop coal mining so as to secure energy resources for government enterprises, and also as a source of government revenue.

From Table 10.7, which shows the government's share in the production of selected metals, we see that the government accounted for more than 50% of gold, silver, and lead production. Needless to say, the government had to invest heavily in mining to achieve this, but it realized a good rate of return as well, as suggested by Table 10.8.

The Ministry of Industry also controlled a number of manufacturing industries: the Nagasaki and Hyogo shipyards, and the Hukagawa, Akabane, and Shinagawa machine factories in Tokyo. In total, the ministry spent ¥2.5 million on these enterprises through the period from 1870 to 1885 (see Table 10.6), and total sales revenue for the same period was about ¥5.66 million. Thus, business performance was satisfactory. Of the total sales of ¥5.66 million, of sales the Nagasaki shipyard accounted for 47.6%, the Hyogo shipyard for 33%, and the Akabane machine factory for 19.4% (see Table 10.9). In 1880, these three factories produced ¥0.9 million (Emi 1971: 102). In general government enterprises engaged in machine production, including shipbuilding, accounted for a very large part of Japan's total production, as this industry remained small in early years.

It should also be noted that the mines and factories under the Ministry of Industry operated in close association, each supplying products to the other, and also supplying products directly to private sectors.

While the nature differs from enterprises under the Ministry of Industry, a large portion of the budget was directed to factories under the control of the ministries of Army and Navy. Although the direct objective of the government in operating machine factories under this ministry was to strengthen the nation's military power, it appears that these factories contributed to the development of the machine industry in general, since they imported the latest technologies from Western nations and diffused them to their subcontractors.

Between 1870 and 1880, the government spent a total of ¥1,074.2 million, of which ¥241.5 million was on military expenditures. In the same period, total expenditure on capital formation by the central government was ¥180.3 million, of which ¥41.3 million was invested in military facilities.[9] (Most of the factories producing weapons were those taken over and expanded by the new

TABLE 10.7. Share of production by government factories in mining

	Composition of government factories	Production unit	1879	1880	1881	1882	1883	1884	1885	1886	1887	1888	1889	1890
Gold	(1) Government and private total	kg	261	312	304	271	300	274	274	464	520	629	768	726
	(2) Government	kg	188	202	180	165	146	132	127	176	225	266	239	253
	(2)/(1)	%	72.1	64.7	59.2	60.9	48.8	48.2	46.5	37.9	43.3	42.3	31.1	34.8
Silver	(1) Government and private total	kg	9,087	10,338	17,861	16,863	24,130	22,901	23,835	33,684	35,617	42,738	42,968	52,844
	(2) Government	kg	4,825	5,904	8,492	8,984	13,622	4,675	4,534	5,851	6,157	6,441	6,193	6,952
	(2)/(1)	%	53.1	57.1	47.5	53.3	56.5	20.4	19.0	17.4	17.3	15.1	14.4	13.2
Copper	(1) Government and private total	tons	4,630	4,669	4,771	5,616	6,774	8,888	10,540	9,774	11,063	13,374	16,254	18,115
	(2) Government	tons	319	254	267	322	681	181	7	19	8	19	32	13
	(2)/(1)	%	6.9	5.4	5.6	5.7	10.1	2.0	0.07	0.20	0.08	0.15	0.20	0.07
Lead	(1) Government and private total	tons	259	269	259	233	289	86	89	235	385	400	601	774
	(2) Government	tons	131	111	104	90	124	—	—	—	—	—	—	—
	(2)/(1)	%	50.7	41.4	40.4	38.8	43.0							
Iron and steel	(1) Government and private total	1,000 tons	13	16	16	12	14	11	6	13	15	18	21	22
	(2) Government	1,000 tons	2	4	4	1	5	4	3	3	3	3	3	3
	(2)/(1)	%	16.3	25.7	27.6	9.0	35.0	40.6	45.8	28.7	21.9	17.4	14.4	16.0
Coal	(1) Government and private total	1,000 tons	863	889	932	936	993	1,149	1,253	1,322	1,683	2,023	2,440	2,618
	(2) Government	1,000 tons	141	174	178	180	171	264	218	342	393	478	187	6
	(2)/(1)	%	16.3	19.7	19.1	19.3	17.3	23.0	17.4	25.9	23.4	23.6	7.7	0.26

Source: MITI (1963).

TABLE 10.8. Profits of government factories and enterprises, 1870–1885
(¥'000)

Railway between Kobe and Otsu	4,672
Railway between Tokyo and Osaka	3,185
Bureau of Telecommunication	1,732
Sado Mine	1,048
Miike Mine	810
Ikuno Mine	650
Aji Mine	399
Kosaka Mine	304
Innai Mine	221
Hyogo Shipbuilding Yard	183
Akabane Manufacture	76
Railway between Tsuruga and Ogaki	58
Okatsu Mine	55
Yudo Mine	30
Hukagawa Manufacture	13
Nagasaki Shipbuilding Yard	4
Petroleum well-drilling	2
Railway between Tokyo and Maebashi	1
Akabane Painting for Wooden products	–2
Nakakosaka Mine	–21
Shinagawa Glass Manufacture	–148
Kamaishi Mine	–318
TOTAL	12,960

Source: Kajinishi (1965).

government.) In contrast, expenditure in this period on operating the enterprises under the Ministry of Industry—the total for mining, railways, factories, and communications in Table 10.6—was only ¥29.7 million, or some three-quarters of the investment in military facilities.

As shown in Table 10.10, between 1876 and 1885, the factories under the ministries of Army and Navy produced machinery and weapons to the value of ¥13,321 million, of which some 76% for the Army and the rest by the Navy (Emi 1971: 108).

The sales of the three factories under the Ministry of Industry and the total production of machines and tools by the factories under the ministries of Army and Navy in the period 1876–85 amounted to ¥18.38 million. According to the Shionoya estimates, however, the cumulative total of all production of machinery and tools until 1885 was ¥70.8 million. On this basis, government enterprises accounted for roughly 26% of total production.

So far we have considered government production only under the ministries of Industry, Army, and Navy, ie. only in the *Z*-sector; the government's *X*-sector enterprises were controlled by the ministries of Internal Affairs and Agriculture

TABLE 10.9. Sales revenue of machinery factories controlled by Ministry of Industry (¥)

	Nagasaki Shipbuilding Yard	Hyogo Shipbuilding Yard	Akabane Manufacture	Total
1870	36,422			36,422
1871	28,958	10,763	359	40,080
1872	11,712	1,563	849	14,124
1873	73,286	28,479	987	102,752
1874	121,785	32,248	7,955	161,988
1875	49,288	15,551	1,217	66,056
	78,251	41,976	7,061	127,288
1876	298,020	72,959	115,012	485,991
1877	131,079	193,585	105,243	429,907
1878	159,877	193,319	97,199	450,395
1879	233,774	251,751	139,278	624,803
1880	416,418	248,930	242,988	908,336
1881	506,077	227,831	167,066	900,974
1882	311,971	165,705	213,528	691,204
1883	233,242	200,201	0	453,443
1884	7,553	160,786	50	168,389
1885	0		137	137
TOTAL	2,697,713	1,865,647	1,098,929	5,662,289
Composition (%)	47.6	33.3	19.4	100.0

Source: Emi (1971).

and Commerce. These two ministries supervised the factories for wool textiles, cotton and silk spinning, tea processing, and sugar processing. They also supervised various experimental laboratories and livestock farms.

Statistics for government enterprises under the Ministry of Internal Affairs became available only from 1875. For the period 1875–85, total expenditure was ¥87,600, of which ¥7,750 was spent on silk and other textile industries. This accounted for 88.5% of its total industrial expenses. In 1880 these industries became the responsibility of the Ministry of Agriculture and Commerce. Between 1880 and 1887 the total expenditure was ¥46,100, of which textile enterprises accounted for 64.7%.[10]

To put this figure in context, we must remember that the Ministry of Industry spent during the period 1870 to 1880 ¥13,746 million on the railway, mining and machine manufacturing sectors,[11] of which ¥1,551 million on activities which could be classified as *Z*-sector. It is clear, therefore, that government investment in the *X*-sector was very much smaller than in the *Z*-sector. (See Table 10.11.) In fact, the government enterprises in the *X*-sector seemed to have played quite a role in the economy: whereas *Z*-sector enterprises were

TABLE 10.10. Production of machinery by government factories (¥'000)

Fiscal year	Ministry of Industry	Ministry of Army	Ministry of Navy	Total	Calendar year production
1876	486	317	361	1,164	645
1877	430	81	453	964	1,064
1878	450	158	542	1,150	1,057
1879	625	411	477	1,513	1,332
1880	908	193	488	1,589	1,551
1881	901	220	457	1,578	1,584
1882	691	221	437	1,349	1,464
1883	453	570	3,846	4,869	3,109
1884	168	423	1,255	1,846	3,358
1885	0	565	1,846	2,411	2,731
TOTAL 1876–85	5,112	3,159	10,162	18,433	18,380
1886		605	1,448	2,053	2,143
1887		964	2,432	3,396	3,060
1888		1,000	2,412	3,412	3,408
1889		803	2,199	3,002	3,105

Source: Emi (1971).

geared to improving Japan's national wealth and military power, the government operated its textile factories as 'model' enterprises, installing modern equipment and inviting foreign engineers to instruct local workers, and employing 'model female workers'. In fact, the total production of these model factories for the period 1882–84 was only ¥1.1 million,[12] in contrast to the total textile production of ¥167.4 million estimated by Shionoya for the same period (Table 10.12).

To sum up, it was in the *Z*-sector that the government invested most heavily, and had the highest share of total production. While there was considerable government investment in mining, government investment in the manufacturing *X*-sector was mostly restricted to establishing 'model factories' to be emulated by private sectors.

10.2.3. Privatization of government enterprises

We now turn to the privatization of government enterprises, which, as will shortly be shown, also favoured the *Z*-sector. Historians have emphasized the importance of privatization in the Meiji era for the industrialization of Japan, but in fact there has been no quantitative analysis. We therefore ignore the politics behind the privatization and concentrate on the quantitative aspects.

Table 10.13 shows the manufacturing companies privatized around 1885 by sector, the cumulative government investment prior to privatization, and the

TABLE 10.11(*a*). Expenditure for promotion of industries by Ministry of Internal Affairs (cumulative from 1875 to 1880 fiscal years)
(¥)

Bureau of Museums	221,474
Geological Survey of Bureau of Geography	10,366
Exhibition of Internal Industry Promotion	106,874
Exhibition in USA	12,683
Bureau for Promotion of Industry, Total	3,124,226
Current expenditure	1,054,225
Other expenditures	2,069,996
Promotion of agriculture	
Naitoshinzyuku Laboratory	318,836
Komaba Agricultural School	394,519
Mita Breeding Laboratory	82,800
Laboratories	
Uchiyamashitamachi Laboratory	52,243
Shin-ogawamachi Laboratory	2,222
Cattle-breeding	
Komaba Cattle-breeding	986
Shimohusa Sheep-breeding	
Shimohusa-Katori Cattle-breeding	780,769
Awa-Mineoka Cattle-breeding	4,143
Selling silk yarn and paper	236,156
Yarn and Spinning	
Senzyu Carpeting	270,022
Sekiguchi Carpeting	43,239
Shinmachi Spinning	232,322
Sakai Spinning	45,891
Aichi Spinning	43,508
Hiroshima Spinning	22,761
Tomioka Silk-spinning	117,112
Tsukiji Tea-processing	488
Sugar-processing	99,741
TOTAL	10,370,215

Source: Kajinishi (1965).

prices at which these enterprises were sold off. (In fact, privatization was carried out not only in the manufacturing sector but also in the mining and transportation sectors.) From the table we see that total government investment in enterprises in the *X*- and *Z*-sectors, including mining, from the first year of Meiji through 1885 was ¥1,533 million, and that these enterprises were then sold off to the private sector for ¥967 million. If we exclude mining sectors from the *X*-sector, however, cumulative government investment in manufacturing to 1885 was only ¥553 million, of which 85% was allocated to the *Z*-sector.

TABLE 10.11(*b*). Expenditure for promotion of industry by Ministry of Agriculture and Commerce (cumulative from fiscal years 1880 to 1887)
(¥)

Current expenditure	
Ordinary expenditure	2,199,033
Komaba Agricultural School	356,517
Cattle-breeding	
Shimohusa Cattle-breeding	186,012
Shimohusa Sheep-breeding	12,671
Spinning	
Senzyu Carpeting	247,942
Shinmachi Spinning	15,724
Aichi Spinning	11,978
Hiroshima Spinning	26,233
Sugar-processing	158,751
Extraordinary expenditure	
Internal Exhibition	168,769
Exhibitions abroad	84,653
Subsidies	
Wine growers	8,000
Sheep-breeding	1,664
Fund for loan to improvement of sugar-processing	60,000
Mita Cattle-breeding	2,100
TOTAL	6,007,166

Source: Kajinishi (1965).

When we look at the prices at which these enterprises were sold, we find that *X*-sector industries were mostly sold at half the cumulative investment value, though mines, understandably, were priced rather higher than the value of cumulative investment. In contrast, government-owned manufacturing and railway enterprises in the *Z*-sector were privatized at about one-third to one-fourth of the cumulative investment values. Thus, the way privatization was implemented clearly favoured the *Z*-sector over the *X*-sector.

Table 10.14 shows the relative significance of this transfer (not a pure transfer, but nevertheless a sale at concessionary prices) to the private sector in the national wealth. The total amount of cumulative investment transferred accounted for about 1.4% of the total productive capital of all industries except agriculture and forestry, and 0.95% if we consider the manufacturing sectors alone. In terms of the generator capacity of manufacturing industries, 2.8% was transferred from the government sector to the private sector (based on 1890). Thus, roughly speaking, privatization resulted in a shift of ownership of some 1%–3% of the total productive capital of manufacturing industries.

TABLE 10.12. Value of textile production by government factories (¥'000)

	1882	1883	1884	1885	1886	1887	1888	1889	1890	1891
Senjyu Carpeting	256	164	74	398	602	536	607	603	590	643
Shinmachi Spinning	47	34	78	73						
Aichi Spinning	17	15	20							
Tomioka Silk-spinning	57	199	115	68	183	214	239	199	399	154
TOTAL	378	413	290	540	785	751	847	802	989	798

Source: Naikaku Tokeikyoku (1881–1939).

TABLE 10.13. Cumulative investment and selling prices of government enterprises (¥)

Enterprise	Cumulative investment by government	Selling price	Installed horsepower
X-sector			
Manufacturing			
Hiroshima Spinning	50,000	12,077	
Aichi Spinning	58,000	—	50
Sapporo Brewing	—	27,672	
Sugar-processing	258,492	994	140
Shinmachi Spinning	130,000	150,000	80
Tomioka Silk-Spinning	310,000	121,460	5
Subtotal	806,492	312,203	275
Mining			
Takashima Coal Mine	393,848	550,000	
Yudo Coal Mine	48,608	27,944	
Nakakosaka Iron Mine	58,507	28,500	
Kosaka Silver Mine	547,476	273,660	
Innai Silver Mine	703,093	108,977	
Aji Copper Mine	1,673,211	337,766	
Okatsu Gold Mine	149,546	117,142	
Miike Coal Mine	757,060	4,555,000	
Sado Gold Mine	1,419,244	1,730,000	
Ikuno Silver Mine	1,760,866		
Subtotal	7,511,459	7,728,989	977
Subtotal	8,317,951	8,041,192	1,252
Z-sector			
Manufacturing			
Hukagawa Cement	101,559	73,964	30
Shinagawa Glass-processing	294,168	79,951	10
Nagasaki Shipbuilding	1,130,949	527,000	335
Hyogo Shipbuilding Yard	816,139	553,660	41
Kamaishi Iron Mine	2,376,625	12,600	
Mita Manufacturing for Agri-tool	—	33,795	
Subtotal	4,719,440	1,280,970	416
Railway			
Horonai Railway for Coal	2,291,500	352,318	
Subtotal	2,291,500	352,318	
Subtotal	7,010,940	1,633,288	
TOTAL	15,328,891	9,674,480	
(in which, total manufacturing)	5,525,932	1,593,173	691

Sources: Kajinishi (1965); Naikaku Tokei Kyoku (1881–1939).

As has been pointed out, the privatization of *Z*-sector factories on favourable terms facilitated capital accumulation in the *Z*-sector thereafter. Moreover, the government had borne the risks of start-up; indeed, most of the firms

TABLE 10.14. Capital stock of privatized government enterprises

Capital Stock	Value at 1934–6 constant prices (¥'000)	Horsepower
(1) Producer's durable equipment	296,986	24,458
(2) Non-residential building	1,368,921	
(3) Structures	1,038,596	
(4) Electric Utilities	1,640	
(5) Railway	369,340	
(6) Capital stock of primary industry	3,075,483	
(7) Direct productive industrial capital in (6)	1,665,907	
(8) Privatized capital	43,672	
(9) Capital of manufacture in (8)	15,743	691
(8)/(6)%	1.42	
(9)/(7)%	0.95	
(9)/(1)%	5.30	2.8

Note: Privatized capital is government investment expenditure deflated by price index of investment.

Sources: Ohkawa *et al.* (1966, 1967); Koyama (1943).

privatized at this time grew to become the bases of today's major conglomerates. This is evidence that it was the direct operation by the government in the early stage that supported the subsequent self-sustaining growth.

10.3. Subsidies to Industry and Investment in Social Overhead Capital

The government influences industrial activities through taxation and fiscal expenditures for administrative services, investment in social overhead capital, and subsidies to industries. Throughout its industrialization processes, whether pre-World War II days or post-World War II days, the Japanese government was far from being the classical 'cheap' government. On the contrary, it actively intervened in the private sector.

We now consider the effects of taxation and government expenditure on capital accumulation in the *X*- and *Z*-sectors, looking first at resource transfer through taxation and subsidies, and then at government investment in infrastructure. In particular, we wish to see how subsidies increased profit, and to what extent production functions shifted upward because of investment in infrastructure. As it is difficult to quantitatively measure such effects, our discussions will remain predominantly qualitative.

Our major findings are that the government favoured the capital-intensive sectors in general and the *Z*-sector in particular, and that such preferential treatment was more significant in the earlier stages of industrialization.

10.3.1. The government budget and revenues

Government revenues comprise the revenue of both central and local governments. As taxes raised by the central government accounted for roughly 50%–60% of total tax revenues (see Table 10.15), we shall focus on this element of government revenues.

TABLE 10.15. Revenues of central and local government (¥m.)

Fiscal year	Revenue of central government	Revenue of local government	Total	Revenue of general account
1880	59.3(64.5)	32.6(35.5)	91.9	63,367
1885	56.4(62.2)	34.3(37.8)	90.7	62,157
1890	74.2(65.1)	39.8(34.9)	114.0	106,469
1895	90.6(61.4)	56.9(38.6)	147.5	118,433
1900	174.2(59.7)	117.5(40.3)	291.7	295,855
1905	375.2(75.8)	119.6(24.2)	494.8	535,256
1910	504.3(68.6)	230.8(31.4)	735.1	672,874
1915	513.0(65.5)	270.0(34.5)	783.0	708,615
1920	1,170.0(59.9)	783.9(40.1)	1,953.9	2,000,652
1925	1,511.0(57.0)	1,142.0(43.0)	2,653.0	2,071,369
1930	1,527.0(52.8)	1,363.6(47.2)	2,890.0	1,596,972

Note: Figures in parentheses are percentages.

Sources: Emi and Shionoya (1965): 174, 178; Toyo Keizai Shinposha (1926).

Central government revenues derived primarily from tax revenues and income from government enterprises and assets (Table 10.16). While the share of tax revenues was larger in the early period, it later decreased to one-half of total revenue. Thus, it is clear that revenues from the government enterprises and government assets were important sources of the budget.

The major taxes in the early Meiji period were land tax (*chiso*), a tax on alcohol, and import tariffs. These were later supplemented by income tax (established in 1887), corporate income tax (1897), inheritance tax (1905), and a tax on sugar consumption (1901). Until 1895 the land tax was the main source of revenue, but subsequently revenues from income tax, the alcohol tax, and import tariffs became more important (Table 10.17). The reason why land tax was so important until 1895 was because Japanese industries were then at an infant stage and agriculture was still the main economic activity. Obviously, the revenue structure changed as the economic structure changed.

In considering the relative burden of taxation one has to compare the ratio of tax to income, or the tax-bearing ratio. This is shown in Fig. 10.2. On the assumption that land tax was borne mostly by the agricultural household and other taxes by the non-primary industry sector, the chart shows that the

TABLE 10.16. Percentage distribution of total revenue of central government (revenues of the General Account)

Fiscal year	Taxes	Stamps	Receipts from government enterprises and properties	Articles sold by the government	Miscellaneous	National bonds	Transferred from previous year	Payments shared by local government
1875	85.2		4.4	3.5	3.1			
1880	87.2		3.3	1.9	6.6			
1885	84.6		3.7	0.6	2.3	4.9		
1890	62.2		8.3	0.7	1.3		18.9	
1895	63.0		13.5	0.5	1.2		16.9	
1900	45.3	4.2	13.5	0.3	2.7	14.8		0.4
1905	46.9	5.7	18.7	0.2	3.0	13.8	9.4	0.1
1910	47.2	4.0	19.1	0.4	1.6	0.5	12.5	0.2
1915	44.1	4.5	23.7	3.9	4.5	0.7	12.1	
1920	34.8	4.2	16.4	0.3	3.8	3.8	31.8	0.5

Source: Toyo Keizai Shinposha (1926).

TABLE 10.17. Tax revenue composition: (percentage distribution of revenues from each tax)

Fiscal year	Land tax	Tax on alcohol	Income tax	Tax on alcohol-making	Import tariffs	Sugar consumption tax	Business tax	Tax on banknote issue	Textile consumption tax	Tobacco tax
1875	85.1			4.3	2.9					0.3
1880	76.6			10.0	4.7					0.5
1885	81.8			2.0	4.0					1.7
1890	60.6		1.7	21.0	6.6					2.7
1895	51.8		2.0	23.8	9.1					3.7
1900	33.4	27.5	4.7		12.7		4.5			
1905	32.0	23.5	9.2		14.6	4.5	7.5	1.0	2.1	
1910	24.0	27.3	10.0		12.6	5.6	8.1	0.3	5.7	
1915	23.5	27.1	12.0		10.3	7.2	6.9	0.3	5.0	
1920	10.6	23.6	27.4		10.0	5.8	8.9	1.2	6.0	
1925	10.6	28.3	28.2		12.4	10.1	8.2	3.9	7.7	
1930	8.2	26.3	24.1		12.6	9.3	6.5	0.9	4.1	
1935	6.3	22.6	24.6		16.3	9.2	6.2	—	4.4	
1940	0.8	7.7	40.2		3.9	3.8	3.5	0.5	2.6	

Sources: Toyo Keizai Shinposha (1926); Toyo Keizai Shinposha (1950); Nippon Hinko (1966).

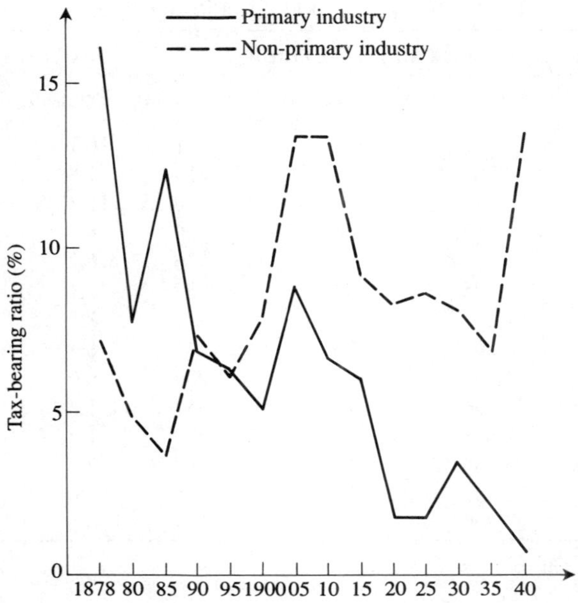

FIG. 10.2 Tax-bearing ratios by industry.

tax-bearing ratio of agriculture in the early years was much higher than that of other industries. But this difference shrank over time, and after 1890 the ratio of tax to income became higher for other industries than for agriculture. Since the agricultural sector also bore some additional taxes in the early years, the alcohol tax-bearing ratio of this sector would have been even higher than our assumed ratio. This implies that the subsistence sector was a major source of revenue for government investment in other sectors until sometime in the 1890s.

Unfortunately, it is almost impossible to classify tax revenues from industrial sources in terms of sectoral origin, as detailed data have not been published. However, there are a number of facts that lead us to infer that tax revenue derived more from the *X*-sector than the *Z*-sector. First, consumption taxes were imposed on several *X*-sector products: alcohol, tobacco, soy sauce, sugar, textiles, and soft drinks. Although taxes were also imposed on gas and electri city, fuel oils, and pharmaceuticals, these *Z*-sector products were a less significant source of revenue than the taxes on alcohol, sugar, and textiles. Therefore, we can conclude that tax revenue came mostly from the *X*-sector.

Secondly, there were various tax reductions and exemptions from corporate income. There was, for example, the Law for the Promotion of the Steel Industry, which provided shelter from corporate income tax and business tax and a law enacted in 1913 which exempted selected industries from corporate income tax in their first three years of operation. The industries which enjoyed such a tax holiday included those producing the following:

1. Gold, silver, lead, zinc, and amalgams of iron or aluminium

TABLE 10.18. Tax-bearing ratios by industry

Fiscal year	Production income (¥m.)		Tax (¥m.)		Tax-bearing ratio (%)	
	Primary	Non-Primary	Land tax	Other tax	Primary	Non-Primary
1878	248	153	40	11	16.1	7.2
1880	536	263	42	13	7.8	4.9
1885	350	271	43	10	12.3	3.7
1890	580	357	40	26	6.9	7.3
1895	630	592	39	36	6.2	6.1
1900	929	1,095	47	87	5.1	7.9
1905	905	1,299	80	171	8.8	13.2
1910	1,133	1,807	76	241	6.7	13.3
1915	1,243	2,639	74	239	6.0	9.1
1920	4,066	7,896	74	657	1.8	8.3
1925	4,094	9,558	75	820	1.8	8.6
1930	2,260	9,338	68	767	3.0	8.2
1935	2,944	12,656	58	868	2.0	6.9
1940	6,154	27,760	29	3,624	0.5	13.1

Source: Nippon Ginko (1966).

2. Various intermediate products of iron and steel
3. Various intermediate products of copper
4. Boiler, engines, and machines operated by generators
5. Soda, ammonium, and various acids as intermediate chemical material
6. Pulp for paper
7. Glass board
8. Condensed milk
9. Silk, linen, and woollen fabrics.

Except for items 8 and 9, the above list suggests that the government gave preferential treatment to the producers of *Z*-sector goods.

A later law, enacted in 1917, awarded firms which had specified equipment a ten-year exemption from corporate income tax and business tax. It can therefore be said that the government's tax policy was to siphon resources from the *X*- and *S*-sectors to the *Z*-sector, through preferential tax treatment and government support for investment.

To sum up: tax was an important source of government revenue. In the early years it was collected mostly from the *S*-sector. Within the industrial sector, the *X*-sector was taxed more heavily than the *Z*-sector, and moreover the tax system was biased to favour capital accumulation in the *Z*-sector. Local taxes heavily depended on additional land tax (Chiso Hukazei) which was proportionate to the land tax of the central government. As the local tax system followed the central one in the pre-World War II period, the overall effect of taxation on resource allocation seems to be as described above.

10.3.2. Industry subsidies

The Meiji government made full use of subsidies to reallocate economic resources and promote investment in heavy industries. Current subsidies increased profit and encouraged capital accumulation indirectly, while capital subsidies directly stimulated investment. The general characteristics of government subsidies prior to World War II can be summarized as follows:

1. Subsidies on capital exceeded subsidies on current accounts.
2. Prior to 1890, the largest subsidies were to construction and investment in transportation.
3. From 1890 to 1910 the share of subsidies to transportation further increased.
4. From 1910 to 1940, subsidies to transportation continued to increase but their relative share declined. Subsidies to construction also increased again in this period. Subsidies to agriculture and fishery, which had previously seen very small, expanded substantially, and became comparable to those to construction.
5. Prior to World War II, subsidies to manufacturing sectors were relatively small. Data limitations make it impossible to classify industrial subsidies by sector, but it may be noted that subsidies to steel industries in the 1920s accounted for some 10% of the total subsidies to mining and manufacturing.

Table 10.19, which shows the distribution of government subsidies to industry through the general account (ie. current expenditures) in more detail,

TABLE 10.19. Distribution of government subsidies to industries (%)

Fiscal year	Agriculture, forestry, and fisheries	Construction	Mining and manufacturing	Electricity with gas utilities	Transport	Other[a]
1875	—	100.0	—	—	—	—
1876–80	—	85.4	—	—	14.6	—
1881–85	0.4	61.6	—	—	12.4	25.6
1886–90	3.2	47.6	—	—	47.6	1.4
1891–95	0.6	59.6	—	—	37.2	2.6
1896–1900	3.8	9.2	—	—	86.2	0.8
1901–05	5.8	0.8	—	2.8	80.8	9.8
1906–10	3.6	1.8	0	0	62.2	32.4
1911–15	11.8	—	0.6	—	85.8	1.6
1916–20	19.4	10.4	11.6	—	54.8	3.8
1921–25	17.6	25.4	9.4	3.4	13.6	30.6
1926–30	33.4	24.8	7.6	2.8	17.4	14.0
1931–35	30.0	47.8	4.0	1.2	12.6	4.6
1931–40	52.2	9.4	15.6	3.0	9.0	10.6

[a] Includes finance, international trade, emigration, exploitation of land, etc.

Source: Emi and Shionoya (1965).

indicates that until 1895 subsidies to construction accounted for more than 50% of all subsidies, and that the next most highly subsidized sectors were transportation and communication. Subsidies to the former covered harbour construction, flood control, and road construction, while the latter were mainly to railway and marine transportation, ship-building, and the manufacture of automobiles and airplanes.

Until 1895, subsidies to construction were directed mostly to the flood control and road-building (Emi 1971: 44–5). The investment in flood control had economic benefits for the *S*-sector, and road construction benefited industries in both sectors. The most important subsidies to transportation and communication were those to shipping—to Nippon Yusen (Japan Line) and Osaka-Yusen (Osaka Line)— and to Nippon Railway (Emi 1971:67).

Between 1900 and 1915 the subsidies to construction stopped and those to transportation and communication were increased, with marine transportation and telecommuniations now receiving the largest subsidies (Emi and Shionoya, 1965: 97). Between 1906 and 1907 the government nationalized seventeen railways, which had formerly received subsidies.

After 1920 subsidies to construction increased again, particularly road construction, and public investment in road construction also expanded, reaching twice the level of investment in flood control. However, subsidies to agriculture, which had so far been relatively small now increased to a level comparable to subsidies to construction. In other words, the government increased its support to the *S*-sector by means of subsidies. As tax-bearing by the *S*-sector decreased in the same period, the government virtually strengthened its support of the sector. Having considered the distribution of subsidies, we need to clarify how subsidies increased profits and how profits were re-invested. As most subsidies after 1915 took the form of capital subsidies (Emi and Shionoya 1965: 260), we investigate how large a part these subsidies played in capital accumulation, assuming that as a first approximation all subsidies were capital subsidies.

Table 10.20 compares levels of subsidy and fixed capital formation by industry. Overall, the ratio of subsidies to fixed capital formation was relatively large until 1885, declined thereafter, but rose again in the years just before World War II. The ratio of subsidies to construction (excluding housing) to total construction investment was also high around 1885, reaching 15%. In the 1920s, the subsidies increased substantially in absolute terms, but the ratio slightly declined. The ratio of subsidies to fixed investment in the private utility sector (transportation and communication) steadily declined over time, but hovered around the 15% level until 1915, after which it declined rapidly. We can thus conclude that subsidies played a very important role in the early years of the Meiji era, especially in construction, transportation, and communication.

To sum up, government subsidies were directed less to industry than to the construction of roads and railways, and the development of communications. However, the latter may be considered social overhead capital, in that its

TABLE 10.20. Industrial subsidies and capital formation (¥'000)

Fiscal year	(1) Total of industrial subsidies	(2) Of which agriculture, forestry, and fisheries	(3) Construction	(4) Mining and manufacturing	(5) Electricity and gas	(6) Transport and communication utilities	(7) Others	(8) Gross domestic fixed formation in Japan	(9) Of which investment of construction capital (except private residential)	(10) Investment in public utilities	(1)/(8)	(3)/(9)	(6)/(10)
1875	1,393	—	1,393	—	—	—	—	10,309	6,199	—	13.5	22.5	—
1876–80	8,656	—	7,391	—	—	1,265	—	66,479	31,837	—	13.0	23.2	—
1881–85	12,167	43	7,500	—	—	1,505	3,119	102,995	50,484	3,106	11.8	14.9	48.5
1886–90	16,569	542	7,895	—	—	7,893	239	279,405	91,678	31,719	5.9	8.6	24.9
1891–95	21,229	148	12,654	—	—	7,876	551	489,770	146,207	38,318	4.3	8.7	20.6
1896–1900	24,025	931	2,207	—	—	20,709	178	1,071,643	404,649	130,957	2.2	0.5	15.8
1901–05	38,006	2,220	289	—	1,036	30,719	3,742	1,072,252	443,981	112,100	3.5	0.1	27.4
1906–10	78,046	2,805	1,399	10	32	48,488	25,312	1,765,217	855,685	246,500	4.4	0.2	19.7
1911–15	64,131	7,564	—	423	—	55,068	1,076	2,721,091	1,148,278	372,980	2.4	—	14.8
1916–20	72,309	14,042	7,495	8,431	—	39,584	2,757	9,533,438	2,734,680	1,174,048	0.8	0.3	3.4
1921–25	254,097	44,629	64,695	23,920	8,740	34,584	77,529	10,966,554	5,438,149	1,755,988	2.3	1.2	2.0
1926–30	302,062	100,710	75,212	22,679	8,492	52,469	42,500	11,484,708	6,271,312	2,163,327	2.6	1.2	2.4
1931–35	677,401	202,781	323,517	26,467	7,843	85,124	31,669	10,892,605	4,632,913	1,033,310	6.2	7.0	8.2
1936–40	1,081,728	564,903	102,433	169,649	32,928	98,017	115,751	33,371,505	6,801,021	2,229,691	3.2	1.5	4.4

Sources: Emi and Shionoya (1965); Emi (1971).

presence can shift production functions upwards. In that sense, subsidies in these factors constituted an indirect subsidy to the development of industry.

10.3.3. Construction of social overhead capital

While social overhead capital may be conceptually divided into two categories—that aimed at industry and that aimed at improving the quality of life—in practice such classification is difficult since many projects have benefits in both spheres. Thus, the construction of roads, harbours, telecommunications and railways has benefits for industry and also improves the quality of life, while the construction of schools, hospitals, and other 'social' facilities has spin-offs for industrial development. Developing the school system improves the quality of labour and thereby it improves the efficiency of production. Developing hospitals and housing have similar indirect effects on production. However, as we deal with these aspects in relation to the labour supply in Section 10.5, we concentrate here on social capital which contributes more directly to production efficiency, focusing on railways as a representative example.

Three stages can be identified in the development of the railways. Until 1880 the railways were developed entirely by the government. This was followed by a boom in the construction of private railways from 1881 to 1906, which was supported by government subsidies. In 1906–7, as we have seen, the private railways were nationalized.

As Table 10.21 shows, the government played a monopolist role in the construction of railways in the early period of the Meiji era. Until 1881 when the first private railway (with government participation in capital ownership),

TABLE 10.21. Length of Japanese railways (km.)

Year	(1) National railways	(2) Local railways	(3) Street railways	(2)+(3)	Total
1883	202	101		101	303
1885	270	217		217	487
1890	886	1,365		1,365	2,251
1895	935	2,731		2,731	3,666
1900	1,325	4,674		4,674	5,999
1905	2,562	5,231		5,231	7,793
1910	7,838	823		823	8,661
1915	9,268	2,806	2,020	4,826	14,094
1920	10,436	3,209	2,126	5,335	15,771
1925	12,593	4,904	2,541	7,445	20,038
1930	14,575	6,902	2,558	9,460	24,035
1935	17,138	7,015	2,386	9,401	26,539
1940	18,400	6,699	2,190	8,889	27,289

Source: Minami (1965).

Nippon Railways, was established, all railways were owned and operated by the central government. We have already reviewed government investment in this area in Section 10.2. After 1881, many private railway companies were established and they invested in this business with government subsidies. While the government screened the entry of private companies, once entry was approved it gave generous subsidies to construction and also subsidized interest payments. The combination of subsidies and favourable profit rates led to a boom in the construction of private railways, and by 1890 the total length of private railways was several times greater than that of national railways. In 1906, however, the government enacted a law on railway nationalization, and thereafter took over private companies one after another; only short, local railways remained private. Japan National Railways (JNR), which covered almost all trunk lines, depended heavily on government revenue.

Although there were some periods in which private enterprises played an important role, most investment in industry-related social capital was undertaken by the government. The major areas of investment were construction of roads, railways, harbours, and flood control facilities; agricultural facilities; electricity and water supply systems; and urban gas supply.

Fig. 10.3 shows the change in government investment in industry related to social capital relative to total government investment in fixed capital formation between 1870 and 1940. The period from 1870 until around 1900 was one of great expansion in government construction of industry-related social capital. (A big drop in 1876 was due to civil war, known as the Seinan War, waged against the government by leaders from Kagoshima, and the subsequent

Source: Takeuchi (1967).

FIG. 10.3 Government investment in industry-related social capital in the total government investment.

termination of railway construction.) By the early 1900s the first round of investment was completed, and investment thereafter stabilized at around 60% while the government concentrated more on repairs than on new construction. In other words, government investment in social capital was almost complete before the heavy industry sector reached the point of take-off.

Government investment in these facilities stimulated *Z*-sector production not only through externalities but also through expanded demand for the *Z*-sector products. As a simple estimation, let us consider the final effects of de mand increases due to investment in the transport and communication sectors, using the inverse matrix of input coefficients from an input–output analysis.

Table 10.22 shows how much the demand for all industrial sectors increased, through indirect as well as direct channels, when investment in the transport and communication sector expanded by one unit: the final effect on the *X*-sector was 0.166 and on the *Z*-sector 0.258. More specifically, investment in the transport and communication sector stimulated demand for metal

TABLE 10.22. Column elements in Leontief inverse matrix of transport and communications industry

Agriculture and fisheries	0.11392
Coal and petroleum	0.15238
Other mining	0.03807
Food	0.03680
Textiles and leather	0.09175
Wood and wood products	0.01767
Pulp and paper	0.00450
Printing and publishing	0.00414
Rubber products	0.00494
Chemicals	0.04691
Coal and petroleum products	0.01918
Nonferrous metals	0.00516
Metals	0.10540
Machinery	0.06789
Transport equipment	0.00912
Other manufactures	0.01070
Construction	0.02430
Electricity and gas	0.00218
Trade	0.02231
Real estate	0.00330
Transport and communications	1.04831
Services	0.08318
Not elsewhere classified	0.10896
Total *X*-sector	0.14574
Total *Z*-sector	0.25816

Source: Mori (1972)

and machine production, which is why development of these facilities also stimulated the Z-sector more than the X-sector.

Our theoretical model focuses on the supply side of development and assumes that there is no shortage of effective demand. In reality this is not always the case, but in a development model the emphasis on the supply capacity nevertheless seem relevant. We should point out, however, that active investment by the government in social overhead capital generated demand large enough to stimulate Z-sector production as well as improving capacity.

10.3.4. The model and the roles of the government

Our basic findings so far are as follows: (*a*) in the early period, the government depended heavily on tax revenues from the S-sector; (*b*) subsidies tended to concentrate on the construction of industry-supporting social overhead capital, especially in the early and intermediate periods; (*c*) government investment in industry-supporting social overhead capital accounted for a large part of total government expenditures; (*d*) investment in transportation, communication, and public utilities such as electricity and gas not only improved the production efficiency of the Z-sector but also stimulated production by generating demand.

Now let us interpret these findings in line with our development model. The investment in social overhead capital plays this role in three ways: (*a*) it makes the $\dot{K}_X$ and $\dot{K}_Z = 0$ curves intersect such that a minimum critical effort curve exists and the economy has a possibility to move above that curve—that is, take-off becomes feasible; (*b*) by shifting the $\dot{K}_X = 0$ and $\dot{K}_Z = 0$ curves downwards through improved efficiency in production it lowers the minimum critical effort curve, or the hurdle for take-off; (*c*) it shifts the actual point of (K_X, K_Z) upward to cross the minimum critical effort curve and then the $\dot{K}_Z = 0$ curve by encouraging investment in the Z-sector.

If the minimum critical effort curve is located infinitely far away, it is impossible for an economy to take off at all. In such a case, social reform is required, through public education and reform of economic institutions. Construction of a minimum level of infrastructure is also considered to be indispensable in this regard. The Japanese government seems to have successfully made such a 'big push' in the late nineteenth and early twentieth centuries, when investment in industry-supporting infrastructures pushed the point (K_X, K_Z) from south to north and also lowered the minimum critical effort curve.

10.4. The Roles of Import Tariffs

10.4.1 Before 1899

As is well known, the Meiji government was forced to conclude treaties with many Western governments that were not to Japan's benefit and violated its national sovereignty. Each prime minister therefore saw it as his responsibility to correct these unequal treaties by eliminating the remaining foreign rights that continued to exist in some districts and establishing a sovereign right to decide import tariffs. The latter was to some extent obtained by a revision of treaties in 1899, but many concessionary rates imposed by Western governments remained.

The basic tariff rates in the Table 10.23 were therefore not always applied. As for Japanese tariffs in those days, it was pointed out:

Originally, there had been tariff agreements between Tokugawa and foreign governments in which basic rates were *ad quantum* [tariff rates per unit quantity] equivalent to a 5% *ad valorem* one [tariff rate per cent value]. After the revision of treaties in 1899, the government became able, for the first time, to decide on Japanese tariff rates on its own account, but Britain, Germany and France demanded to apply *ad valorem* tariff rates not higher than about 10% on their important exportables. The Japanese government after all accepted this request and it had to apply such concessionare rates to other countries due to the most favoured nation treatment. (Nippon Gakujutsu Shinkokai 1951: 27)

TABLE 10.23. Basic tariff rates of Customs Tariff Law (promulgated 1887, effective 1899)

Classification of dutiable goods		Commodities	Rate of *ad valorem* tax (%)
Part 1	Grains	Grains	5
	Beverages	Tea	25
	Yarn and cloth	Cotton textiles	10
		Other yarns	15
		Raw silk	15
	Chemicals		10
	Dyes, pigments, coating matter		10
	Metal and metal products	Iron, soft steel	5
		iron bar, plate	10
	Ordnance, Machinery	Agricultural implements	5
		Machinery	10

Notes: Part 1 covers dutiable goods.
Part 2 covers tax-exempt goods, including raw and waste cotton and rice.
Source: Ohkurasho Shuzeikyoku (n.d.).

Although only partial sovereignty was obtained with regard to tariff decisions in 1899, from then till 1911, when a second revision of unequal treaties was made, tariffs became increasingly important policy instruments. Thus, before 1911 import tariffs were not important measures to protect domestic industries.

Table 10.24 shows the average ratio of tariff revenues to the total value of taxable goods imported and the shares of tariff revenues in total government revenues.

TABLE 10.24(*a*). Average customs tariff rates and customs revenue, 1880–1900 (¥'000)

	(1) Revenue from export duties	(2) Value of dutiable exports	(3) Value of total exports	Dutiable tariff-bearing ratio (exports) (1)/(2)%	Total tariff-bearing ratio (exports) (1)/(3)%	(4) Revenue from import duties	(5) Value of dutiable imports	(6) Value of total imports	Dutiable tariff-bearing ratio (imports) (4)/(5)%	Total tariff-bearing ratio (imports) (4)/(6)%	(7) Total tariff revenue (including other items)	(8) Total financial revenue[a]	Composition of tariff revenue (7)/(8)%
1880	904	24,154	28,395	3.74	3.18	1,688	34,817	36,626	4.85	4.61	2,636	55,262	4.77
1881	984	26,169	31,058	3.76	3.17	1,473	30,141	31,191	4.89	4.72	2,511	61,675	4.07
1882	1,262	32,287	37,721	3.91	3.35	1,387	28,546	29,446	4.86	4.71	2,704	67,738	3.99
1883	1,324	30,552	36,268	4.33	3.65	1,342	27,753	28,444	4.84	4.72	2,717	67,659	4.02
1884	1,136	26,385	33,871	4.31	3.35	1,356	27,442	29,672	4.94	4.57	2,547	67,203	3.79
1885	1,313	30,109	37,146	4.36	3.53	1,367	27,353	29,356	5.00	4.66	2,734	52,581	5.20
1886	1,481	37,590	48,876	3.94	3.03	1,556	31,158	32,168	5.00	4.84	3,103	64,371	4.82
1887	1,562	38,984	52,407	4.01	2.98	2,116	42,171	44,304	5.02	4.78	3,763	66,255	5.68
1888	1,995	46,058	65,705	4.33	3.04	2,725	63,208	65,455	4.31	4.16	4,825	64,727	7.45
1889	1,747	45,121	70,060	3.87	2.49	2,863	63,403	66,103	4.52	4.33	4,720	71,294	6.62
1890	1,433	35,962	56,603	3.98	2.53	2,932	65,795	81,728	4.46	3.59	4,488	66,114	6.79
1891	2,202	51,073	79,527	4.31	2.77	2,401	54,822	62,927	4.38	3.82	4,723	64,423	7.33
1892	2,204	59,184	91,102	3.72	2.42	2,745	64,299	71,326	4.27	3.85	5,069	67,167	7.55
1893	1,772	51,531	89,712	3.44	1.98	3,098	79,462	88,257	3.90	3.51	5,006	70,004	7.15
1894	2,217	62,860	113,246	3.53	1.96	3,506	103,195	117,481	3.40	2.98	5,881	71,286	8.25
1895	2,273	72,003	136,112	3.16	1.67	4,238	118,680	129,260	3.57	3.28	6,684	74,697	8.95
1896	1,777	50,595	117,842	3.51	1.51	4,915	130,431	171,674	3.77	2.86	6,904	76,387	9.04
1897	2,540	80,738	163,135	3.15	1.56	5,295	139,975	219,300	3.78	2.41	8,094	94,912	8.53
1898	2,080	68,067	165,753	3.06	1.25	6,280	165,522	277,502	3.79	2.26	8,674	97,629	8.88
1899	971	32,490	214,929	2.99	0.45	13,251	136,489	220,401	9.71	6.01	14,664	126,034	11.63
1900	0	0	204,429	0		16,764	200,458	287,261	8.36	5.84	17,333	133,926	12.94

Note: Export duty was abolished in 1899.

Source: Toyo Keizai Shinposha (1935); Data of financial revenue based on Nippon Ginko (1966).

TABLE 10.24(*b*). Average customs tariff rates and customs revenue, 1901–1930 (¥'000)

	(1) Revenue from import duties ¥'000	(2) Value of dutiable imports ¥m	(3) Value of total imports ¥m	Dutiable tariff-bearing ratio (imports) (1)/(2)%	Total tariff-bearing ratio (imports) (1)/(3)%	(4) Total tariff revenue (including other items)	(5) Total financial revenue	Composition of tariff revenue (4)/(5)%
1901	14,457	165	255	8.76	5.67	15,122	139,574	10.83
1902	14,727	151	271	9.75	5.43	15,313	151,084	10.14
1903	16,372	168	317	9.75	5.16	17,047	146,163	11.66
1904	20,519	208	371	9.86	5.53	21,230	194,362	10.92
1905	34,298	294	488	11.67	7.03	35,181	251,275	14.00
1906	41,230	281	418	14.67	9.86	42,148	283,468	14.87
1907	46,959	307	494	15.30	9.51	47,944	315,983	15.17
1908	44,817	281	436	15.95	10.28	45,777	322,636	14.19
1909	35,438	221	394	16.03	8.99	36,431	323,407	11.26
1910	36,072	232	464	15.55	7.78	37,100	317,285	11.69
1911	42,007	278	513	15.11	8.19	43,169	329,071	13.12
1912	58,242	312	618	18.67	9.42	59,611	360,969	16.51
1913	73,580	368	729	20.00	10.09	75,090	369,479	20.32
1914	50,512	255	595	19.81	8.49	51,828	343,708	15.08
1915	30,195	174	532	17.35	5.68	31,335	312,744	10.02
1916	33,832	294	756	11.51	4.48	35,393	348,672	10.15
1917	38,019	448	1,035	8.49	3.67	40,441	430,604	9.39
1918	62,696	759	1,668	8.26	3.76	66,296	600,882	11.03
1919	69,435	810	2,173	8.57	3.20	73,208	834,649	8.77

Table 10.24b (*contd.*)

	(1) Revenue from import duties ¥'000	(2) Value of dutiable imports ¥m	(3) Value of total imports ¥m	Dutiable tariff-bearing ratio (imports) (1)/(2)%	Total tariff-bearing ratio (imports) (1)/(3)%	(4) Total tariff revenue (including other items)	(5) Total financial revenue	Composition of tariff revenue (4)/(5)%
1920	74,413	899	2,336	8.28	3.19	79,780	730,553	10.92
1921	89,172	758	1,614	11.76	5.52	93,684	790,938	11.84
1922	117,295	888	1,890	13.21	6.21	121,855	897,320	13.58
1923	100,686	754	1,982	13.35	5.08	104,376[a]	787,337	13.26
1924	113,904	827	2,453	13.77	4.64	118,664	887,364	13.37
1925	108,688	863	2,572	12.59	4.23	112,332	894,895	12.55
1926	147,969	937	2,377	15.79	6.23	151,951	887,020	17.13
1927	144,645	827	2,179	17.49	6.64	148,664	898,690	16.54
1928	155,147	876	2,196	17.71	7.06	159,385	915,937	17.40
1929	147,336	854	2,216	17.25	6.65	152,407	893,505	17.06
1930	113,173	584	1,546	19.38	7.32	118,120	835,041	14.15

[a] 8 days' revenue of Yokohama district excluded due to the 'Great Earthquake'.

As the import of goods on which there are low tariffs increases while the import of goods on which there are high tariffs is held down, the former ratio does not reflect the real tariff burden, as is well known. None the less, because of forced low tariffs, average tariff rates were in fact low before 1880. But this is not shown in the table. The table shows the following points:

1. In the interests of government revenue, export taxes amounted to 3%–4.5% of the total value of taxable exports.
2. The ratio of import tariffs to the total value of taxable imports remained stable until 1899 at around 4%–5%.
3. Import tariffs and export taxes together accounted for about 4%–9% of total government revenues and thus constituted important financial resources. While tariff rates were held fairly constant, the revenue they generated increased as foreign trade steadily expanded. Tariffs therefore became an increasingly important source of government revenue.

In those days there were two important policy issues as far as tariffs were concerned. First, industrialists argued that export taxes should be abolished, while others advocated their continuation in the interests of government revenues. As Table 10.24 shows, export taxes were abolished in 1900, so the industrialists won the debates. The most earnest campaigners in this respect were the textile manufacturers, who argued that the tariffs made Japanese textile production unable to compete with that of India.

In fact, the production of cotton yarn in Japan in 1890 almost exceeded imports, meaning that Japanese cotton production was half way to import substitution. In 1891, the Japan Association of Cotton Spinning Industry submitted a request for the abolition of export tax on cotton yarn.[13] At the same time, the League for the Complete Abolition of Export Taxes, which was directed by Taguchi Ukichi, campaigned strongly to abolish all export taxes, but in vain. This latter attempt was unsuccessful at that time, but the tax on the export of cotton yarn was abolished in 1894.

A second conflict related to the imposition of import tariffs on raw cotton. Since Japanese farmers produced cotton, landowners were in favour of tariff protection; textile manufacturers obviously protested against this, and in 1891 they issued a request to abolish the import tariffs on raw cotton. After five years of heated debates, the increased political power of textile manufacturers won the day and in 1896 the government exempted raw cotton from import tariffs. An economic historian would call this a monumental victory of manufacturers over landowners.

Although tariff rates thus existed until 1899, both on imports and exports, they were low, and the terms of trade faced by the *X*- and *Z*-sectors were not distorted from the assumptions of free trade to any significant degree. Thus, the basic assumptions of our model largely held in this period.

10.4.2. 1899–1911

Although the government achieved the first revision of unequal treaties with West European countries in 1899, there remained a number of unequal tariff rates which were non-reciprocal. The government was still forced to accept a 10% *ad valorem* tariff on cotton fabric, wool fabric, and steel from the UK, on dyestuffs and medicine from Germany, and cosmetics and wine from France. In all, one-way tariffs were applied to some seventy-eight items, comprising about 42% of total Japanese imports from 1890 to 1892 (Nippon Gakujutsu Shinkokai 1951: 3).

Therefore, although the government could resort to protectionist tariffs a little more than it could before 1899, the possibilities were fairly limited until a second revision was made in 1911. The ratio of the tariff revenue to the total value of taxable imports jumped from 4% before 1899 to about 9% in 1901. Export taxes were mostly abolished. Despite this, the share of import tariff revenue in the total government revenue increased to 10%.

Table 10.25 shows that the tariffs protected the *X*-sector rather than the *Z*-sector. As there were many concessionary tariff rates based on treaties rather than on the basic rates in this table, however, the dominant rate was virtually about 10%. Thus there was no conspicuous discrimination against either the *X*- or the *Z*-sector as the concessionary rates were applied. As Japan had no alternative to importing producer durables and intermediate products, import tariffs on *Z*-sector products would only have led to higher costs of production for both sectors.

None the less, if we look at expenditure, we see that tariff revenue was an important source of financial support for the *Z*-sector. In other words, the

TABLE 10.25. Basic tariff rates of Customs Law (promulgated 1910, effective 1911)

Classification		Commodity	Basic tariff rate	
Group 2	Grains	Rice	per 100 kin	1.00¥
		Barley	per 100 kin	0.55¥
Group 3	Beverages	Black tea	per 100 kin	22.60¥
Group 8	Yarns	Cotton yarn	per 100 kin	5.80–30.00¥
		Raw silk	*ad valorem*	30%
Group 9	Cloth	Cotton textiles	per 100 kin	30–40¥
			ad valorem	20%
Group 6	Drugs, chemicals		*ad valorem*	20%
Group 14	Ores and metals	Iron	per 100 kin	0.10¥
Group 15	Metal products	Metal net	per 100 kin	13.5–34.1¥
			ad valorem	25%
Group 16	Machinery	Clocks	*ad valorem*	40%
		Cars	*ad valorem*	50%

Note: 1 kin = 600g.

Source: Gaimusho (1951).

tariff policy promoted the take-off of the *Z*-sector—not through the price effects of tariffs on *Z*-sector goods, but through the totality of tariffs imposed on both sectors which were used mostly to subsidize the *Z*-sector.

As the Meiji government tried to keep the prices of agricultural products as low as the prices prevailing in the world markets, the prices of *X*- and *Z*-sector goods relative to *S*-sector goods can be measured at the international prices. As industry-protecting tariffs were raised in 1899, however, the domestic relative prices became a little more favourable to industrial sectors than international prices. Therefore, profits must have risen in terms of the prices of *S*-sector goods. In 1905, however, at the time of the Russo-Japanese War, the government imposed a special tax on rice, of ¥0.641 per kin (1 kin = 0.6 kg), which was equivalent to a 15% *ad valorem* tax. A similar 15% *ad valorem* tax was imposed on other cereals. The taxes were imposed to finance the war but resulted in a deterioration of the prices of industrial goods relative to agricultural goods.

10.4.3. After 1911

The Tariff Law of 1910 was one of the government's major achievements at the second revision of unequal treaties *vis-à-vis* Western countries. It was at this time that the policy of protecting domestic industries was established. Table 10.26 shows the tariff rates of selected items corresponding as far as the data permit to our *S*-, *X*-, and *Z*-sector classification. According to data provided by the Japanese Association for Science Promotion (1951), the overall tariff structure may be summarized as follows:

TABLE 10.26. Basic tariff rates of Customs Law (promulgated 1926)

Classification		Commodity	Basic tariff rate	
Group 2	Grains	Rice	per 100 kin	1.00¥
		Barley	per 100 kin	0.60¥
Group 3	Beverages	Tea	per 100 kin	22.60¥
Group 8	Yarns	Cotton yarn	per 100 kin	5.80–41.00¥
		Raw silk	*ad valorem*	25%
Group 9	Cloth	Cotton textiles	per 100 kin	29–83¥
			ad valorem	20%
Group 6	Drugs, chemicals		*ad valorem*	20–40%
Group 14	Ores and Metals	Iron	per 100 kin	0.10¥
Group 15	Metal products	Metal net	per 100 kin	14.8–74.9¥
			ad valorem	25%
Group 16	Machinery	Table clocks	*ad valorem*	40%
		Cars	*ad valorem*	50%

Note: 1 kin = 600g.

Source: Gaimusho (1951).

1. Tariffs on industrial materials which could not be produced domestically were set at zero.
2. On raw materials for which there was an alternative domestic supply, such as silk cocoons, furs, and skins, 5% tariffs were imposed.
3. Relatively high tariffs were set on agricultural products—¥1 per 100 kin, which was equivalent to a 23% *ad valorem* tariff, but this rate could be reduced to ¥0.64 at the time of a poor crop; 20% *ad valorem* tariffs on other cereals.
4. Semi-finished goods:
   - 5.0%: Metal ingots, coke, pulp, and other minimally processed goods
   - 10.0%: Textile yarns of various kinds
   - 7.5%: Steel bars and sheets
   - 15.0%: Steel wires and poles
   - 20.0%: Sheets, rods, and wires of non-ferrous metals.
5. Completely finished goods:
   - *Fabrics*: Cotton 20%, wool 25%, silk 40%
   - *Machines*: Mostly *ad quantum* tariffs of 20% *ad valorem* tariff equivalent, but reductions for specified machinery used for manufacturing exportable goods
   - *Chemicals*: Using dyestuffs as an example—on non-competing imports of dyestuffs, 10%; on dyestuffs competing with domestic products, such as indigo, 18%.
   - *Others*: Rubber and leather goods and celluloid: 20–25%; consumer luxuries 40%.

It can be concluded that tariffs were high for consumption goods, especially for so-called luxuries, but that there was no discrimination between *X*- and *Z*-sector goods. While textiles were an important export sector for Japan at this time, import tariffs on these items were extremely high. Since Japan could not supply machines for producing *X*-sector goods from domestic sources only, and since the government could not adopt tariff policies which were unfavourable to *X*-sector producers, as an increasingly influential interest group, tariffs on *Z*-sector goods (producer goods especially) were kept low.

As we saw in Chapter 9, market forces generated expansion only in the *X*-sector. Some argued for the protection of the domestic steel industry by tariffs, but this group was defeated as most other industrial sectors needed cheaper industrial materials to survive. The steel industry grew rapidly during World War I, but once Europe's steel industries recovered from war damage it soon stagnated, as did other *Z*-sector industries. Thus, when the government revised the Tariff Law in 1920, *ad quantum* tariffs on steel were changed to *ad valorem* tariffs thereby increasing the effective tariff rate as steel prices rose substantially. Even so this was not enough to protect the Japanese steel industry, partly because there were still concessionary tariffs on some industrial goods. The government therefore had no alternative to tax subsidies to support the *Z*-sector.

The next revision of the tariff structure was made in 1926 when the Japan–UK tariff treaty expired. The government avoided drastic increases on steel products but decided instead to provide a production subsidy of 3–6 yen per ton of iron to firms that were fully vertically integrated. Skipping the details we can confirm that tariff rates on iron and steel products were not raised by any substantial degree.

The fact that some sectors argued for the abolition of tariffs on cotton yarn in 1926 implies that Japan's *X*-sector had been fairly competitive in the international market. If Japan had not entered into a war economy, the tariff structure would have been protecting the *Z*-sector, leaving the *X*-sector to the market forces. But the international economy at that time was more interested in military alliances than in tariff negotiations.

10.4.4. Summary

Japan's tariff policy was not aimed at protecting the *Z*-sector prior to World War II. This was because *X*-sector industrialists were a stronger interest group at that time, and because the development of the *Z*-sector also depended on taxes on *X*-sector growth. It would indeed have been dangerous to protect the *Z*-sector at such an early stage: as discussed in Part I, if the economy had failed to take off, the result would have been worse than without protection of the *Z*-sector.

The policy combination that the government adopted can be summarized as follows: (*a*) protection of the *X*-sector by import tariffs; (*b*) protection of the *Z*-sector by tariffs, but with no preferential treatment; (*c*) tariff revenues to subsidize the *Z*-sector, ie. the *Z*-sector enjoyed preferential tax treatment; (*d*) prices of *S*-sector goods were originally kept low by promoting free trade, but were ultimately raised substantially to finance massive military expenditures. This caused a deterioration in the terms of trade of the industrial sectors, but generated compulsory savings by workers.

As long as savings were used to finance military expenditures, however, compulsory savings did not contribute to capital accumulation. Higher prices of grains would create compulsory savings, which might be suggested by the tariff schedule for grains. Since, however, Japan had two important colonies, namely Korea and Taiwan, the high tariffs on imported grains may not have raised the prices of food too much if trade with them was regarded as domestic trade. In these circumstances tariffs on food did not influence domestic prices significantly.

10.5. Food Supply, Labour Supply, and Wage Rates

The supply of food, labour, and subsistence wage rates, while not directly related to the minimum critical efforts of government, play critical roles in economic development in our model.

There is significant disguised unemployment in contemporary LDCs. But it is another matter whether the labourers who receive subsistence wages when their marginal productivity is even below that level can be as efficient as industrial workers without additional investment in education and training. Therefore, it is possible that LDCs have a shortage of industrial workers despite the existence of significant disguised unemployment.

It seems that the government efforts to spread skills through investment in primary education contributed greatly to the unlimited supply of 'industrial workers' in the Meiji (1868–1911) and Taisho (1912–25) eras. Therefore, a more important question is if the economy faced a general shortage of labour prior to World War I, as discussed in Section 4.3, and if so when. We then consider whether there was a shortage of industrial workers, as analysed in Section 4.2.[14]

Obviously, in the first stages of economic development it is the subsistence sector which supplies industrial workers. The supply of industrial workers, therefore, depends heavily on technological progress in agriculture. With appropriate technology, this sector can supply labour without increasing the prices of foodstuffs. It is assumed that such labour outflows will not reduce agricultural production as long as marginal productivity remains below subsistence wage rates, and the terms of trade of the industrial sectors thus remain constant.

Following our theoretical model, we found that the prices of *X*-sector goods declined relative to *S*-sector goods as the production of *X*-sector goods expanded. This is because Japan, in those years, an LDC, was inclined to specialize in specific *X*-sector products which occupy a significant share in the world market, despite stable prices of *S*-sector goods. On the other hand the relative prices of *Z*-sector goods are assumed to be independent of production because Japan's share of the international market for these products is negligible.

One problem is that the labour force needed by the *Z*-sector may not be congruent with that used in the *X*-sector. More specifically, the *Z*-sector might employ male labour who are prime breadwinners while the *X*-sector uses mostly women whose contribution to the household income is only secondary. In the analyses below, however, we do not go into details with regard to the quality of labour and differentials in wage rates between the sectors since the behaviour of wage rates were not so different as to require the modification of the model. (As a matter of fact, male workers in the Z-sector were mostly the second or third sons of farmers, while female workers in the *X*-sector were mostly daughters of the same family.) Even though wage differentials existed between the two sectors, they were not significant enough to change the results of the analysis.

Agricultural productivity affects both the demand for and the supply of industrial workers. In this context we look at population growth, which determines the demand for *S*-sector goods, and the increase in the productivity of agriculture, which affects the supply of industrial labour. Further, since the economy can depend on imports for food supply if there are no budget constraints, we trace the import of foodstuffs as well as production.

To state the major conclusion first: although agricultural productivity increased rapidly, it was in fact imports from Korea and Taiwan that provided the food supply in the mid-1920s and beyond when the *Z*-sector was taking off. Labour productivity in agriculture did not increase as rapidly as after World War II, but this was partly a reflection of the slower expansion of demand for labour in industrial sectors in the pre-war period. While there were occasional increases in food prices when war or abnormal weather resulted in food shortages, the terms of trade between the agricultural and the industrial sectors were largely those that the model assumed, as described in Chapter 8.

According to Minami (1970), the labour supply only really became limited in 1960. We did not test this finding because our main concern was not finding the 'turning point' as such but finding the point of take-off of the *Z*-sector. If Minami is correct, then it follows that the test on the 1945–60 period can be carried out on our basic model, not expanding it to cover the case of a general labour shortage.

10.5.1. Supply of foodstuffs

Tables 10.27 and 10.28 show the long-term supply of, and demand for, foodstuffs from 1868 to 1940. From the growth rates in Table 10.27, we see that while population expanded at an annual rate of 1.4% between 1920 and 1930, the growth rate was as low as 0.9% between 1880 and 1900, and about 1.2% annually for the rest of the period. In this respect, Japan differed substantially from contemporary LDCs, which commonly have annual population increases of 2%–3% or more, in that the burden of population pressure on take-off was relatively small. This relatively low population growth in turn contributed to increased surpluses in agriculture and promoted the supply of industrial labour.

Although population growth picked up after 1900, partly due to government encouragement with an eye to military purposes, the long-run average growth rate of 1%–1.2% was not especially high compared with the growth of agricultural production at constant prices. Indeed, the latter exceeded the former. After 1930, however, agricultural production began to stagnate, and in fact production was slightly lower in 1940 than in 1930. This was not due to climatic factors but was clearly the result of a trend of domestic agricultural production, as was confirmed by Umemura (1961). Although Table 10.28

TABLE 10.27. Long-term trend of population and agricultural production

	(1) Population ('000s)	(2) Employed in agricultural sector ('000s)	(3) Employed in secondary industry ('000s)	(4) Agricultural production (¥m. at 1934–6) prices	(5) Price index of agricultural produce 1934–6 = 100
1880	36,649	14,660	1,105	1,515	36.5
	(0.9)	(–0.2)	(6.2)	(2.1)	(–1.2)
1890	39,902	14,298	2,013	1,874	30.4
	(0.9)	(–0.1)	(3.8)	(0.8)	(3.7)
1900	43,847	14,191	2,929	2,036	43.7
	(1.2)	(–0.1)	(2.9)	(1.4)	(1.8)
1910	49,184	14,017	3,887	2,330	52.1
	(1.2)	(–0.1)	(1.7)	(2.8)	(10.4)
1920	55,391	13,939	4,593	3,062	140.5
	(1.4)	(–0.0)	(1.0)	(1.0)	(–6.5)
1930	63,872	13,911	5,067	3,390	71.8
	(1.1)	(–0.3)	(4.2)	(–0.0)	(9.6)
1940	71,400	13,504	7,659	3,388	179.6

Note: Ratios of annual increase are in parentheses.

Sources: Columns (1), (3) Nippon Ginko (1966); (2), (4) Umemura *et al.* (1966).

gives no information about imports of foodstuffs from the colonies, it is known that increased imports of rice and sugar from Korea and Taiwan relieved shortages in domestic supply and kept prices more stable than otherwise. Prices did in fact rise in the 1920s and 1930s, but for a long period preceding this they had been stable as Japan's agricultural output steadily grew, which in turn contributed to industrial growth.[15]

Between 1880 and 1920, agricultural production at constant prices expanded faster than population growth, allowing an outflow of labour to the industrial sectors at an annual rate of 0.1%–0.2%. Furthermore, as discussed in Section 10.4, Japanese trade was close to free trade until 1910, due to pressures from the Western industrialized countries, and the government was forced to keep the domestic markets for agricultural products open too. A situation whereby food shortages deteriorated the terms of trade of industries *vis-à-vis* the *S*-sector was thus avoided.

Since the rate of labour outflow from the agricultural sector was only 0.1%–0.2% annually, there was no shortage in the supply of foodstuffs. The price index for cultivated agricultural products rose rapidly between 1890 and 1900, but as we have seen (Chapter 8), relative prices did not change sufficiently to lead to a deterioration in the terms of trade of the industrial sectors. This was due to general inflation which was triggered by the increased money supply. In a sense, this was a reflection of the government policies.

In fact, agricultural production expanded with a gradual saving of labour in the same sector. It will be shown later (in Table 10.29) that this was because

TABLE 10.28. Japan's food supply

	(1) Agricultural production (¥m. at 1934–6 prices)	(2) Rice production (¥m. at 1934–6 prices)	(3) Exports of rice (current prices)	(4) Imports of rice (current prices)	(5) Rice imports from Taiwan	(6) Rice imports from Korea	(7) Total (5)+(6)	(8) Price index of agricultural products (1934–6 = 100)	(9) Price index of rice (1934–6 = 100)
1880	1,429	952	x	x	—	—	—	39.9	34.4
1885	1,530	1,006	1	1	—	—	—	28.0	21.4
1890	1,773	1,184	1	12	—	—	—	34.4	29.3
1895	1,754	1,098	7	4	—	—	—	37.9	29.9
1900	1,845	1,139	4	9	x	—	—	49.8	41.6
1905	1,748	1,049	3	48	5	—	5	56.9	44.4
1910	2,069	1,281	6	9	7	4	11	57.4	46.1
1915	2,472	1,537	10	5	8	21	29	55.8	45.2
1920	2,675	1,737	6	18	17	73	90	145.6	135.2
1925	2,577	1,641	4	120	72	173	245	148.8	130.0
1930	2,768	1,838	7	20	39	110	149	75.3	60.8
1935	2,582	1,579	n.a.	n.a.	n.a.	n.a.	n.a.	101.0	102.0
1940	2,781	1,673	n.a.	n.a.	n.a.	n.a.	n.a.	190.2	152.6

Note: x means less than ¥1m.

Source: Columns (1),(2),(8),(9) are based on Umemura (1966); (3) and (7) on Toyo Keizai Shinposha (1935).

labour productivity improved as the use of intermediate inputs, such as chemical fertilizer, increased. While agricultural production stagnated in the 1930s, this reflected the changing division of labour between Japan and its colonies, the latter being forced to specialize in agricultural production.

According to Table 10.28, rice imports from the colonies equalled 5.1% of domestic production in 1920, but 14.9% in 1925. Prices of rice had increased heavily in the world market during World War I, but price inflation was controlled by increasing the import of rice from the colonies. A further inflation of food prices took place just before World War II, by which time Japan was already in a war-time economy. In fact, a law by which the government purchased all rice at specified prices and distributed it to consumers was enacted in 1933, and as early as 1928 the government had controlled the quantity of rice imported.

Thus, in comparing Japan with contemporary LDCs, one must remember that Japan avoided the food shortage which many LDCs suffered in the 1950s and 1960s before the Green Revolution, when they frequently had to use scarce foreign exchange to import foodstuffs to feed a rapidly growing population. It is true that this was partly due to Japan having colonies, unlike today's LDCs, but on the other hand it should be emphasized that Japan did not receive any foreign assistance.

10.5.2. Labour supply and wage rates

The labour supply to the industrial sector has received special attention in development economics. Lewis (1954) and Ranis and Fei (1961) are representative examples of development models that focus on the 'turning point' of labour supply.[16] The outline of the theory is that marginal productivity of labour in the subsistence sector is lower than the institutionally determined subsistence wage rate due to disguised unemployment. Therefore, for a given demand for labour from the industrial sector at a given institutional wage rate, the labour supply responds with infinite elasticity. Here, the marginal productivity of labour of the *S*-sector need not be zero, as discussed in Part I.

As economic development proceeds (ie. employment in traditional subsistence sectors declines as the industrial sector expands), the marginal productivity of labour in the *S*-sector increases to match the subsistence wage rate. At that point, the labour supply to the industrial sector stops being infinitely elastic with regard to wage rates. If the industrial sector wants to fill the demand for labour, wage rates will have to rise. Minami (1970) showed empirically that this transition occurred in 1960. Prior to that Ranis and Fei (1961) had concluded that the turning point was some time between 1916 and 1919, while Lewis (1958), Ohkawa (1965*a*), and Tsujimura (1963) independently argued that the actual turning point was sometime after World War II.

However, as the turning point itself is no more than a minor issue in our development theory, we simply adopt Minami's conclusion as given. In what follows, we concentrate on confirming that the Japanese economy had not reached the turning point before World War II, ie. that in our terminology it had never experienced a 'general shortage of labour', and investigate whether it had nevertheless experienced a shortage of skilled labour in the industrial sectors.

Table 10.29 shows the wage rates, in terms of cultivated agricultural products, for agricultural workers employed on a daily basis. The figures indicate that the real wage rate of agricultural workers was fairly stable until 1905 and then rose substantially through World War I till 1930. This is why Ranis and Fei (1961) identified the 1920s as a turning point. But as Minami (1970) correctly pointed out, a temporary rise in real wages is not a sufficient condition for a turning point, because the institutional subsistence wage rate can rise over time. From the point of view of development, one has to see a sustained wage rise in order to conclude that the economy has passed the turning point. The fact that the real agricultural wage rate dropped in the latter half of the 1930s suggests that the economy had not yet reached the turning point.[17] Whereas other investigators made only rough empirical tests, Minami measured agricultural production functions and estimated the marginal productiv lity of labour to obtain real wage rates and it was on this basis that he concluded that only in 1960 did real wage rates achieve parity with the marginal productivity of labour for the first time.

In order to investigate whether the economy experienced a shortage of skilled labour in industrial sectors, let us look at Table 10.30, which shows real wage rates in terms of cultivated agricultural products for agricultural workers (male), manufacturing workers (average for male and female), female workers in silk-spinning, and male workers in machine industries, and expresses these wage rates relative to those of agricultural workers. The wage differentials which existed from the initial stage of industrialization are regarded as expressing the costs of skill acquisition, regional movement of workers and so on.

If the differentials did not change over time, we can conclude that industrial sectors paid $w^* = w(1 + \alpha)$, where α is the cost of acquiring the skills for agricultural workers to become industrial workers, w is the wage rate of agricultural workers, and w^* is that of industrial workers. If α increases at a given point of time, we understand this to mean a shortage in skilled labour for industrial work. In fact, the average of male and female wage rates in manufacturing sectors remained equal to that of agricultural workers, namely $\alpha = 0$, except for 1930 to 1935. The wage rate of female workers at silk-spinning factories, which represented that of X-sector workers, remained steady at 0.5 to 0.7 of the wage rate of agricultural workers.

The value of α for the wage rate of male machinists varied between 0.5 and 1.70, reflecting skill requirements, but remained stable at around 0:50 to 0.80 except for 1930 to 1935. It is to be noted, however, that the wage rate for both male and female workers averaged over manufacturing sectors stayed at a

TABLE 10.29. Real wages of agricultural workers

	Farm value of production (¥m.)	Agricultural workers ('000s)	Arable land '00 cho[a]	Per Worker: Farm value of production (¥)	Arable land (tan)	Current inputs (¥)	in which agricultural origin (¥)	Gross capital stock (¥)	Agricultural wages (male) (sen/day)[b]	Index of agricultural prices (1934–6 = 100)	Real wages
1880	1,462	14,660	47,769	100	3.26	17.9	5.9	315	22	36.5	60
	(2.2)	(–0.3)	(0.4)	(2.5)	(0.6)	(0.8)	(1.3)	(0.4)	(–7.4)	(–7.5)	(0)
1885	1,632	14,473	48,640	113	3.36	18.6	6.3	321	15	24.7	61
	(1.6)	(–0.2)	(0.4)	(1.9)	(0.7)	(0.3)	(0.9)	(0.4)	(—)	(4.2)	(—)
1890	1,771	14,298	49,635	124	3.47	18.9	6.6	328	n.a.	30.4	—
	(1.0)	(–0.1)	(0.5)	(1.1)	(0.6)	(1.3)	(2.3)	(0.7)	(—)	(1.2)	(—)
1895	1,865	14,209	50,772	131	3.57	20.2	7.4	339	19	32.3	59
	(1.5)	(–0.0)	(0.6)	(1.5)	(0.6)	(1.6)	(3.3)	(0.4)	(10.3)	(6.2)	(3.8)
1900	2,005	14,191	52,281	141	3.68	21.9	8.7	346	31	43.7	71
	(1.8)	(–0.1)	(0.5)	(2.0)	(0.7)	(1.5)	(3.6)	(0.7)	(0)	(2.7)	(–2.7)
1905	2,195	14,091	53,626	156	3.81	23.6	10.4	358	31	49.9	62
	(2.3)	(–0.1)	(0.9)	(2.4)	(1.0)	(3.8)	(7.0)	(1.1)	(5.8)	(0.9)	(5.0)
1910	2,465	14,017	56,140	176	4.01	28.4	14.6	378	41	52	79
	(2.1)	(–0.1)	(0.8)	(2.2)	(0.9)	(2.4)	(4.0)	(0.7)	(2.3)	(–0.6)	(2.9)
1915	2,739	13,952	58,411	196	4.19	32.0	17.8	392	46	5.05	91
	(1.1)	(–0.0)	(0.6)	(1.2)	(0.6)	(2.3)	(4.1)	(0.5)	(25.0)	(20.0)	(1.7)

Table 10.29 (*contd.*)

	Farm value of production (¥m.)	Agricultural workers ('000s)	Arable land '00 cho[a]	Per Worker					Agricultural wages (male) (sen/day)[b]	Index of agricultural prices (1934–6 = 100)	Real wages
				Farm value of production (¥)	Arable land (tan)	Current inputs (¥)	in which agricultural origin (¥)	Gross capital stock (¥)			
1920	2,893	13,939	60,155	208	4.32	35.8	21.8	401	139	140.5	99
	(0.6)	(0.1)	(–0.1)	(0.6)	(–0.1)	(2.3)	(4.2)	(0.5)	(3.5)	(–0.9)	(4.4)
1925	2,980	13,941	59,896	214	4.30	40.2	26.8	411	165	134.3	123
	(1.5)	(–0.0)	(0.1)	(1.5)	(0.2)	(2.1)	(3.1)	(0.9)	(–7.5)	(–11.8)	(4.9)
1930	3,217	13,911	60,310	231	4.34	44.5	31.3	430	112	71.8	156
	(0.6)	(–0.2)	(0.4)	(0.8)	(0.6)	(2.1)	(2.7)	(0.7)	(–4.0)	(6.8)	(–10.2)
1935	3,318	13,750	61,501	241	4.47	49.3	35.7	444	91	99.7	91
	(0.1)	(–0.4)	(0.1)	(0.4)	(0.4)	(0.7)	(0.7)	(0.2)		(12.5)	
1940	3,322	13,504	61,548	246	4.56	51.1	37.0	449		179.6	

Notes: Farm values of production are seven-year moving averages in terms of 1934–6 constant prices.

[a]1cho = 10 tan = 9,917m^2, [b]1sen = 1/100 yen.

Source: Umemura (1966).

TABLE 10.30. Real wages in terms of index of agricultural prices (sen per day)

	(1) Agriculture (daily contract workers, male)	(2) Manufacturing (average of male and female)	(3) Textile (cotton-spinners, female)	(4) Machinery (blacksmiths and iron workers)	(2)/(1)	(3)/(1)	(4)/(1)
1880	60	65.8	n.a.	96	1.09	n.a.	1.59
1885	61	56.7	44.5	85	0.93	0.73	1.40
1890	n.a.	42.8	n.a.	n.a.	n.a.	n.a.	n.a.
1895	59	52.6	40.2	90	0.89	0.68	1.53
1900	71	59.5	45.8	110	0.84	0.65	1.55
1905	62	58.1	44.1	110	0.94	0.71	1.77
1910	79	78.7	59.5	132	1.00	0.76	1.68
1915	91	91.1	65.3	137	1.00	0.72	1.50
1920	99	99.6	63.3	155	1.01	0.64	1.57
1925	123	115.4	72.2	167	0.85	0.59	1.36
1930	156	197.8	119.8	299	1.27	0.77	1.92
1935	91	135.4	64.2	245	1.48	0.70	2.68
1939		104.9	49.4	195.6	0.99	0.47	1.85

Note: For index of agricultural prices see Table 10.29. 1 sen = 1/100 yen.
Source: Ohkawa *et al.* (1967).

high level in 1939, suggesting that there was a shortage of industrial workers. Therefore, in the late 1930s, there might be a shortage of Z-sector labourers. This corresponded to the fact that expansion of the war in China increased military enrolments and caused a shortage of male workers, while demand for Z-sector goods expanded drastically. As Japan entered into a war-time economy, however, it became difficult to make a rigorous investigation.

10.5.3. Public investment in education

We showed in Sec. 4.2 that the $\dot{K}_z = 0$ curve and the $\dot{K}_x = 0$ curve will not intersect, ie. take-off will not be possible, if labour does not fill the skill requirements of the Z-sector. In the late 1930s, there was a danger of this happening when the Z-sector was almost squeezed out of the labour market by the militarization of the economy. In the long run, however, the supply of skilled workers for the Z-sector was maintained. Since the marginal wage rate in the Z-sector *vis-à-vis* that of the S-sector had remained stable for a long period, the supply of skilled workers to the Z-sector had been unlimited. This fact has an important policy implication for contemporary LDCs. The importance of manpower policy in promoting economic development has only recently been realized. Theoretically, the aim of this strategy is to make the $\dot{K}_x = 0$ curve and the $\dot{K}_z = 0$ curve intersect so as to make take-off feasible.

Table 10.31 shows school enrolment and the ratio of government investment in public education to national income from 1875. Even in 1875, enrolment in compulsory education, measured as the ratio of those enrolled to the population of the same age groups, reached 35%. Since Japan had very few factories in the *X*-sector at that time, having only a few silk factories, this spread of primary education was significant. The corresponding ratios for secondary and higher education, however, were less than 1%, and the skill level of Japanese workers when the privatization of government factories began in 1880 was as low as this figure suggests. This situation was compatible with government strategy, however, which was to ensure a small number of middle-class engineers and managers and a large number of factory workers. Investment in higher education was promoted only much later.

TABLE 10.31. School enrolment and expenditure on education

	School enrolment			Ratio of government investment in public education to national income		
	Compulsory (6–9-year-olds)	Secondary (12–16-year-olds)	Higher (17–22-year-olds)	Compulsory	Secondary	Higher
1875	35.2	0.1	0.4	n.a.	n.a.	n.a.
1880	41.1	0.2	0.4	n.a.	n.a.	n.a.
1885	49.6	0.5	0.5	1.34	—	0.08
1890	48.9	0.4	0.5	0.80	0.03	0.20
1895	61.2	1.0	0.5	0.88	0.09	0.15
1900	81.5	2.9	0.9	1.26	0.32	0.33
1905	95.6	4.0	1.3	1.23	0.37	0.32
1910	98.1	5.2	1.5	1.95	0.49	0.44
1915	98.5	6.2	1.5	1.49	0.42	0.34
1920	99.0	8.3	1.8	1.59	0.46	0.38
1925	99.4	13.4	2.9	1.92	0.72	0.59
1930	99.5	15.7	3.1	2.23	0.95	0.74
1935	99.6	16.3	2.9	1.95	0.76	0.49
1940	99.6	20.3	4.0	1.32	0.67	0.39

Sources: Monbu-sho (1969): 377, 381.

In 1910, school enrolment in compulsory education was as high as 98%, indicating that almost all children aged 6–13 attended primary schools. The ratios for secondary and higher education in the same year were 5.3% and 1.5%, respectively. The ratio between public investment in education and national income was 1.95% for compulsory education, 0.49% for secondary education, and 0.44% for higher education, giving a total of 2.88%. As the British counterpart ratio for 1920 was only 1.6%, it can be said that Japan expended a large portion of its wealth on public education. This in turn enhanced the supply of skilled labour to industrial sectors. Thus, investment in education facilitated the take-off of the *Z*-sector.

Notes

1. We have not investigated the regulation of private borrowing in this period, but we imagine that it was restricted by the government's nationalist stance.
2. According to a more recent work by Emi (1971), however, gross private capital formation in 1933 at current prices was ¥1,272 million, giving a ratio of 2.8%.
3. 'Government enterprises' includes enterprises owned and operated by local governments as well as the central government. In Japanese all these are called *kan-ei*, which means 'government-operating'. Most of the public utility sectors were operated by local governments. Enterprises which produced manufactured goods and mineral products were operated by the central government. As we are concerned with the manufacturing sectors, our focus is on central government activities.
4. These were branches of the Machine and Tool Office in Shinagawa, Akabane, and Hukagawa (all in Tokyo) and they were engaged in machinery production. See also Table 10.4–6.
5. Asahishimbun (1930) gives 1,100 hp instead of 2,509 hp for the same year. It is conjectured that factories under the ministries of Army and Navy were excluded.
6. In 1869 these 4 factories ceased operation; see Nakajima (1967: Chronological Table: 94). The factory in Kagoshima was privatized as early as 1872; see Iwanami Shoten (1968). The same *han* owned another factory for cotton spinning in Sakai, near Osaka, far from its own territory. It used four modern spinning machines but was bought by the Meiji government in 1872; see Kajinishi (1965: 51).
7. The generator capacity of factories under the mining sector of the Ministry of Industry was estimated by subtracting those which were sold for privatization, taking the data for 1881 as a benchmark. Generator capacity for the factories under the Ministry of Army and Navy was estimated from Naikaku Tokeikyoku (1881–1939).
8. The Law on Mining was enacted in 1873. One of the largest mines operated by a *daimyo* was a coal mine, Takashima-Tanko in Kyushyu, operated by Saga-Han, which started operation in 1868. This mine was reputedly equipped with modern facilities.
9. See Emi and Shionoya (1965: 162, 186). The last figure is our estimate from Emi (1971: 225).
10. Calculated by us from Kajinishi (1965: 53, 62).
11. Calculated by us from Kajinishi (1965: Appendix Table, 46).
12. Calculated from Naikaku Tokeikyoku (1881–1919) which gives data only for after 1882. After 1885, data on the production of individual factories became unavailable because of privatization.
13. For details of the manufacturers' movement for the abolition of export taxes, see Matsui (1963: vol. ii), ch. 4.
14. For a discussion of this 'turning point' in the labour supply in Japan, see Minami (1970).
15. There was a controversy between Ohkawa (1956) and Nakamura (1966) on the growth rate of Japanese agriculture, due to differences in evaluation of statistics for the initial point of time. We do not join this debate because for us it is sufficient for our model to confirm that agriculture had grown just fast enough to create a surplus for industrial development.
16. These models consist of two sectors, ie. subsistence and capitalist sectors only,

whereas our model has three, i.e., the *S*-, *X*-, and *Z*-sectors. The Ranis and Fei model has not dealt explicitly with the mechanism of capital accumulation.

17. On the other hand, the real wage rate of the *S*-sector can drop even after the economy has passed the turning point if there is large-scale shrinkage of industrial activities as at the time of the Depression. The Depression might have brought about a temporary shift of industrial employment to agriculture which would have led to a drop in the real wage rate in the *S*-sector.

PART III

Economic Development of Japan Post-World War II

11

Introduction and Summary

11.1. Introduction

The path of economic development that Japan was following was interrupted by World War II. Thus while we showed in Part II that the Z-sector, by which we mean the heavy industry sector, was on a path to self-sustaining growth by the late 1920s, we could not trace the path of development any further because the Japanese economy was devastated by the war. National wealth dropped to three-fourths of the pre-war level, and there was enormous damage to social overhead capital as well as to X- and Z-sector capital. The damage to the Z-sector was especially serious because it was the major target of bombing. It seems that (K_X, K_z) dropped below the minimum critical effort curve at this time, and far below the $\dot{K}_z = 0$ curve. Thus, after the war, Japan had once again to construct its industrial base as it had once done in the Meiji period. But the initial conditions differed significantly from those of the earlier period, in the following respects:

1. Population pressure was far heavier in the post-war period relative to food supply because a greater number of people now had to be fed from a smaller territory. The pressure was further aggravated by the return of Japanese soldiers and other overseas Japanese from abroad.
2. The quality of labour was far higher in 1945, in terms of education, experience of industrial production, and so forth. Thus, industrial workers and engineers were available in abundant supply after the war.
3. Foreign aid and foreign capital had played very minor roles in the Meiji period. Now, assistance in foodstuffs saved the nation from starvation, and foreign loans to purchase raw cotton contributed to economic restoration.
4. While import tariffs did not play a significant role in the earlier part of the Meiji era, it now became possible for the nation to use tariff policy to protect domestic industries. In fact, however, the government imposed relatively low tariffs on Z-sector goods in order to ensure supply of machines at lower prices. Import policy depended more on quantitative restraints than tariffs.
5. Government support for industries was now implemented through fiscal and monetary policies rather than by establishing government enterprises. Investment in the Z-sector was promoted by means of fiscal investment and loan programmes sponsored by the government and preferential loans through government financial institutions.

The initial conditions for industrialization in 1945 thus differed significantly from those of the pre-war period. Accordingly, although it had taken some sixty years for the economy to take-off in the pre-war period, the Z-sector started to attain self-sustaining growth only ten to fifteen years after the defeat, i.e. sometime between 1955 and 1960.

We must make it clear, however, that our analysis of the post-war period is still tentative in nature. First, when we disaggregated industries to test our assumptions on terms of trade and production technology, some did not satisfy the conditions; even the aggregated Z-sector tended to be short of the terms of trade assumptions. This was probably because it was so highly protected that flooding of the domestic market due to increased supply capacity caused it to decline until exports picked up. Thus the prices of Z-sector goods relative to S-sector goods tended to decrease after 1955. At the same time, it should be noted that the protection of domestic agriculture at this time kept the prices of S-sector goods high, which led to a substantial deterioration in the terms of trade of Z-sector goods.

The statistics that we used to estimate production functions did not substantiate the existence of economies of scale in the chemical and metal industries. Therefore, there remains the task of preparing consistent data among production, capital stocks, and labour input. Our judgement at the moment is that the frequent business fluctuations associated with stop-and-go policies caused wild fluctuation of operation ratios. Since Japanese industries were protected from foreign competition, domestic industries had to lower operation ratios to adjust to tight demand control policies. Thus, it seems that our data did not fully reflect the real changes in labour inputs, operation of capital stocks, and other factors.

Data constraints were another serious problem. Reliable data for estimating production functions only became available after 1953, whereas the most critical period for testing Z-sector take-off and minimum critical efforts was before 1953, when the government used whatever policy instruments it could to raise Z-sector productivity. This defect was a serious weakness of our analysis when we used national accounts statistics instead of Census of Manufacturing data.

Finally, it is to be noted that we confine our test to the period before 1960, in line with Minami's assertion that this is when the Japanese economy passed the turning point.

11.2. Summary of Findings

Since the correspondence between theory and empirical test was discussed in Chapter 7, here we merely summarize the major findings:

1. *Classification of industries.* As domestic markets were separated from overseas markets, the empirical data did not so consistently support our

assumptions on the terms of trade for some industries. This is partly attributable to poor classification of the original data. On the other hand, artificial promotion of capital accumulation in the *Z*-sector under strict protection by import quota shifted the supply curve for *Z*-sector goods east. In contrast, the upward shift of the demand curve due to the growth of population and income was not large enough to raise the price of *Z*-sector goods, P_z, and in fact towards 1960, which was the last year of our testing, P_z actually declined. Generally speaking, it seems that the terms of trade conditions hold in the post-war period too, at least until 1960.

On the other hand, classification based on production technology resulted in the following:

X-*sector*: textile industries, food-processing, miscellaneous manufacturing.
Z-*sector*: paper and pulp, chemicals, primary metals, machines (transportation and communication, utilities)

The estimation results, however, revealed that the food-processing industry did not satisfy the terms of trade assumption, and that the chemical and metal industries, remarkably, did not satisfy the economies of scale assumption. We assume that the latter was because the operation ratios and labour inputs which we used did not sufficiently reflect the real-life situation, which fluctuated considerably due to drastic cut-backs at the time of excess supply in the domestic market which was protected from foreign competition in earlier years. Industries in parentheses are those that were not included in the classification when we estimated profit rates and the path of capital accumulation. In the post-war period, 'miscellaneous manufacturing' included various industries which should be classified into either the *X*- or *Z*-sectors. So we excluded 'miscellaneous manufacturing' from our testing for the post-war period.

2. *The path of capital accumulation.* Having classified industries into *X*- and *Z*-sectors as shown above, in order to maintain consistency among data we investigated the relations among profit rates, capital stocks and net investment using two kinds of data: Mitsubishi Keizai Kenkyusho (1930–), and Keizai Kikakucho (1969). The latter provides industry statistics based on national accounting. The result from the former confirmed that except during the Korean War (1950–1), the *X*-sector profit rate was negatively related to capital stock, whereas the *Z*-sector profit rate increased as capital stock expanded. On the other hand, net investment in the *Z*-sector always exceeded that in the *X*-sector regardless of the profit rate differential, due to strong government support for investment in the *Z*-sector.

Measuring profit rates in terms of the fixed assets of large corporations, the *Z*-sector began to overtake the *X*-sector in around 1956, but net investment in the *Z*-sector was already much larger than the realized profit value from 1951. The ratio of net investment to realized profit was also higher in the *Z*-sector.

When we used national account statistics, or more precisely the data estimated from national account statistics, we were unable to derive the expected results. The results of the latter test are therefore excluded from the body of the text and included in the statistical appendix.

3. *Critical efforts by the government.* Corresponding to Chapter 10 our analysis focuses as the following: (*a*) the role of foreign capital and foreign aid; (*b*) industry subsidization and investment in social overhead capital; (*c*) import restriction and tariffs; and (*d*) the supply of food and labour.

Foreign capital and foreign aid were far more important in the early post-war period than in the Meiji era. Food shortages were critical, due to the huge numbers of Japanese who returned from abroad and the extremely low agricultural production at this time. Aid from the United States, through GARIOA (Government Appropriation for Relief in Occupied Areas) and EROA (Economic Rehabilitation in Occupied Areas) contributed enormously to alleviating the crisis. Japanese imports funded by these programmes accounted for 60%–70% of all imports between 1945 and 1950.

As for capital importation, the government revised the Law of Foreign Investment in April 1951 to encourage more investment from abroad. Due to the uncertain prospects of the Japanese economy, however, foreign investors were not attracted; and while the government was anxious to import capital, it was cautious of allowing foreign control of management, so direct investment was subject to strict screening. Thus it was only after 1960, when the government published its *Grand Design for the Liberalization of Foreign Trade and Exchange*, that capital imports began to increase. While the ratio of capital imports to the formation of fixed capital had never been very high, it seems that foreign capital nevertheless played an important role in financing investment in public utilities, transportation, and communication.

In the economic restoration process after World War II, the Japanese government protected important industries by allocating resources under a Priority Production Scheme (*Keisha Seisan Hosiki*) through preferential financing by the Restoration Bank which operated from 1946 to 1952 as a government financial institution. Selected industries, mostly in the *Z*-sector, were offered low interest funds through a Fiscal Investment and Loans Programme which used postal savings as financial sources. Postal savings have played an important role in collecting small-scale deposits made by households at post offices all over the country. The fund has been lent to government financial institutions under that programme. Subsidies were also available to ensure that the supply of important commodities could be secured.

Furthermore, there were various tax incentives to promote investment in the *Z*-sector: special allowances for the depreciation of production equipment, tax exemptions for specified important production activities, and exemption from import tariffs for specified important machines. There was also a special depreciation allowance for textile industries, but this was a temporary measure designed to promote exports and earn foreign currency. The government investment in social overhead capital which supported productive activities may also be considered to have encouraged the *Z*-sector, both by creating a demand for Z-sector goods and making production in the Z-sector more efficient.

Import policy, which consisted of quantitative restraint and tariff policies, played several roles in capital accumulation in the post-war period. First, import quotas were a dominant policy until 1960 when the government published its *Grand Design for the Liberalization of Foreign Trade and Exchanges*. In fact, it was only after 1960 that import tariffs became an important policy instrument. Our development model, which is based on free trade assumptions, must therefore be significantly modified.

Import quotas for machines and intermediate goods needed for investment in the Z-sector were more generous, although import tariffs did not necessarily favour the domestic Z-sector. In this respect, the tariff structure was similar to that in the pre-war period, and other policy instruments, such as subsidies, tax incentives, and preferential finance, were used to promote investment in the Z-sector. However tariff exemptions for specified important machines encouraged the importation of machines needed for the production of Z-sector goods. It was necessary for Japan to secure machines and intermediate goods at low cost in order to reduce the cost of production in all industrial sectors. For this reason, tariffs on Z-sector goods were not necessarily higher than those on X-sector goods. The criterion for determining tariff rates was whether the goods in question were producer goods or consumer goods: passenger cars, for example, were considered consumer goods and were subject to a high tariff.

Finally, let us summarize the supply of food and labour. For several years after the war, there were serious food shortages: only in 1955 did the per capita food supply reach the pre-war level. However, as the diet pattern changed, with a notable decrease in the relative weight of rice in total food intake, it is difficult to make comparisons simply on the basis of rice consumption. Roughly speaking, however, it seems safe to say that the supply of foodstuffs almost matched demand in 1955 and tended to exceed demand thereafter. Prior to 1955 the government raised the price of rice each year in order to encourage an increase in domestic supply.

This policy disfavoured industrial sectors because the relative prices of industrial goods declined against those of agricultural goods. Over time, however, an increase in productivity in the agricultural sector generated an outflow of labour to the industrial sectors. As we have seen, according to Minami (1970), there began to be a problem with the labour supply to industries in around 1960. We also concluded that there was no shortage of skilled labour for the Z-sector before 1960; we reached this conclusion because there was no significant rise in the wage rate differential between the Z-sector and other sectors until 1960.

12

Classification of Industrial Sectors

We apply the method used in Chapter 8 to the post-war period too. The period of investigation is from 1945 to 1960 as it has already been confirmed that the Japanese economy passed the turning point sometime between 1955 and 1960 with respect to labour supply (Minami 1970: 181–6). Thus, after 1960, industrial sectors could not expand employment any more without raising real wage rates. Although our development model can be easily modified to cover development at a time of general labour shortage, as explained in Part I, we terminate our analysis at this point of time because the main focus of our study is the situation before this.

As in the pre-war period, agriculture, forestry, and fishery are classified as *S*-sector. Industries are classified as *X*- or *Z*-sector according to their terms of trade and production technology. Occasionally, public utilities (transportation, gas, and electricity) are included in the *Z*-sector. In making the industrial classification, it was found that some industries satisfied neither the terms of trade nor the production technology assumptions, so we had to relax our criterion for classification; that is, our test was made on aggregated *X*- and *Z*-sectors. If the aggregated sectors satisfied the two conditions, we regarded the classification as acceptable.

12.1. Terms of Trade

As already shown in Table 8.1, Japan's market shares in the *X*- and *Z*-sectors in world trade right after the war were similar to those in the early Meiji period. Exports of metals, machines, and chemicals accounted for a slightly higher proportion of world trade, but were still as low as 2%–3%. Exports of textile products and miscellaneous manufactures, however, increased their shares of the market to around 20%–25%. In other words, Japan's export structure reverted to that of the earlier development period, with increasing specialization in light industries. This structure did not change until 1955.

It follows that the Japanese *X*-sector, represented by textile industries, faced a downward-sloping demand curve in the international market, whereas the *Z*-sector faced a horizontal demand curve. We can therefore skip some steps in testing terms of trade conditions. That is, applying the classification made for the pre-war period, we shall test the relationship between production and relative prices in the *X*- and *Z*-sectors against agricultural products.

Fig. 12.1 shows production and relative prices by industry from 1953 till

1965.[1] Many industries show different patterns before and after 1960. From 1953 to 1960, relative prices reacted in various ways to production in the individual industries, but after 1960 almost all *X*-sector industries showed declines in relative prices as production expanded in the *X*- and *Z*-sectors against agricultural products.

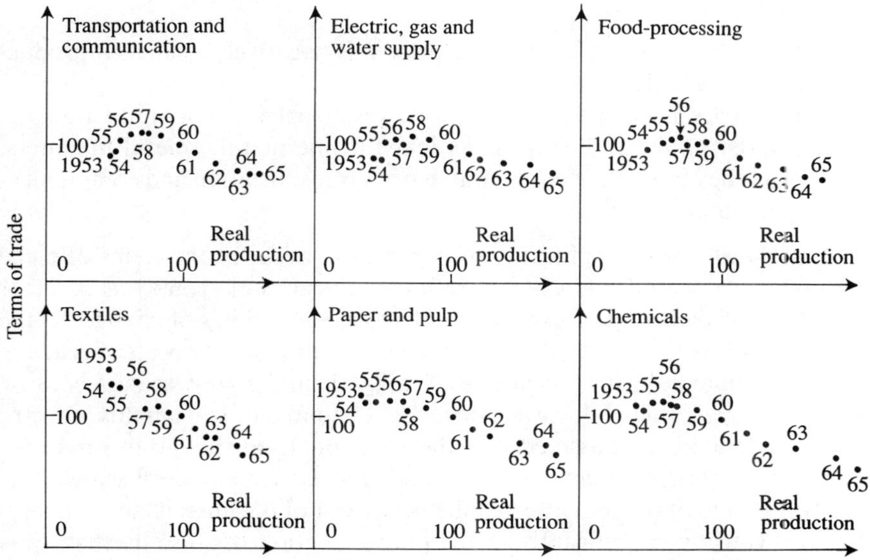

FIG. 12.1(i) Terms of trade in post-war period (by industry).

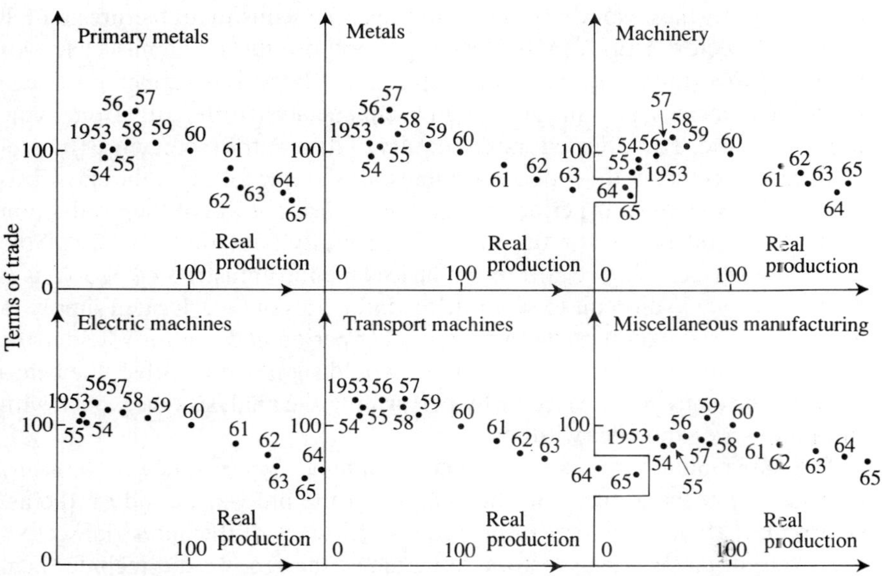

FIG. 12.1(ii) Terms of trade in post-war period (by industry).

Relative prices in many *Z*-sector industries also tended to decline as production grew after 1960. This probably reflects the fact that the labour supply became limited and that price support for agriculture deteriorated the terms of trade. However, our investigation is restricted to the period before 1960.

From observation of the data shown in Fig. 12.1 we can classify industries into two types:

1. Industries in which the relative price was negatively related to production: textiles.
2. Industries in which the relative price was independent of production: Foods, paper and pulp, chemicals, prime metals, metal products, machines, miscellaneous manufactures, transportation and communication, public utilities.

A straightforward application of our criteria as far as the terms of trade are concerned would lead us to classify the industries of group 1 as *X*-sector and those of group 2 as *Z*-sector. As there were a number of changes in the statistics, however, we modified our methods in some respects. First, the paper and pulp industry was not listed explicitly in the pre-war data because it was extremely small. However, since the economies of scale in this industry are considerable, we classified it in the *Z*-sector. Even though this industry had faced a horizontal demand curve and enjoyed economies of scale in the pre-war period, it had not affected the behaviour of the miscellaneous manufactures category significantly in the pre-war period because its share was too small.

When we classified industries on the basis of domestic markets, food-processing, textiles, wood products, and miscellaneous manufactures all fell into the *X*-sector. Thus, food-processing, wood products, and miscellaneous manufactures (including glass and ceramics, rubber and rubber products, precision machines, printing and binding, etc.) behaved differently from in the pre-war period. This affected the continuity of data in the sector since the pre-war years. As regards the food-processing industry, we followed the classification used in the pre-war period, partly from consideration of the production technology and partly for the sake of continuity with the analysis of the preceding period. With regard to the miscellaneous industries category, as it proved extremely difficult to separate the industries out we decided simply to adopt the classification made for the pre-war period and test for whether the inclusion of miscellaneous manufactures would significantly affect the behaviour of the aggregated *X*-sector by performing the analysis twice, once with these industries and once without.

The public utility, transportation and communication sectors were basically excluded from the *Z*-sector in the analysis of the pre-war period on the assumption that although they improved the efficiency of the industrial sectors they were in fact external factors. Sometimes, though, we did include them into a broadly defined *Z*-sector. For the post-war period, we included them in the Z-sector more positively, in the light of the economies of scale in these industries and their independent terms of trade.

Thus, as a final product of the above considerations, we adopted the following classification:

X-*sector*: textiles, food-processing, (miscellaneous manufactures);
Z-*sector*: transport and communication, public utilities, paper and pulp, chemicals, metals, machinery.

Using aggregated data for the X- and Z-sectors thus defined, we obtained the relations between relative prices and production shown in Fig. 12.2. As the number of samples was limited, however, it was difficult to find a definite relation. Looking at the period 1953–60, however, it can be said that P_x was negatively related to production, X, and that P_z was rather independent of production, Z. An estimation of a simple regression equation of X on P_x and Z on P_z for the period 1953–60 gave us:[2]

$$P_x = 124.7 - 0.2535x$$

$$P_z = 107.3 - 0.0011z.$$

12.2 Production Technology

To test for the existence of economies of scale for the Z-sector and constant returns to scale for the X-sector, we estimated production functions for the

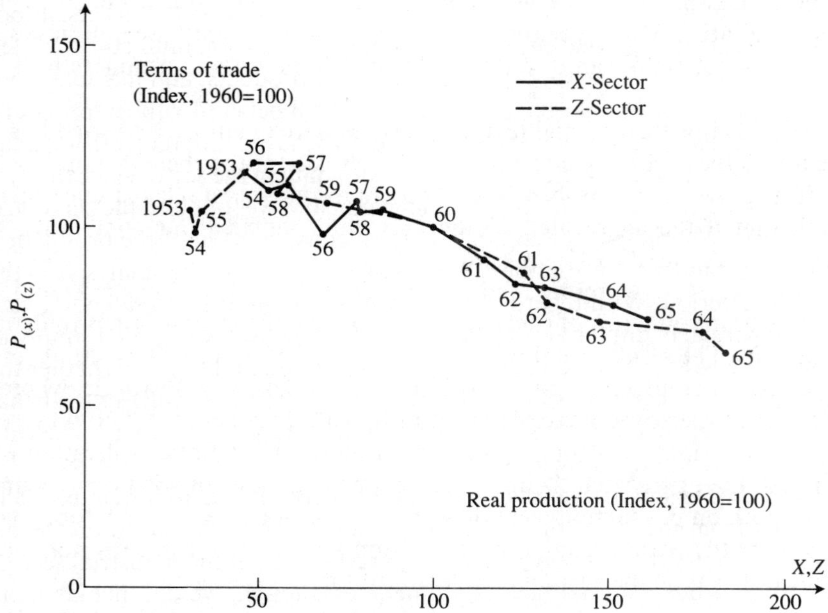

FIG. 12.2 Terms of trade in post-war period (by sector).

two sectors as we had done for the pre-war period. Although we confined our empirical test to the period until 1960, data from after 1960 were used in estimating the production functions because the sample sizes would otherwise have been too small since the annual data we used were available for 1953 on. Moreover, there were no significant changes in production technology in the later period.

The nature of the production functions was essentially the same as in the pre-war period, except for some elaborations to remove cyclical variations. More specifically, capital stock and labour input variables were modified by multiplying them with operating ratios of capital stock[3] and a man-hour index,[4] respectively.[5] Letting V, K, L, ρ, and h stand for real production, capital stock, number of employees, operating ratios, and man-hours, respectively, the following equation was estimated:

$$\log \frac{V}{hL} = A_t + B \log \frac{\rho K}{hL} + C \log hL.$$

The estimated results are given in Table 12.1. When we applied our test to individual industries, the parameter C, which indicated the existence of economies of scale if $C > 0$, was positive except for food-processing and chemicals. Thus, it appears that all other industries enjoy increasing returns to scale. When we applied a stricter criterion for the acceptance of the hypothesis at a 1% level, in line with the improved data reliability in the post-war period, t-test on C led us to identify an absence of increasing returns to scale in the food-processing, textiles, chemicals, metal and metal products industries and public utilities. More precisely, the production technology used in these sectors was not significantly different from that offering constant returns to scale.

The food-processing and textile industries were in any case classified as X-sector on the terms of trade criterion, but the case of the chemicals and metal industries was more problematic. As a compromise, we relaxed our criteria such that if the aggregated X- and Z-sectors satisfied our conditions, we regarded the classification as valid.

Our tests were therefore made in two steps: first for aggregate X- and Z-sectors based on terms of trade, and then for aggregates based on production functions. The results are shown in Table 12.1. The X-sector (excluding 'miscellaneous manufacturing') was subject to constant returns to scale, whereas the Z-sector, enjoyed increasing returns to scale, whether public utilities were included or not. Thus, although the criterion for industrial classification was relaxed, the aggregated X- and Z-sectors nevertheless satisfied our assumptions both on production technology and terms of trade. We therefore go on to review the path of capital accumulation and the government's minimum critical efforts on the basis of this classification.

TABLE 12.1. Economies of scale, 1945–1965

	Production elasticities to capital (B) (*t*-value)	Economies of scale (C) (*t*-value)	$\bar{R}$	Production elasticities to capital (B) (*t*-value)	Economies of scale (C) (*t*-value)	Technical progress (A) (*t*-value)	$\bar{R}$
Food processing	0.986*	–1.216	0.978	0.094	–0.332	0.037*	0.998
	(10.040)	(–1.472)		(1.019)	(–1.328)	(10.176)	
Textiles	1.361*	2.599	0.976	0.504	1.055	0.025*	0.992
	(7.403)	(2.784)		(2.265)	(1.640)	(4.374)	
Miscellaneous manufactures	0.598*	0.735*	0.995	0.450*	0.003	0.018	0.997
	(11.783)	(4.182)		(5.640)	(0.008)	(2.188)	
X-sector (excluding miscellaneous manufactures)	1.141*	1.219	0.977	0.177	0.008	0.035*	0.997
	(7.491)	(1.150)		(1.221)	(0.002)	(7.281)	
X-sector (including miscellaneous manufactures)	0.626*	1.846*	0.996	0.396*	0.687	0.019*	0.999
	(9.178)	(6.402)		(7.463)	(2.725)	(5.522)	
Paper and pulp	0.669*	0.869*	0.995	0.331	0.299	0.025*	0.998
	(8.433)	(4.297)		(2.964)	(1.393)	(3.438)	
Chemicals	1.018*	–0.670	0.993				
	(6.366)	(–1.064)		**			
Primary metals	0.699*	0.704*	0.983	0.690*	0.680	0.001	0.980
	(8.436)	(8.528)		(4.503)	(1.869)	(0.069)	
Metal products	0.880*	0.019	0.988	0.767*	–0.264	0.016	0.988
	(7.856)	(0.104)		(4.247)	(–0.662)	(0.804)	
General machinery	0.492*	0.642*	0.985	0.580*	1.144*	–0.023	0.987
	(5.433)	(5.641)		(5.756)	(3.407)	(–1.574)	
Electric machinery	0.671*	0.407*	0.985	0.693*	0.445	–0.003	0.983

Table 12.1 (*contd.*)

	Capital elasticities to production (B) (*t*-value)	Economies of scale (C) (*t*-value)	$\bar{R}$	Production elasticities to capital (B) (*t*-value)	Economies of scale (C) (*t*-value)	Technical progress (A) (*t*-value)	$\bar{R}$
	(7.031)	(3.768)		(4.270)	(1.885)	(–0.183)	
Transport machinery	0.465*	1.189*	0.991	0.292	0.598	0.026	0.994
	(5.863)	(9.065)		(2.898)	(2.098)	(2.248)	
Transport and	0.452*	0.907*	0.990				
communications	(5.504)	(8.783)		**			
Electricity, gas and	0.884*	0.400	0.996	0.354	–0.639	0.033*	0.998
water supply	(10.810)	(0.743)		(2.167)	(–1.356)	(3.452)	
Z-sector							
(excluding public	0.664*	0.649*	0.996	0.640*	0.588	0.028	0.995
utilities)	(10.891)	(8.586)		(6.035)	(2.524)	(0.280)	
Z-sector							
(including public	0.699*	0.909*	0.996	0.874*	1.362	–0.017	0.996
utilities)	(9.751)	(11.347)		(6.112)	(4.080)	(–1.393)	

Notes: * significant at 1% level.

**: multicolinearities between log K/L and time (technical progress).

Log-linear estimations are used.

Variables of technical progress are excluded in left-hand equations and included in right-hand equations. Equation: $\log(V/hL) = B \log(\rho K/hL) + C \log(hL) + At$

Source: Keizai Kikakucho (1969).

Notes

1. Both relative prices and real production were estimated from the Keizai Kikakucho (1959–63). For the price index of the *S*-sector, we adopted the EPA price index of agriculture, forestry, and fisheries (1960 = 100). Indices of real production were also calculated from the data.
2. As these equations were estimated by hand, statistics such as *t*-values were unavailable. It seems, however, that the parameters of production variables are significant. While the parameter of the *Z*-variable is negative, its size is extremely small. That is, the negative effects of production on terms of trade are very slight.
3. MITI published its own index of the operating ratio which had an upward trend, ie. an absolute level of rising indices. After correcting for this upward trend and plotting the points on a line which was drawn by connecting peaks as 100, actual operating ratios were calculated as percentages of this 100% line (a so-called 'Warton index').
4. The man-hour index was based on average man-hours and also had an upward trend, but, in this case we did not revise the series to exclude the trend elements.
5. The data for production, capital stock, and employment were obtained from Keizai Kikakucho (1969). Operating ratios are taken from the materials compiled by the Econometrics Department of the Japan Economic Research Centre, and man-hour data from the Ministry of Labour's *Monthly Survey of Employment Statistics*.

13

The Path of Capital Accumulation

Immediately after World War II Japanese industries faced the mass destruction of their production facilities and massive unemployment due to the return of soldiers stationed abroad and people who had formally lived in the colonies. In other words, there was a conspicuous capital shortage and an equally conspicuous excess supply of labour. However, there were two institutional factors that mitigated the situation: land reform, which determined the institutional framework in which the supply both of food and of labour were secured; and the establishment of the Bank for Restoration which between 1946 and 1952 promoted investment in specified important industries. As we have seen, US aid was also important in the early period of restoration. In this chapter, however, we concentrate on the relationship between profit rates and capital accumulation, leaving a consideration of the other factors to Chapter 14.

As was made clear in the preceding chapter, *X*- and *Z*-sectors in the post-war period, which were aggregated on the terms of trade and production technology conditions, were as follows:

X-*sector*: food-processing, textiles;
Z-*sector*: paper and pulp, chemicals, prime metals, machines (including electric and transport equipment).

Miscellaneous manufacturing was excluded because some components, such as ceramic and other industries, were difficult to classify.

Data on capital stock were taken from Mitsubishi Keizai Kenkyusho (1930–) and Keizai Kikakucho (1969).

We then performed the following tests. First, is the *X*-sector profit rate in terms of *S*-sector goods negatively related to its own capital stock (i.e. is $\frac{dR_X}{dK_x} < 0$)? Likewise, is the *Z*-sector profit rate in terms of *S*-sector goods non-negatively related to its own capital stock (i.e. is $\frac{dR_Z}{dK_Z} \geqq 0$)? Is the entire *X*-sector profit and some portion of *Z*-sector profit re-invested in the *X*-sector if $R_X - R_Z > 0$? Finally is the proportion of *Z*-sector profit which is re-invested in the *X*-sector a non-decreasing function of $R_X - R_Z$? Because of the many simplifying assumptions that were made, however, as discussed in Chapter 9, we were unable to make a strict test of our hypotheses.

Our findings may be summarized as follows. First, the relationship between profit rates and capital stock showed a random variation until around 1953, but thereafter the distribution of the points (K_i, R_i) $(i = x, z)$ behaved as the hypothesis suggested—namely, the K_X–R_X curve was downward sloping as K_x expanded, and the K_Z–R_Z curve was upward sloping.

On the other hand, it turned out that it was only around 1955–6 that R_z exceeded R_x for the first time after the war. Thus, for the greater part of the period under investigation, the *X*-sector profit rate was higher than that of the *Z*-sector. The fact that *Z*-sector investment was overwhelmingly larger than that in the *X*-sector despite this implies either that the government strongly intervened in investment allocation or that investors had a longer-term perspective in making investment decisions, such as that described in Section 4.8.

The above summary is based on the empirical test using the Mitsubishi data for large corporations. The results obtained through the test on the EPA (Economic Planning Agency: Keizai Kikakucho) data (i.e. national account data), are more ambiguous, as summarized in the Appendix to this chapter (see Tables 13-A.4 and 13-A.5). While the data for large corporations confirmed our hypotheses the results achieved with the EPA data suggest that our investigation of capital accumulation in the post-war period is still tentative.

13.1. The Period of Post-War Restoration

Since there is a series of studies in, and documentation of, the post-war restoration process (e.g. Keizai Kikakucho 1959–63), we shall not go into details here.

After Japan's defeat on 15 August 1945, various emergency programmes were initiated: a memorandum on land reform was issued in December 1945, emergency measures for financing were introduced in February 1946, price controls were instituted in March 1946 and a committee for the dissolution of holding companies was set up in April 1946. These were all important steps in determining the institutional framework of post-war economic development. Additionally, in June 1946 foreign trade was put under the control of the occupation forces.

In October 1946, the Bank for Restoration (Hukko Kinyu Kinko) was set up to finance investment in important industries. This also had the effect of triggering inflation, (the so-called 'Hukkin' inflation). In reality, prices rose as money supply increased through this bank, since there were bottlenecks in some important intermediate industries. Thus the government had to reduce the number of industries to be encouraged. Thus, at the end of the year the cabinet decided to adopt a priority production strategy (*keisha seisan hosiki*) to induce investment in strictly specified industries such as coal-mining, electricity generation, and steel production.

Table 13.1 compares the production of mining and manufacturing immediately after the war i.e. in 1946, 1950 and 1955 with that before the war. The extent of war damage to Japanese industries is immediately apparent. While the index of production in 1940 was 47.1 (1960 = 100), in 1946 it dropped as low as 8.5. As a result of the priority investment programme it recovered in

TABLE 13.1. Index of production (1960 = 100)

	1940	1946	1950	1955
Total industries	44.9	9.1	23.0	47.4
Public utilities	31.2	26.0	38.8	54.6
Mining and Mfg.	47.1	8.5	22.3	47.0
Manufacturing	44.3	7.4	20.4	45.7
Iron and steel	32.4	4.3	22.9	43.8
Machinery	25.0	5.0	11.9	28.5
Chemicals	37.3	6.0	20.8	49.5
Textiles	63.7	9.1	26.4	62.0

Source: Nippon Ginko (1966).

1950 to 22.3. Among the manufacturing industries, steel production re covered faster than chemicals and machinery, reaching 22.9—clear evidence of the priority production strategy.

As the economy was still in chaos due to supply shortages, various direct controls were also adopted. In March 1947, the compulsory sale of foods to the government was reinforced, in June 1947 private foreign trade came under the control of the occupation forces, and in December 1947 the government took over the management of the coal mines. Thus, until 1951 when Japan regained independence, the Japanese economy was strictly controlled by the occupying forces.

The preparation for industrial take-off, or recovery to the pre-war level, was thus not carried out under a laissez-faire strategy. Rather, in the critical period of 1945–50, the government intervened heavily to promote investment in the *Z*-sector.

There is no doubt that the Korean War (June 1950–July 1953) stimulated investment in heavy industries in Japan. Because of the extra demand, even the *Z*-sector realized a high profit rate over this period. In fact, the influence of this war was so strong that it distorted the K_Z–R_Z graph; if we want to neutralize the path of capital accumulation from the effects of war, we have to concentrate on the period after 1953.

Before concluding this section, let us look at Table 13.2, which shows the distribution of capital stock by industry by five-year periods since 1950. According to the Mitsubishi data, in the early post-war years the majority of capital stock in the *X*-sector was in the textile industry, while that of the *Z*-sector was in paper and pulp, chemicals, and primary metals. As time went on, the share of the textile industry in the *X*-sector declined, whereas that of machinery and metals in the *Z*-sector rose significantly.

The same trend emerged from the analysis of the economy-wide EPA data, although the shares of food-processing and metal manufacturing industries increased since these industries are based mostly on small and medium-scale

TABLE 13.2. Percentage composition of fixed assets by industries

	1950	1955	1960	1953	1955	1960	1965
X-sector							
Food processing	8.3	5.1	4.7	11.6	12.4	10.9	10.8
Textiles	33.1	22.0	15.6	27.1	26.9	21.0	13.9
Subtotal	41.5	27.1	20.2	38.7	39.3	31.9	24.7
Z-sector							
Paper	10.5	7.9	8.0	4.9	5.4	6.0	5.8
Chemicals	23.8	16.4	16.3	12.1	13.0	17.5	18.1
Primary metals	12.4	25.6	29.5	26.2	23.7	21.0	22.0
Other metals	0.2	0.6	0.7	1.6	1.8	2.9	3.8
General machinery	2.6	3.5	2.8	4.9	4.6	4.9	6.0
Electrical machinery	4.5	9.6	13.1	3.4	4.4	7.5	8.8
Transport machinery	4.5	9.3	9.3	8.1	7.8	8.2	10.7
Subtotal	58.5	72.9	79.8	61.2	60.7	68.0	75.2
TOTAL VALUE[a]	1,059	9,262	22,910	333,330	374,040	674,590	1,310,300

[a] ¥100m.

Source: Mitsubishi Kezai Kenkyusho (1930—), Keizai Kikakucho (1969).

enterprises had increased in the EPA data, as EPA data presumably covered all of the production units.

13.2. Profit Rates, Capital Stock, and Net Investment

Table 13.3 shows the estimates of the ratios of profits to fixed assets and total assets in the *X*- and *Z*-sectors, measured at current prices. The table makes it clear that there was temporary extra demand in 1950–3 due to the Korean War. More importantly, $R_X - R_Z$, or the *X*-sector profit rate minus that of the *Z*-sector, when measured in terms of fixed assets, was negative only in 1957. When measured in terms of total assets, however, $R_X - R_Z$ became negative only in 1958 (see cols. 5 and 6). Even so, the investment allocation was much more strongly oriented to *Z*-sector.

Fig. 13.1 compares $R_X - R_Z$ with the ratios of net investment to total profit in the respective sectors in the preceding period. In other words, this figure compares $R_X - R_Z$ with $\frac{dK_X}{(R_X K_X)_{-1}}$. Let us call the latter the *X*-sector reinvestment ratio. Although it was only after 1957 that *Rz* exceeded *Rx*, the *Z*-sector re-investment ratio, or $\frac{dK_Z}{(R_z K_z)_{-1}}$, had been higher than the *X*-sector ratio throughout the period 1951–60. Since the data were for large corporations only, a re-investment ratio of 200%–400% implied that the investment of large corporations was financed either by the savings of small firms and personal

TABLE 13.3. Ratios of profit to capital and flow of investment

		Ratios of profits(%)				Difference in ratio between two sectors		Increment in fixed capital assets (¥m)		Increment in total assets (¥m)		Profits (¥m)	
		to fixed capital assets		to total assets		$R_X' - R_Z'$	$R_X - R_Z$	dK_X'	dK_Z'	dK_X	dK_Z	$R_X K_{X-1}$ X- sector	$R_X K_{Z-1}$ Z-sector
		X-sector	Z-sector	X-sector	Z-sector								
1950	FH	9.2	2.7	2.5	0.6	6.5	1.9					4,029	1,702
	SH	16.8	3.6	5.4	1.1	13.2	4.3	32,023	88,484	73,493	197,862	12,752	5,364
1951	FH	37.3	10.5	9.3	2.9	26.8	6.4	15,518	16,944	133,834	121,290	34,100	17,534
	SH	13.3	10.3	4.2	2.9	3.0	1.3	6,326	75,877	–61,646	241,846	12,969	25,070
1952	FH	8.4	7.2	2.7	2.2	1.2	0.5	49,151	54,034	148,194	136,787	12,319	21,412
	SH	6.1	6.2	2.2	1.9	–0.1	0.3	13,152	42,714	–19,393	101,108	9,686	21,005
1953	FH	6.1	5.8	2.3	1.9	0.3	0.4	19,884	60,287	37,354	107,509	11,004	23,202
	SH	7.3	5.2	3.1	1.9	2.1	1.2	34,246	98,625	34,050	153,690	15,680	25,909
1954	FH	5.5	3.0	2.3	1.2	2.5	1.1	18,657	71,335	47,843	97,953	12,875	17,153
	SH	4.0	2.2	1.7	0.9	1.8	0.8	9,663	79,770	15,823	124,939	9,735	14,204
1955	FH	3.7	3.1	1.6	1.3	0.6	0.3	8,352	26,309	–5,809	85,178	9,261	20,776
	SH	4.6	3.9	2.1	1.5	0.7	0.6	13,638	26,016	1,163	115,898	12,119	27,214
1956	FH	5.0	4.8	2.4	1.8	0.2	0.6	22,228	44,519	36,549	152,681	14,354	35,469
	SH	5.8	5.5	2.9	2.0	0.3	0.9	31,912	94,184	47,032	323,935	18,522	45,986
1957	FH	5.0	5.2	2.4	1.9	–0.2	0.5	29,037	135,846	80,737	393,642	17,381	51,053
	SH	3.4	4.0	1.7	1.5	–0.6	0.2	22,116	120,211	28,655	173,718	12,638	43,529
1958	FH	2.2	3.3	1.1	1.3	–1.1	–0.2	14,335	106,683	31,176	86,896	8,600	39,196
	SH	1.8	3.4	0.9	1.5	–1.6	–0.6	10,067	98,999	–11,385	138,583	7,038	44,727
1959	FH	2.7	4.8	1.4	2.0	–2.1	–0.6	16,696	112,757	27,832	276,211	11,041	67,197
	SH	5.2	6.1	2.7	2.6	–0.9	0.1	18,380	175,621	35,757	451,676	22,367	96,363
1960	FH	5.4	5.8	2.7	2.5	–0.4	0.2	33,870	237,756	93,223	466,291	25,124	106,444
	SH	4.5	5.3	2.2	2.3	–0.8	–0.1	43,254	302,242	70,465	706,748	22,577	113,929

Note: FH = first half of year; SH = second half.

Source: Mitsubishi Keizai Kenkyusho (1930–).

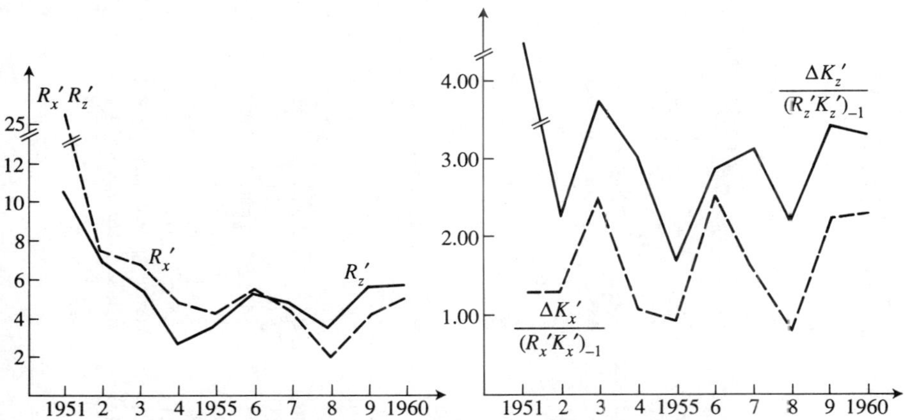

Note: ΔK_i stands for changes in K_i. ($)_{-1}$ stands for the value of the previous period.

FIG. 13.1 Difference in profit ratios and re-investment ratios: Ratio of Increase in Capital to Profit.

businesses, or by the savings of those industrial sectors which were not covered by these statistics (such as mining, construction, services, and distribution). (Supporting statistical data are given in Appendix 1).

Why was *Z*-sector investment significantly higher than that in the *X*-sector even around 1955 when R_x was much higher than R_z? One reason, as we have seen, was the subsidization of the Z-sector by the government. Another reason might be that the market structure of the *X*-sector had already become fairly oligopolistic. As early as February 1953, the Fair Trade Commission of Japan had approved the establishment of the Association for the Co-ordination of Export of Cotton Products, and in November 1954, the MITI published regulations on productive equipment for cotton and cotton staple fibres and restrictions on the use of unregistered productive equipment. This situation, whereby the sector was thus pricing higher than in a competitive situation and producing less, caused investment to be lower than the high profit rates would have suggested.

We next consider the relationship between capital stock and profit rates. Fig. 13.2 shows the point (K_i, R_i), where $i = x, z$, plotting R_i on the vertical axis in terms of profit measured in *S*-sector goods prices over fixed capital assets as the denominator, and K_i on the horizontal axis. If we ignore the points for 1952 and 1953, which reflect the windfall profits of the Korean War, we can conclude that $\mathrm{d}R_X'/\mathrm{d}K_X' < 0$ and $\mathrm{d}R_z'/\mathrm{d}K_Z' = 0$, where R_i' is the profit rate in terms of fixed capital assets.

A similar feature can be observed from Fig. 13.3, which plots R_i, the profit over the total asset of corporations K_i. That is, we changed the denominator into total assets, not limiting it to fixed assets. It also suggests that $\frac{\mathrm{d}R_X}{\mathrm{d}K_X} < 0$

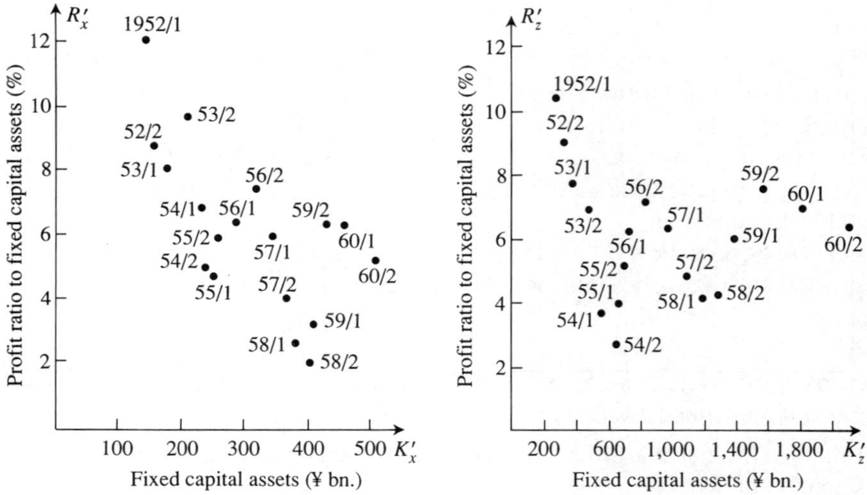

Note: 53/1 stands for the first half of 1953,
53/2 stands for the second half of 1953.

FIG. 13.2 Profit ratios measured in *S*-goods prices and fixed capital assets.

and $\dfrac{dR_Z}{dK_Z} > 0$ hold even if we control business cycles. Supporting statistical data are given in Tables 13-A.2 and 13-A.3 in the Statistical Appendix.

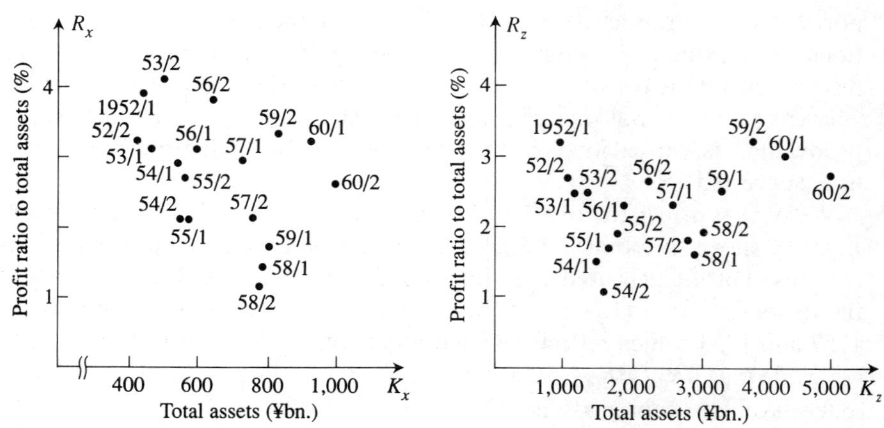

Note: 53/1 stands for the first half of 1953,
53/2 stands for the second half of 1953.

FIG. 13.3 Profit ratios measured in *S*-goods prices and total assets.

13.3. The Path of Capital Accumulation

The two series of capital stocks for the *X*- and *Z*-sectors are shown in Fig. 13.4 (fixed capital assets) and Fig. 13.5 (total capital stocks). The data are those for large corporations estimated by the Mitsubishi Research Institute (Mitsubishi Keizai Kenkyusho, 1930–). As we accept Minami's (1970) proposition that the supply of labour had already become limited by 1960, our graph does not go beyond this point.

These figures show no tendency for capital stocks (K_x, K_z) to fall to (K_X^*, 0) after 1950—that is, the poverty trap was averted. Even though R_x had been much higher than R_z until 1957, the actual accumulation of capital was significantly larger in the *Z*-sector. This suggests that the government had actively encouraged investment.

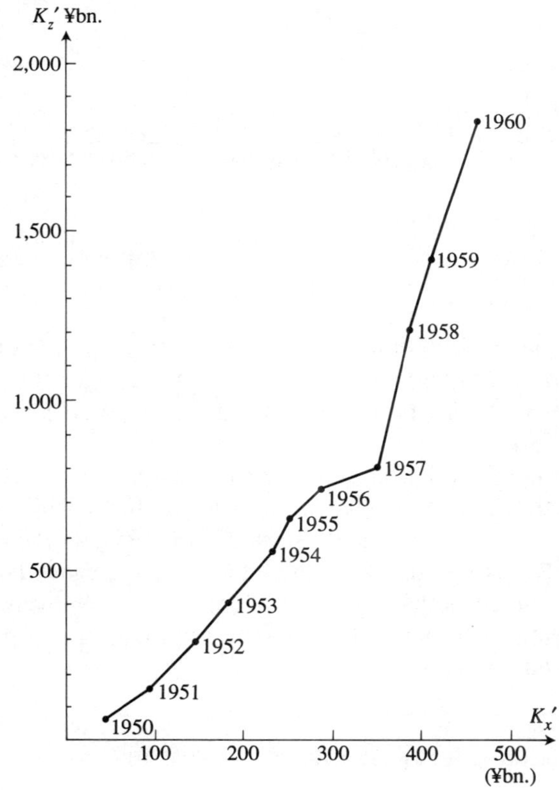

Note: Figures are for the first half of the year.

FIG. 13.4 Capital accumulation path: fixed assets in Mitsubishi Keizai Kenkyusho (1930–).

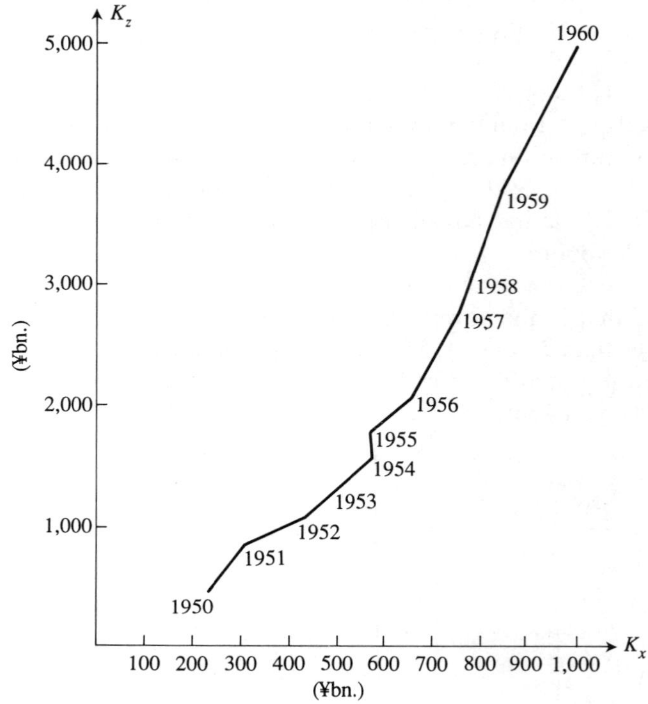

Note: Figures are for the second half of the year.

FIG. 13.5 Capital accumulation path: total assets in the Mitsubishi Keizai Kenkyusho (1930–).

We did not make use of data on capital stock per capita in our analysis of the post-war period, but doing so would not have changed the picture to any significant degree because population increase was always a stable 1% in the period in question.

If we add in the period after 1960, we can say that the point moved more quickly north, that is, K_Z expanded more rapidly than K_X. This acceleration of investment in the Z-sector actually started after 1957, as indicated in Figs. 13.4 and 13.5. Retesting this finding with the EPA aggregate data we get Fig. 13.6. This also indicates that the point of capital assets moved more rapidly north in the 1960s, that there was i.e. faster accumulation of capital in the Z-sector in the 1960s than the 1950s.

13.4. Some Reservations

The results we obtained using data for large corporations support our hypotheses. When we used data based on national accounts, however, the data did not support our hypotheses, as shown in the Statistical Appendix to this chapter.

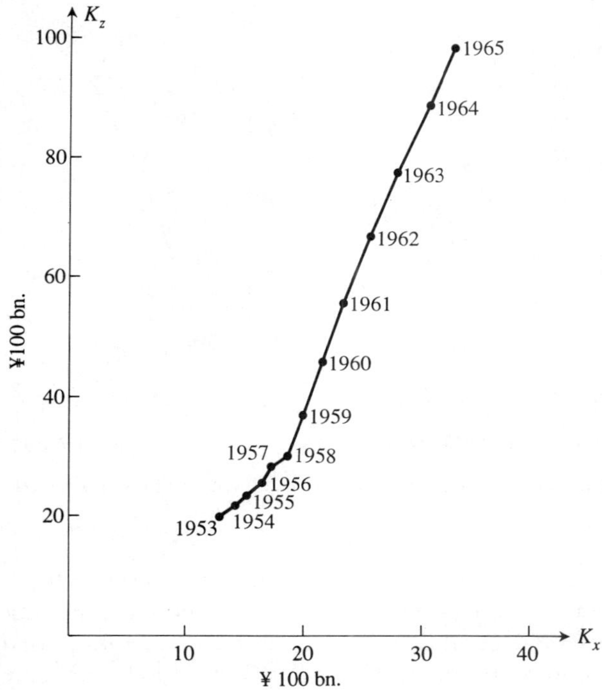

Source: Keizai KikaKucho (1969).

FIG. 13.6 Capital accumulation path, 1953–65.

The way in which variables were estimated from the national account data is as follows. For each sector, after calculating GDP of *X*- and *Z*-sectors at current prices, the proxies of profits were estimated by subtracting wage bills (wage rate multiplied by number of employees) published in Kezai Kikakucho (1969) from the calculated GDP. The profit at current prices was then converted to profit measured in terms of *S*-sector goods. This figure was then divided by capital stocks to obtain an estimated proxy of profit rates. In the absence of reliable deflators for capital stocks by sector, our capital stock data have not been deflated.

The paths of R_X–K_X and R_Z–K_Z shown in Fig. 13.7 for these data sets deny such relations as $\frac{dR_X}{dK_X} < 0$ and $\frac{dR_Z}{dK_Z} \geqq 0$. In particular, the data seem to nullify our hypothesis about the relationship between R_Z and K_Z after 1960. On the other hand, R_Z was almost always higher than R_X, with no identifiable critical point when R_Z caught up with and exceeded R_X. This may be because the data for value added, wage bills, and capital stocks were inconsistent. Or the reason may be that R_Z had always been higher than R_X since 1953 if we include small firms in our data coverage. Investment in the *Z*-sector was always encouraged,

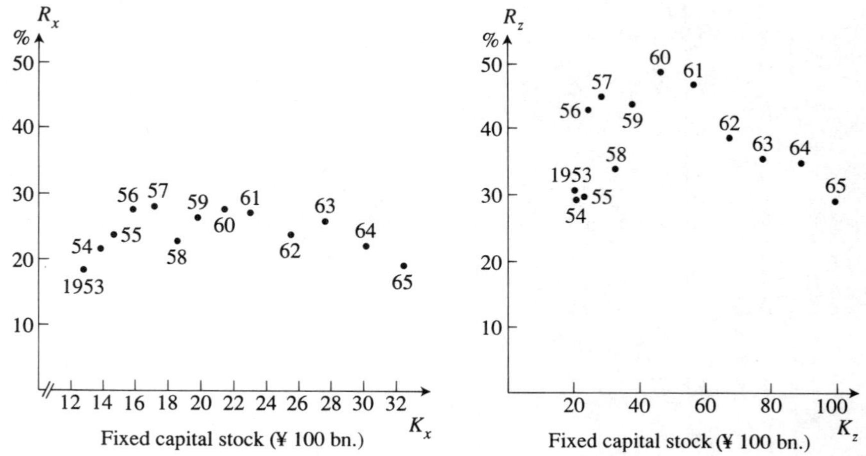

FIG. 13.7 Relationship between profit ratios and fixed capital stock based on the national account statistics.

moving the point (K_X, K_Z) north. The reduction of protection for the *Z*-sector after 1960 might also be a powerful contributor to the negative relation between R_Z and K_Z since 1960, but this period is really beyond the scope of our study.

The application of our model to the post-war period is not complete, and there are many points in our empirical tests that could be refined. None the less, from the data for large corporations we can conclude that it was around 1957 that R_Z exceeded R_X, and that the *Z*-sector was on a path of self-sustaining capital accumulation. The take-off of the *Z*-sector, in our terminology, was thus around 1957–60, with acclerated industrialization having taken place since 1955.

13.5. Statistical Appendix

TABLE 13.A.1. Profit rate differentials and the ratios of investment in own sector to total profit of own sector

		Increase in fixed assets / Profit of previous period (%)		Profit rate differential $R_X' - R_Z'$	Increase in fixed assets / Profit (annual rate) (%)		Profit ratios (annual average, %)		Profit rate differential
		X-sector	Z-sector	(% point)	X-sector	Z-sector	R_X'	R_Z'	$R_X' - R_Z'$ (% points)
1950	FH			6.5			13.0	3.2	9.8
	SH	7.95	51.99	13.2					
1951	FH	1.22	3.16	26.8	1.30	13.14	25.3	10.4	14.9
	SH	0.19	4.33	3.0					
1952	FH	3.79	2.12	1.2	1.32	2.25	7.3	6.7	0.6
	SH	1.07	1.99	–0.1					
1953	FH	2.05	2.87	0.3	2.46	3.75	6.7	5.5	1.2
	SH	3.11	4.25	2.1					
1954	FH	1.19	2.75	2.5	1.06	3.08	4.8	2.6	2.2
	SH	0.75	4.65	1.8					
1955	FH	0.86	1.85	0.6	0.97	1.67	4.2	3.5	0.7
	SH	1.47	1.25	0.7					
1956	FH	1.83	1.64	0.2	2.53	2.89	5.4	5.2	0.2
	SH	2.22	2.66	0.3					
1957	FH	1.57	2.95	–0.2	1.56	3.14	4.2	4.6	–0.4
	SH	1.27	2.35	–0.6					
1958	FH	1.13	2.45	–1.1	0.81	2.17	2.0	3.4	–1.4
	SH	1.17	2.53	–1.6					
1959	FH	2.37	2.52	–2.1	2.24	3.44	4.0	5.5	–1.5
	SH	1.66	2.61	–0.9					
1960	FH	1.51	2.47	–0.4	2.31	3.30	5.0	5.6	–0.6
	SH	1.72	2.84	–0.8					

Note: FH = first half of year; SH = second half of year.

TABLE 13.A.2. Capital stocks of *X*- and *Z*-sectors

		(1) 'Analysis of Japanese Business' (¥m.) Fixed capital assets		Total assets		(2) 'Economic Analysis No.27' (¥bn.) Fixed capital stock	
		X-sector	*Z*-sector	*X*-sector	*Z*-sector	*X*-sector	*Z*-sector
1950	FH	43,955	61,964	160,489	290,166		
	SH	75,978	150,448	233,982	488,028		
1951	FH	91,496	167,392	367,816	609,318		
	SH	97,822	243,269	306,170	851,164		
1952	FH	146,973	296,303	454,364	987,951		
	SH	160,025	339,017	434,971	1,089,059		
1953	FH	179,909	399,304	472,325	1,196,568	1,291	2,042
	SH	214,155	497,929	506,375	1,350,258		
1954	FH	232,812	569,264	554,218	1,448,211	1,400	2,166
	SH	242,475	649,034	570,041	1,573,150		
1955	FH	250,827	675,343	564,232	1,658,328	1,470	2,269
	SH	264,465	701,359	565,395	1,774,226		
1956	FH	286,693	745,878	601,944	1,926,907	1,589	2,448
	SH	318,605	840,062	648,976	2,250,842		
1957	FH	347,642	975,908	729,713	2,644,484	1,705	2,843
	SH	369,758	1,096,119	758,368	2,818,202		
1958	FH	384,093	1,202,802	789,544	2,905,098	1,853	3,250
	SH	394,160	1,301,801	778,159	3,043,681		
1959	FH	410,856	1,414,558	805,991	3,319,892	1,983	3,695
	SH	429,236	1,590,179	841,748	3,771,568		
1960	FH	463,106	1,827,935	934,971	4,237,859	2,149	4,596
	SH	506,360	2,130,177	1,005,436	4,944,607		

Sources: Column (1) Mitsubishi Keizai Kenkyusho (1930–); Column (2) Keizai Kikakucho (1969).

TABLE 13.A.3. Profit rate in terms of *S*-sector prices (%)

		Profit measured in *S*-sector prices: Fixed capital assets		Profit measured in *S*-sector prices: Total assets		*S*-sector prices index
		X-sector	*Z*-sector	*X*-sector	*Z*-sector	1965=100
1952	FH	12.2	10.4	3.9	3.2	69.1
	SH	8.8	9.0	3.2	2.7	
1953	FH	8.1	7.7	3.1	2.5	75.3
	SH	9.7	6.9	4.1	2.5	
1954	FH	6.9	3.7	2.9	1.5	80.1
	SH	5.0	2.7	2.1	1.1	
1955	FH	4.8	4.0	2.1	1.7	77.2
	SH	6.0	5.1	2.7	1.9	
1956	FH	6.5	6.2	3.1	2.3	77.2
	SH	7.5	7.1	3.8	2.6	
1957	FH	6.0	6.3	2.9	2.3	82.8
	SH	4.1	4.8	2.1	1.8	
1958	FH	2.7	4.1	1.4	1.6	80.6
	SH	2.2	4.2	1.1	1.9	
1959	FH	3.3	5.9	1.7	2.5	81.5
	SH	6.4	7.5	3.3	3.2	
1960	FH	6.4	6.9	3.2	3.0	84.3
	SH	5.3	6.3	2.6	2.7	

TABLE 13.A.4. Estimated profit
(¥bn.)

	X-sector						*Z*-sector				
	Nominal GDP	Total pay-roll	Nominal profit	Agricultural price index (*S*-sector prices) (1960 = 100)	Profit in terms of *S*-sector prices	Fixed capital stock	Nominal GDP	Total pay-roll	Nominal profit	Profit in terms of *S*-sector prices	Fixed capital stock
1953	561	324	237	98.2	241.3	1,291	1,063	443	620	631.4	2,042
1954	648	348	300	98.1	305.8	1,400	1,126	488	638	650.4	2,166
1955	690	359	331	94.2	351.4	1,470	1,171	526	645	684.7	2,269
1956	772	371	401	91.4	438.7	1,589	1,638	673	965	1055.8	2,448
1957	839	385	454	96.0	472.9	1,705	2,006	775	1,231	1282.3	2,843
1958	810	412	398	93.3	426.6	1,853	1,907	876	1,031	1105.0	3,250
1959	927	436	491	94.3	520.7	1,983	2,530	988	1,542	1635.2	3,695
1960	1,080	482	598	100.0	598.0	2,149	3,423	1,175	2,248	2248.0	4,596
1961	1,280	577	703	111.6	629.9	2,313	4,352	1,413	2,939	2633.5	5,568
1962	1,434	692	742	120.1	617.8	2,549	4,738	1,625	3,113	2592.0	6,699
1963	1,701	790	911	127.7	713.4	2,777	5,425	1,899	3,526	2761.2	7,703
1964	1,789	883	906	134.7	672.6	3,030	6,379	2,184	4,195	3114.3	8,882
1965	1,902	994	908	145.8	622.8	3,243	6,663	2,433	4,230	2901.2	9,858

Note: *X*-sector: Food processing, textiles; *Z*-sector: Paper and pulp, chemicals, primary metals, machinery, electrical machinery, transport machinery.

Source: data based on Keizai Kikakucho (1969).

TABLE 13.A.5. Estimated profit rate
(¥bn.)

	Nominal profit		Fixed capital		Nominal profit (%)		Agricultural price index	Profit rate (%)		Differential of profit	Increase in fixed capital	
	X-sector	*Z*-sector	K_x	K_z	R_x'	R_z'	(1960 = 100)	R_x	R_z	$R_x' - R_z'$	dK_x	dK_z
1953	237	620	1,291	2,042	18.4	30.4	98.2	18.7	31.0	–12.0		
1954	300	638	1,400	2,166	21.4	29.5	98.1	21.8	30.1	–8.1	109	124
1955	331	645	1,470	2,269	22.5	28.4	94.2	23.9	30.1	–5.9	70	103
1956	401	965	1,589	2,448	25.2	39.4	91.4	27.6	43.1	–14.2	119	179
1957	454	1,231	1,705	2,843	26.6	43.3	96.0	27.7	45.1	–16.7	116	395
1958	398	1,031	1,853	3,250	21.5	31.7	93.3	23.0	34.0	–10.2	148	407
1959	491	1,542	1,983	3,695	24.8	41.7	94.3	26.3	44.2	–16.9	130	445
1960	598	2,248	2,149	4,596	27.8	48.9	100.0	27.8	48.9	–21.1	166	901
1961	703	2,939	2,313	5,568	30.4	52.8	111.6	27.2	47.3	–22.4	164	972
1962	742	3,113	2,549	6,699	29.1	46.5	120.1	24.2	38.7	–17.4	236	1,131
1963	911	3,526	2,777	7,703	32.8	45.8	127.7	25.7	35.9	–13.0	228	1,004
1964	906	4,195	3,030	8,882	29.9	47.2	134.7	22.2	35.0	–17.3	253	1,179
1965	908	4,230	3,243	9,858	28.0	42.9	145.8	19.2	29.4	–14.9	213	976

Source: data based on Kezai Kikakucho (1969).

14

Critical Efforts behind Japan's Post-War Economic Take-off

14.1. Foreign Capital and Aid

14.1.1. The balance of payments in the immediate post-war period

After the end of the war, Japan's foreign trade was put under the control of the occupation force, and private trade was restarted, albeit with many restrictions, in June 1947. Production was at extremely low levels, both in the manufacturing and agricultural sectors, and to meet the level of local demand, swelled by the enormous numbers of soldiers and overseas residents who repatriated after the war, Japan had to import large quantities of goods, both for consumption and production.

Table 14.1 shows Japan's balance of payments from 1946 to 1960. From 1946 till 1949, deficits were large—in fact the annual deficit in the current overseas account reached some \$400–\$500 million. As Japan's exports, at FOB, remained at only several hundred million dollars in those years, imports had to be financed largely by capital imports and foreign aid. In 1947, for instance, aid to Japan stood at \$400 million, whereas the current account deficit was \$360 million. This situation lasted until 1951 when Japan regained political independence and aid was terminated.

The outbreak of the Korean War in June 1950 stimulated Japanese exports both directly and indirectly, which decreased the deficit on the external current account. With the termination of foreign aid in 1951, however, Japan had to maintain its balance of trade through capital imports by government financial institutions. As the government was cautious about foreign domination of industry, it aimed to promote capital imports without management participation. This can be seen from cols. 6 and 7 of Table 14.1. As already explained, by the end of the 1950s Japan's *Z*-sector had taken off, thanks to vigorous investment in fixed assets from the mid-1950s on. A surplus was achieved on the trade balance, and even the current account became balanced by the early 1960s. By the late 1960s, Japan had even become a net capital exporter.

Since our concern is with the critical efforts needed to recover from the war damage, however, we concentrate below on the role of capital imports and foreign aid in fostering the development of heavy industry. More specifically, we focus on US aid from 1945 to 1951 and Japanese capital imports thereafter, as well as on the effects of the Korean War.

TABLE 14.1. Balance of payments ($USm.)

	Balance of goods trade	Balance of investment income	Balance of goods and services	Balance of grants	Balance of private long-term capital (net)	Balance of long-term capital government and financial institutions (net)	Balance of short-term capital government financial institutions (net)
1946	–238.0	—	–272.6	194.5	0.0	–22.4	–59.8
1947	–267.4	–3.4	–358.4	404.8	0.0	–5.5	63.7
1948	–284.3	2.2	–386.8	461.6	0.0	–3.6	102.7
1949	–194.8	–1.3	–306.4	513.3	0.0	17.5	175.8
1950	34	–6	47	429	–4	98	430
1951	–292	–5	159	171	–24	–4	365
1952	–413	–5	190	34	–16	77	198
1953	–792	–23	–226	21	–19	157	–381
1954	–429	–39	–80	29	–23	–27	–1
1955	–54	–41	205	22	–9	7	284
1956	–131	–39	–59	25	–48	–26	0
1957	–402	–49	–584	–36	–8	–534	
1958	369	–40	468	–204	–41	337	
1959	361	–37	393	–31	20	339	
1960	268	–33	168	–25	83	77	

Source: Nippon Ginko (various years).

14.1.2. Aid and war-related demand

US aid to Japan immediately after the war consisted of two elements: a loan to cover the cost of importing raw cotton, and aid provided through the GARIOA and EROA funds. GARIOA aid, which aimed at stabilizing US-occupied countries by supplying food, energy, medicines, fertilizers, and so on.[1] This fund operated in US fiscal years 1947/51.

The commodities supplied under EROA funding were mostly industrial raw materials such as cotton and minerals, and the machinery needed for economic reconstruction. The value of aid from these sources, according to US data, is shown in Table 14.2.

The relative importance of this aid is clear from Table 14.3: Japan's imports from 1945 through December 1946 totalled $304 million, of which 63.2% was financed by aid. Dependence on aid was even higher during 1947–8, and still accounted for 37% of imports even in 1950, just before independence.

Data on the scale of aid-funded imports became available only after April 1949, when a special account was established for managing the funds released by sales of aid in kind. Table 14.4 summarizes these aid-funded imports by commodities from FY 1949 on. It is clear from this table that aid was used

TABLE 14.2. American aid to Japan in the budget (¥m.)

US fiscal year	GARIOA	EROA	Administrative cost	Total
1946	93	—	—	93
1947	287	—	13	300
1948	351	—	20	371
1949	426	97	26	549
1950	237	188	19	444
1951	183	—	15	198
Cumulative total	1,577	285	93	1,955

Note: Original data are based on the SCAP documents. The values of aid are based on US data.

Source: Keizai Kikakucho (1959–63).

TABLE 14.3. Comparison of imports based on aid with commercial imports

	Imports based on aid		Commercial imports		Total	
	Value ($m)	%	Value ($m)	%	Value ($m)	%
1945 (Sept.–1946)	192	63.2	112	36.8	304	100.0
1947	404	77.2	119	22.7	523	100.0
1948	461	67.4	223	32.6	684	100.0
1949	519	58.4	370	41.6	889	100.0
1950	357	36.9	613	63.1	970	100.0
1951	151	6.9	2,037	93.1	2,188	100.0
Cumulative total	2,084	37.5	3,474	62.5	5,558	100.0

Note: Survey by Ministry of International Trade and Industry.

Source: Keizai Kikakucho (1959–63).

principally to guarantee the supply of food and to purchase cotton to encourage the textile industry; in fact, US aid from 1945 to 1951 was vital in preventing starvation and financing recovery from war damage. Without it Japan could not have supplied even enough rice to feed its population, let alone the industrial materials needed to rebuild the shattered economy.

A meaningful question in the context of contemporary LDCs is how Japan achieved take-off within just a few years of the devastation of all its production facilities. The most important factor seems to have been that Japan had already accumulated the human resources and institutions for industrialization, which most of the countries trying to develop in the 1950s–60s lacked. In other words, Japan had a high absorptive capacity for foreign capital.[2] A

TABLE 14.4(*a*). Japanese imports with GARIOA by US Aid (after 1949 fiscal year)

Commodities	Value($US'000)	Percentage
Main foodstuffs	352,143	41
Foods	26,588	3
Oils and fats	63,480	7
Raw cotton	241,598	28
Jute	15,217	2
Fertilizers and feeding stuffs	53,126	7
Mineral oil	72,517	8
Other industrial raw materials	44,985	5
TOTAL	869,654	100

Note: Survey by Ministry of International Trade and industry.

Source: Keizai Kikakucho (1959–63).

TABLE 14.4(*b*). Imports based on US aid to Japan ($USm.)

Fiscal year	Aid imports	Military reliefs	Total
1945–46	139	54	193
1947	360	9	404
1948	458	3	461
1949	516	3	519
1950	355	2	357
1951	150	1	151
Cumulative total	2,014	72	2,086

Note: Data through March 1949 are based on ESS statistics, data after April are based on the survey by Ministry of International Trade and Industry.

Source: Keizai Kikakucho (1959–63).

second factor, though it also required a high production capacity, was the demand generated by the Korean War.

Most studies of the Korean War–related demand do not distinguish between demand directly generated by the war and demand for goods and services for the US forces stationed in Japan after Japan's independence in 1951; sometimes, even the demand for goods for the reconstruction of Korea after the Korean War is included in this category. In this sense it is difficult to isolate the true demand effect of the Korean War on the Japanese economy.

The Japanese government's own estimate of the Korean War-related demand is shown in Table 14.5. According to this, the average monthly value of sales contracts related to the Korean War from July through December 1950, half a year after the beginning of the war, was \$32 million. As Japan's total export in 1950 was \$829 million FOB, the war-related demand accounted for 46% of total exports on an annual base ((32 × 12) ÷ (829 × 100) = 46.3%). In 1951, war-related demand totalled \$367 million or 27% of total exports (\$1,354 million). (See Tables 14.5, 14.6, and 14.7.)

TABLE 14.5. Sales contracts of war-related demands (\$USm.)

	1950	1951		1952	1950–1952	
	July–Dec. monthly average	Jan.–June monthly average	July–Dec. monthly average	Jan.–May monthly average	Cumulative total	Percentage (%)
Total	31.9	22.9	36.8	16.0	634.1	100.0
Goods	21.2	17.1	26.0	13.6	457.1	72.1
Services	10.7	5.8	10.8	2.4	177.0	27.9

Source: Keizai Kikakucho (1959–63).

TABLE 14.6. Sales contracts of war-related demands (\$USm.)

	1950	1951		
		First half	Second half	Total
1. Goods	120.2	25.5	156.1	269.1
Textiles	19.9	44.6	31.7	76.3
Machinery	34.4	21.8	26.3	48.1
Metals and their products	31.8	22.5	36.4	58.9
Paper	15.8	6.4	14.8	21.2
Chemicals	5.8	5.8	14.5	20.3
Non-ferrous metals etc.	12.3	14.4	32.4	46.8
2. Services	62.3	23.9	65.3	97.6
3. TOTAL	182.5	149.4	221.4	366.7

Source: Keizai Kikakucho (1959–63).

During the years 1945 to 1952 Japan managed to maintain the balance of payments equilibrium, depending on foreign aid. Therefore war-related demand played a significant role in the take-off of heavy industries. In the meantime, Japan accepted a fixed exchange rate of \$1.00 = ¥360 in 1949, which was an overvaluation of the yen for many tradable goods. Thus, the Japanese economy was in a serious recession due to the tight anti-inflationary policy

TABLE 14.7. Balance of payments in pre- and post-war
($USm.)

	1936	1947	1948	1949	1950	1951
Current account						
Exports (f.o.b)	1,034.9	181.6	262.3	533.3	829.4	1,354.0
Imports (f.o.b)	–1,049.3	–449.0	–546.6	–728.1	–885.9	–1,641.0
Net exports	(–14.4)	(–267.4)	(–284.3)	(–194.8)	(–56.6)	(–287.1)
Freight and insurance	68.4	–88.0	–120.2	–164.0	–90.5	–233.8
Investment income	50.8	–3.4	–2.2	–1.3	–5.6	–5.3
Others	–34.3	–1.4	17.3	50.8	195.7	673.9
Total current account (1)	70.5	–358.4	–389.4	–309.3	43.1	147.7
Private grants and investments	—	—	—	—	—	—
Private grants	36.9	0.4	0.6	—	69.2	15.8
Independent investments	–93.8	—	—	–35.9	4.8	24.4
Total (2)	–56.9	0.4	0.6	–35.9	74.0	40.2
Errors and omissions (3)	–6.7	13.6	26.9	–10.9	49.7	9.9
Above the line (1) + (2) + (3)	+6.9	–336.2	–361.9	–356.1	+166.8	+197.8
Compensation balance						
US aid to Japan	—	404.4	461.0	531.9	360.3	159.9
Net foreign loan	—	–17.6	–58.1	19.2	–98.2	–2.5
Increase (–), Decrease (+)	15.7	–40.6	–41.0	–195.0	–428.9	–355.2
Gold, outflow (+), inflow (–)	–22.6	—	—	—	—	—
Total compensation balance	–6.9	+336.2	+361.9	+356.1	–166.8	–197.0

Notes: 'Others' in current account contains war-related demand and consumption by the US forces.
Net foreign loan covers transaction of net long-term capital.
The values in this table are preliminary because of occupation by US.

Source: Keizai Kikakucho (1959–63).

directed by a US adviser. An additional export demand equivalent to 30%–40% of total exports must have been a significant encouragement to Japanese industries that would otherwise not have been competitive in the international market.

14.1.3. Capital imports

The basic laws regulating capital imports in post-war Japan were the Law of Management of Foreign Exchange and Foreign Trade (Gaitameho) enacted in 1949 and the Law Concerning Foreign Investment (Gaishiho) enacted in May 1950.

The Gaishiho law described the basic conditions under which capital imports were allowed, the regulations governing the repatriation of profit and interest, and the protection of foreign capital. It was important because the government wanted to promote capital import so as to compensate for the deficit on the current external account. The Gaitameho law regulated all foreign exchange transactions not covered by the former. More generally, Gaishiho regulated transactions of capital assets, contracts, and/or payment periods exceeding one year, while Gaitameho covered all short-run current transactions.[3]

Capital imports were not encouraged indiscriminately. Projects encouraged under the Gaishiho were those which (*a*) contributed to the balance of payments; (*b*) promoted important industries, and especially public utilities; or (*c*) facilitated technology imports. Projects in certain categories were prohibited—those which (*a*) seriously damaged competing domestic industries; or (*b*) endangered Japanese control of management. It is well known that the principle of majority Japanese ownership in joint ventures was maintained for many years.

Until 1950, foreign investors were rather concerned about the prospects of the Japanese economy, and in particular about the chronic balance of payments difficulties and the business performance of Japanese enterprises. In consequence, in April 1951 the government revised the Gaishiho to somewhat relax its control over capital imports, and similar measures were made again in 1952 and 1956. Liberalization was only made a long-term goal in June 1960, when the government announced its Grand Design of Liberalization of Trade and Foreign Exchange (Boeki Kawase Jiyuka Keikaku Taiko), but in fact many reservations about liberalization still remained.

Government efforts to promote imports of capital and technology were tempered somewhat by its grave concerns about foreign control on management and the displacement of local firms due to pressure from foreign subsidiaries. Since the government therefore intervened in the import of capital and technology, the real extent to which they actually promoted capital imports is unclear.

Table 14.8 shows the evolution of the import of capital and technology

TABLE 14.8. Evolution of import of capital and technology in Japan ($US'000)

	Number of the issue	Type of foreign capital import					
		Stock	Bonds	Securities	Loan and investment	Foreign bonds	Total
1950	27	3,150					3,150
1951	101	13,326			4,026		17,352
1952	133	10,123	25	146	34,457		44,751
1953	103	5,002		562	49,362		54,926
1954	82	3,970		58	15,279		19,307
1955	72	5,101	7	52	47,054		52,214
1956	144	9,520	15	115	93,652		103,302
1957	118	11,490		128	123,979		135,597
1958	90	11,350	28	116	231,473	30,000	272,967
1959	153	27,031	30	214	127,615		154,890
1960	327	74,151	20	555	127,132	9,800	211,658

Source: Gaishi Donyu Nenkan Hensyu Iinkai (1962–7).

from 1950 to 1960 on the basis of authorizations granted. Obviously, these figures do not correspond to the balance of payments figures, but the trends are nevertheless apparent. (Balance of payment tables based on IMF formats only became available in Japan after 1960.) However, it is clear from Table 14.1 that imports of long-term capital between 1945 and 1955 were never significant. Although foreign capital was badly needed around 1951–5, for the take-off of heavy industries, the Japanese economy does not seem to have been so attractive to foreign investors. One can assume that many foreign enterprises wanting to invest in Japan would have been deterred by the restrictions on foreign management controls.

How important were capital imports in total domestic investment in 1955? Capital imports in this year (approved amounts, not real flows), totalled $52 million, or only 1.9% of total private investment in production equipment. In 1960, capital import expanded to $212 million but this was still only 2.4% of total private investment in production equipment ($8,807 million). In other words, the contribution of foreign capital to Japan's capital accumulation from 1945 to 1960 was rather small.

Despite the relatively small contribution of capital imports to total private investment in productive equipment, we are interested in knowing to which sectors foreign capital was principally directed. Table 14.9 gives a partial answer to this question, showing cumulative foreign loans by industries for the period April 1949–September 1960: the *Z*-sector absorbed $476 million, or 45.6% of total foreign loans, whereas the *X*-sector accounted for only $1.7 million. Other industrial sectors which we could not classify received $566 million, or 54.2% of the total.

TABLE 14.9. Foreign capital (loan) import by industry (cumulative total from 1949.4 to 1960.9)

		Value ($000)	Percentage
X-sector	Subtotal	1,711	0.2
	Spinning	1,431	0.1
	Food processing	280	0.0
Z-sector	Subtotal	476,038	45.6
	Machinery	68,139	6.5
	Metals	243,159	23.3
	Chemicals	36,458	3.5
	Petroleum refining	125,782	12.1
	Paper and pulp	100	0.0
	Ceramics	2,400	0.2
Others	Subtotal	565,611	54.2
	Mining	5,447	0.5
	Electric and gas supply	267,269	25.6
	Construction	70,300	6.7
	Transport and communications	213,771	20.5
	Commerce, storing, finance	8,824	0.8
TOTAL		1,043,360	100.0

Source: Gaishi Donyu Nenkan Hensu Iinkai (1962–7).

The gross capital stock of the *X*- and *Z*-sectors in 1960 was ¥2,149 billion and ¥4,596 billion, respectively, equivalent to $5,969 million and $12,766 million. Cumulative foreign lending thus accounted for only 0.1% of gross capital stock in the *X*-sector, 3.7% in the *Z*-sector. If we were to include the utilities in the *Z*-sector, this ratio might have been much higher. In fact, these sectors—electricity, gas, transportation, and communication—were those which borrowed most heavily from abroad, with electricity and gas taking 25.6% of total foreign loans and transportation and communication 20.5%.

More generally, we would argue that technology imports, which were often tied up with capital imports, played a more important role than the funds themselves. Many empirical studies on technical progress in the post-war Japan support this argument (see e.g. Watanabe and Egaitsu, 1967).

14.2. Industrial Policies and the Role of Social Overhead Capital

As discussed in Chapter 13, the process of capital accumulation stabilized after 1953. As the war damage was so enormous, however, the level of economic activity for several years after the war dropped to that of 1935–6 or lower, and the peak of the pre-war level was not attained until 1955. There were thus some critical years before 1953 when the economy was not yet on the threshold of take-off and was still in danger of falling into the poverty trap.

Our main concerns are, first, how far the economy had dropped below the minimum critical effort curve because of the war, and whether it was located below the $\dot{K}_z = 0$ curve; and, secondly, how productive facilities were actually reconstructed. As it seems that the point (K_X, K_z) was already on the path leading to the autonomous expansion of the *Z*-sector by the mid-1950s, the critical period for which we must analyse government policies towards resource allocation is the late 1940s and early 1950s.

The remainder of Section 14.2 will deal with the government's industrial policies—more specifically, subsidies to industries and the supply of funds for fixed investment; tax policies; and the construction of social overhead capital as a factor contributing to industrial efficiency.

Section 14.2.1 reviews the destruction of productive facilities and tries to estimate the war damage in Japan's *X*- and *Z*-sectors. Section 14.2.2 traces Japanese capital accumulation for the decade from 1945. (The path of capital accumulation after 1955 was considered in Chapter 13.) In the following three sections, from 14.2.3 through 14.2.5, we examine subsidies and the supply of investment funds, tax policies, and the construction of social overhead capital.

The major findings are that the *Z*-sector showed an occasional tendency to shrink in the early years after 1945, and that in consequence the government intervened heavily to encourage investment. Thus, the roles of the post-war Japanese government in developing modern industries was as important as that of the Meiji government had been.

14.2.1. War damage in the *X*- and *Z*-sectors

An estimate of war damage was published by the government (see Keizai Antei Honbu 1949), while estimates of national wealth for the years 1935 and 1945 are also available (Nakayama 1959, 1961: 64). Taken together the figures show that war damage amounted to some 25% of Japan's national wealth at its pre-war highest.

The rates of asset damage, defined as the value of direct and indirect damage relative to the value of assets under a no-damage assumption,[4] by type of asset were as follows: 80%, the highest, for ships; 34% for machines and equipment for industry; 25% for construction; 11% for facilities for electricity and gas supply; 15% for communication facilities; 7.5% for harbours and canals; 3.5% for bridges; 7% for railways; and 17% for the water supply system. Where the damage exceeded the average rate, which was 25%, was thus in ships and machines and equipment for industry. When all infrastructure facilities are considered together, however, it is clear that the implications for industry were even more serious than the raw data suggest.

Table 14.10, which examines war damaged buildings and machines by

TABLE 14.10. War damage to industry by kind of stock (¥m.)

	Damage to private factory buildings			Direct damage to machines and equipment in private industries		
Z-sector	Metal	231		Metal	761	
	Machinery	764		Machinery	1736	
	Chemicals	194		Chemicals	617	
	Ceramics and stone	30		Ceramics and stone	71	
	Electric and gas supply	7				
	Total	1,226	(67.4%)		3,185	(71.4%)
X-sector	Mining	96		Textile	356	
	Textile	159		Printing and publishing	94	
	Printing and publishing	25		Food processing	153	
	Food processing	70		Wood and wood products	108	
	Wood and wood products	28		Construction	95	
	Construction	48		Others	472	
	Others	166				
	Total	592	(32.6%)		1,278	(28.6%)
TOTAL		1,818			4,463	

Source: Keizai Antei Honbu (1949): 65, 71.

industry, indicates that damage in the *Z*-sector was much higher than in the *X*-sector, accounting for some 70% of the total. Thus, while overall capital stock in 1945 was roughly comparable to that in 1935, the *Z*-sector was far worse off than in 1935 because it had suffered much greater war damage.

14.2.2. Capital accumulation a decade after the war

While the point (K_X, K_Z) was located above the $\dot{K}_z = 0$ curve, i.e. K_Z was growing in a self-sustaining way in the 1930s, the war pushed the point of capital stock in a generally south-westerly direction, though the actual direction was more south than west. It is likely that the war pushed the point below the $\dot{K}_z = 0$ curve and further below the minimum critical effort curve.

Given that despite the very severe war damage, the economy had by 1955 recovered to the level of 1935, it would be most interesting to shed light on the process of capital allocation and accumulation for the entire period between 1945 and 1955, but due to data constraints we can examine only the period after 1949. Fig. 14.1 plots the point (K_X, K_Z) as derived from the Survey of National Wealth of 1955, which was the first survey since 1945. Capital stock here refers to net capital stock after depreciation, and thus corresponds directly to K_X and K_Z in our model. K_z in the left-hand panel does not include public utilities but in the right-hand panel it does. The figure shows that K_z declined absolutely for four or five years from 1949, though K_x steadily expanded during the same period. This is clearest in the left-hand panel, namely

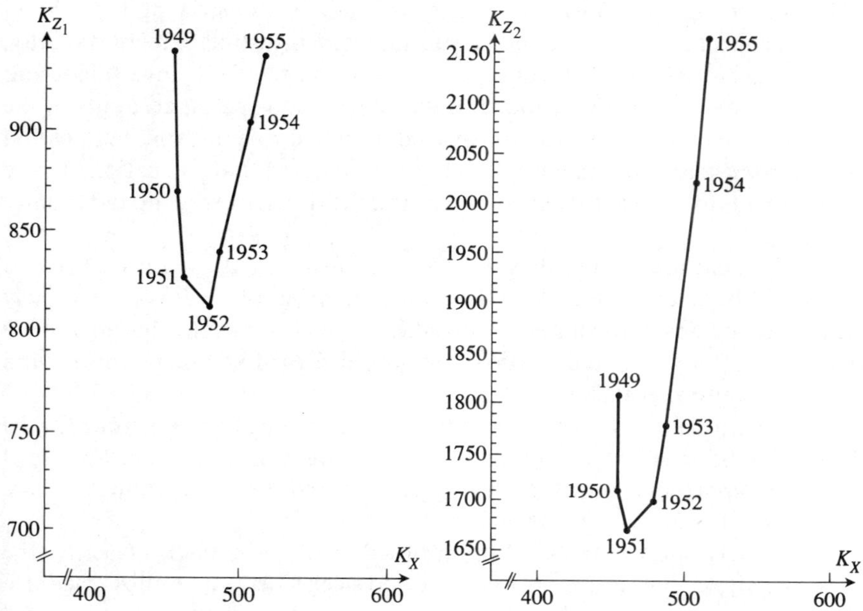

Source: Nakayama (1959).

FIG. 14.1 Capital accumulation before 1955 (¥bn. at 1955 constant prices).

where K_z does not include public utilities. This suggests that the decline in K_z was not a reflection of business cycles but more a structural mechanism of movement to a lower equilibrium. Since K_x expanded in the same period, the absolute decline in K_z in the above years indicates that capital was somehow reallocated from the Z- to the X-sector: over the years 1949–52, the Z-sector could not even raise enough net positive investment to cover capital depreciation. The point (K_x, K_z) began to move north, indicating net positive investment in the Z-sector, only around 1952. Recovery to the level of 1949 was achieved sometime between 1953 and 1955.

From the above, we can conclude that in the period 1949–51 the points (K_x, K_z) were below the $\dot{K}_z = 0$ curve, and seemed set on a path where the Z-sector would be forced to vanish,[5] while the X-sector expanded. After 1951–2, the point moved north-east, indicating expansion in both sectors. Through vigorous investment thereafter, total capital assets recovered to the pre-war peak sometime between 1953 and 1955.

14.2.3. Subsidies and supply of investment funds

We now investigate the government's industrial policies during the critical period between the end of the war and 1955, by which time Japan was no longer in danger of falling into the poverty trap. As there are a variety of policies that could be considered under the heading 'industrial policies', however, we concentrate on government subsidies and investment. Institutionally, this was achieved largely through a new account, the fiscal investment and finance account, which was added to the general and special accounts of the pre-war period. This new account played a critical role in capital allocation among industries. We therefore start by tracing capital allocation by the government through fiscal expenditures and the fiscal investment and finance account.

Government use of subsidies to encourage investment by private firms in the period between 1946 and 1960 is shown in Table 14.11. It is immediately clear that the absolute values of subsidies were larger in the period 1946–9 than in the later years, so the ratio of the subsidies to GNP and private capital formation were also higher.

Subsidies in the period 1946–55 were mostly provided to compensate for the differential between production costs and the government-controlled prices of specified important products such as steel, coal, and chemical products. These subsidies were appropriated from the general account.

Table 14.12 shows the distribution of subsidies by industry. Clearly, the most heavily subsidized sectors were coal and steel. Chemical fertilizer was also classified as a priority good because increasing food production was of crucial importance. As the poor state of Japan's agriculture made food imports inevitable, the government subsidized the imported grains to hold prices stable.

TABLE 14.11. Evolution of government subsidies in post-war period (¥bn.)

	(1) GNP	(2) Gross domestic capital formation	(3) Private capital formation	(4) Subsidies	(4)/(1) %	(4)/(3) %
1946	474.0	105.4	75.7	26.4	5.6	34.9
1947	1,308.7	345.2	199.1	36.7	2.8	18.4
1948	2,666.1	752.2	494.2	109.8	4.1	22.2
1949	3,375.2	830.6	532.0	211.1	6.3	39.7
1950	3,946.7	1,007.3	818.1	68.0	1.7	8.3
1951	5,444.2	1,663.8	1,252.1	36.5	0.7	2.9
1952	6,118.0	1,678.0	1,204.0	50.7	0.8	4.2
1953	7,084.8	1,978.0	1,335.2	46.2	0.7	3.5
1954	7,465.7	1,749.5	1,156.9	15.2	0.2	1.3
1955	8,235.5	2,086.6	1,371.6	7.5	0.1	0.5
1956	9,292.9	2,955.2	2,263.5	7.4	0.1	0.3
1957	10,149.8	3,210.2	2,401.6	36.3	0.4	1.5
1958	10,394.7	2,831.7	1,903.7	4.8	0.0	0.3
1959	12,577.8	4,404.7	3,289.4	7.2	0.1	0.2
1960	14,678.9	5,607.2	4,267.1	34.9	0.2	0.8

Source: Nippon Ginko (1966).

TABLE 14.12. Subsidies filling the gap between production costs and selling prices
(¥bn.)

Specified products		Actual expenditure of subsidies (1946–9 fiscal year total)
Coal		59.2
Iron and steel		76.0
Non-ferrous metals		5.3
Fertilizers		57.5
Sodium hydroxide		4.0
Foodstuffs	Domestic	12.5
	Imported	45.3
Others		12.5
TOTAL		272.4

Source: Keizai Kikakucho (1959–63).

Basically, however, the subsidies were addressed to the heavy industries: from 1946 to 1949 ¥2,724 million out of ¥3,840 million, or some 70% of the total. This clearly stimulated investment in the *Z*-sector.

Another important channel of investment was the fiscal investment and finance programme which started anew in the post-war period. The financial sources for this fund comprised postal savings by households and firms, bonds of governmental financial institutions, and transfers from the general and special accounts of the budget. These resources were used by government enterprises and lent through government financial institutions to private enterprises. Funds from the sales of US aid goods, allocated through the special account for industrial investment (*sangyo toshi tokubetsu kaikei*), also played a major role in capital allocation in the earlier years of economic restoration. This aid-based fund, existed from 1949 till 1953 and was mostly used to support investment in the *Z*-sector.

Before this fund was established, it was the Bank for Restoration that was responsible for industrial investment. This government institution borrowed partly from the government budget and partly from the private sectors by issuing bonds, and lent the capital thus raised to industrial enterprises. As most of the bonds issued were purchased by the Bank of Japan, this was tantamount to the government printing money and caused drastic inflation. According to Table 14.13, 85% of all loans to industry between 1949 and 1950 were Restoration Bank loans. By sector, the major recipients were the steel, coal, fertilizer, electricity, and shipping industries. However, this might be somewhat of an exaggeration since some of the funds lent by commercial banks under the title of current funds were in fact used to finance fixed investment and were not repaid for many years. Thus, commercial banks supported fixed investment by productive enterprises much more than Table 14.13 suggests. None

TABLE 14.13. Size of the restoration finance loan and total loan at end of March 1949 (¥m.)

	Total loan (A)	Restoration finance loan (B)	(B)/(A)%
Coal			
Fixed capital expenditure	33,877	33,424	98.1
Current expenditure	33,373	13,903	41.7
Total	67,251	47,328	70.4
Iron and Steel			
Fixed capital expenditure	2,821	2,071	73.4
Current expenditure	19,110	1,608	8.4
Total	21,931	3,680	16.8
Fertilizers			
Fixed capital expenditure	7,113	4,555	64.0
Current expenditure	9,030	1,654	17.3
Total	16,143	6,119	37.9
Electricity supply			
Fixed capital expenditure	20,580	19,125	92.9
Current expenditure	4,842	3,270	67.5
Total	25,422	22,395	88.1
Sea transport			
Fixed capital expenditure	15,569	13,079	84.0
Current expenditure	5,009	156	3.1
Total	20,578	13,235	64.3
Textiles			
Fixed capital expenditure	11,088	4,975	44.9
Current expenditure	58,778	20	0.0
Total	69,866	4,995	7.1
Above total			
Fixed capital expenditure	91,048	77,299	84.5
Current expenditure	130,142	20,521	16.1
Total	221,191	97,750	44.0
Total loan			
Fixed capital expenditure	127,380	94,341	74.1
Current expenditure	438,738	37,623	8.6
TOTAL	566,118	131,965	23.3

Source: Keizai Kikakucho (1959–63).

the less, it is clear that the Bank for Restoration played a critical role in industrial investment in the period immediately after the war, accounting for 44% of all loans to industries by all financial institutions.

We now consider where (K_x, K_z) would have been located without such government support to fixed investment in the *X*- and *Z*-sectors. Fig. 14.2 shows what happens when the new investment funds supplied by the government and government financial institutions (see Table 14.14) are subtracted from K_x and K_z (depicted in Fig. 14.1). In other words, it indicates the capital accumulation path in the absence of government intervention in capital allocation.[6]

The dotted lines indicate the path of capital accumulation under the assumption of no government support. They indicate that the effect of government loans in stimulating investment was much larger in the *Z*-sector than in the *X*-sector. Thus, the figure suggests that, without government intervention, Z-sector capital stock after 1949 would have been much smaller, and that there would have been no upturn of the point (K_x, K_z) until 1953. As our estimates

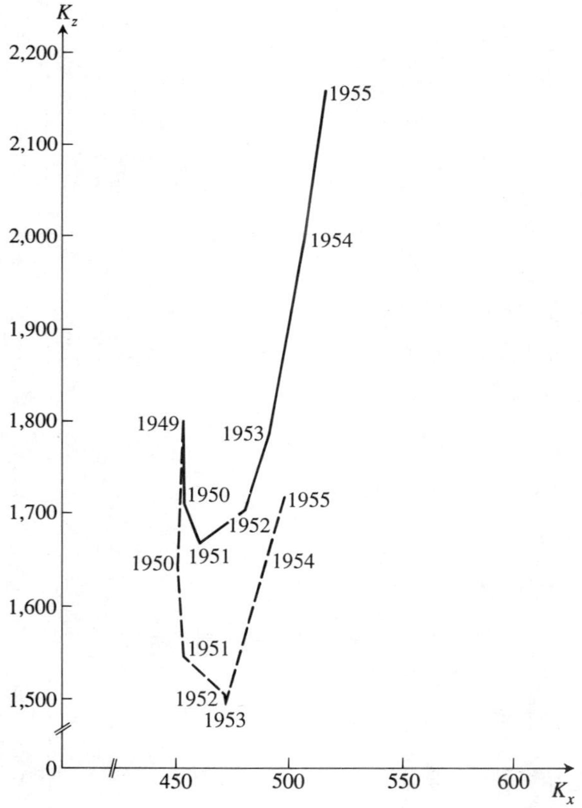

——— Actual path of capital accumulation;
------- Estimated path of capital accumulation without government support;
Z-sector includes public utilities.

FIG. 14.2 Effects of government loan on capital accumulation (¥bn. at 1955 constant prices).

TABLE 14.14. New supply of investment funds by the government and the governmental financial institutions (¥bn.)

	Z-sector						*X*-sector		
	excluding marine transport and electricity supply			including marine transport and electricity supply					
	(1) New supply of investment funds Total	(2) Government	(3) (2)/(1) (%)	(4) Total	(5) Government	(6) (5)/(4) (%)	(7) New supply of investment funds Total	(8) Government	(9) (8)/(7) (%)
1950	26.3	0.9	3.4	73.4	23.8	32.4	43.2	0.7	1.6
1951	57.4	5.4	9.4	144.7	55.1	38.1	59.4	0.8	1.3
1952	74.8	9.3	12.4	241.7	68.6	28.4	23.5	2.0	8.5
1953	82.1	8.0	9.7	274.7	90.6	33.0	41.1	4.5	10.9
1954	72.3	6.6	9.1	230.4	83.5	36.2	18.1	3.2	17.7
1955	69.5	16.4	23.6	234.1	85.6	36.6	15.1	3.0	19.9
1956	131.7	10.1	7.7	393.5	80.2	20.4	51.3	4.5	8.8
1957	267.1	21.0	7.9	605.9	118.8	19.6	44.3	6.4	14.4
1958	234.8	20.4	8.7	578.6	119.4	20.6	37.2	5.2	14.0
1959	346.8	27.5	7.9	660.6	116.9	17.7	46.7	8.3	17.8
1960	624.6	36.3	5.8	986.6	121.8	12.3	73.5	9.2	12.5

Notes: Z-sector is iron and steel, machinery, and chemicals: *X*-sector is textiles.

Sources: Nakayama (1961); Nippon Ginko (various years).

of the effects of government support are rather low because we neglect the indirect effect that capital was attracted to the Z-sector by the improved prospects of profits, K_z was probably even lower than depicted in Fig. 14.2.

In other words, (K_x, K_z) probably located further south, and perhaps below the minimum critical effort curve. That is, if investment had been left to market forces only, the Z-sector might have vanished. Or, to put it in another way, it was government intervention through subsidies and loans that pushed (K_x, K_z) across the minimum critical effort curve.

14.2.4. Tax policies

Producers of Z-sector goods were further assisted by preferential taxation. Corporate income taxes were reduced through favourable treatment of depreciation and other factors, and the government also encouraged fixed investment in important industries by tax exemptions and/or by the alleviation of commodity taxes. Thus, one channel of encouragement was to improve the profit prospects of the Z-sector and the other was to stimulate demand for Z-sector goods.

Policy instruments applied in the industrial sectors in the early post-war period included a special allowance for capital depreciation, tax exemptions for specified goods, and exemption of import tariffs on specified machines (see MITI 1957*a*).

The special depreciation allowance consisted of the following measures:

1. Depreciation of capital equipment at a rate 50% above the normal rate was allowed for three years from purchase.
2. Tax-free depreciation 50% of the value of the newly purchased equipment.
3. Special allowances for depreciation of equipment for research and development.
4. Special allowances for depreciation of equipment for use by co-operatives.

As these allowances enabled firms to pay less tax if they invested in equipment, they encouraged the renewal of capital equipment and encouraged the retention of earnings for future investment. As for the actual sizes of allowances: item 1 was applied to investment totalling ¥96 bn. over the period from September 1954 to fiscal year 1955. Item 2 was applied to investment totalling ¥110.8 bn. for the period 1952–7. Item 3 was applied to a total of ¥9.15 million.

The industries that benefited from item 2 were largely Z-sector industries, centring around the production of steel, petroleum, machinery, and cement (Table 14.15). The largest single beneficiary of item 1 was the textile industry, accounting for some 30% of the total. Obviously, such treatment encouraged firms to renew their production capacity by purchasing those machines which were eligible, and thereby stimulated demand for these Z-sector products.[7]

The tax exemptions for specified goods aimed at reducing the corporate

TABLE 14.15. Acquisition of machines and equipment on benefit of 50 per cent special depreciation allowance in first year (¥m.)

		1952–6 Cumulative total
Dyeing		6,877
Chemical fertilizers		1,268
Petroleum refineries		9,160
Cement		6,289
Iron casting		35,815
Iron and steel wire rods		1,492
Steel cable		1,621
Power generator for electricity and ships		1,433
Metal-working machines		1,108
Spindle and steel bearings		1,138
Cable and wire for electricity		4,087
Electric communication machines		1,208
Cars and parts		4,688
Coalmining	(machines)	8,820
	(tunnel)	6,031
Construction and Engineering		3,194
TOTAL	(machines)	97,188
	(tunnel)	6,031

Source: MITI (1957*a*).

income taxes on the producers of those goods.[8] Most of the specified goods belonged to the *Z*-sector, including petrochemical products, synthetic fibres,

TABLE 14.16. Acquisition of machines on benefit of extra 50 per cent depreciation in three years

	1951.9–1956 Cumulative total	
	number	value (¥m.)
Machinery	6,332	14,690
Iron and steel	961	6,643
Chemicals	2,931	9,795
Miscellaneous articles	316	1,035
Textiles	20,165	39,022
Paper	902	9,998
Mining	750	1,799
Coal	314	725
Others	2,342	12,297
TOTAL	35,013	96,009

Source: MITI (1957*a*).

and chemical fertilizer. The upper limit of tax exemption was 40% of the fixed investment required for the production of the specified goods. The system, therefore, favoured those firms that took greater risks by producing new products and those that had to invest on a large scale for the production of new goods.

An exemption of import tariffs on specified important machines was introduced in 1951, and the number of items to which this system was applied reached a peak of 316 in 1954. This system aimed at tariff exemption for machines that Japan was unable to produce and yet were indispensable for improving efficiency. As renewing machinery and equipment for modernization and increasing domestic production of these machines were both urgent tasks, the government had to specify those which could be imported and those which must be produced locally. The system undoubtedly contributed to the modernization of equipment in the sectors where these imported machines were used and it is clear from Table 14.17 that the main beneficiary was the *Z*-sector.

TABLE 14.17. Import tax exemption

Industries	1951–6 Cumulative total (¥m.)	
	Imports	Amount of tax exemption
Iron and steel	18,561	2,806
Electricity generation and transmission	11,439	1,717
Mining and engineering	5,345	805
Pulp and paper-making	4,023	640
Textiles	3,360	522
Dyeing of textile products	3,163	491
Artificial textile fibre	2,902	434
Petroleum refinery	1,493	230

Source: MITI (1957*a*).

To sum up: tax policies in the early period after 1945 favoured investment in the *Z*-sector more than in the *X*-sector, both directly and indirectly.

14.2.5. Social overhead capital

The economic impact of social overhead capital has two aspects: there is a productive effect, which is reflected in the upward shift of production functions due to increased social capital; and there is a demand effect, which increases the incomes accruing to the sectors that produce the goods to be used in investment in social capital.[9] The former is referred to as the external effects of social capital while the latter comprises a multiplier effect of investment under some factor redundancy.[10]

The war damage to social overhead capital was slightly less than that to private productive industrial capital. At the start of the period of economic restoration the government invested heavily in infrastructure for industries, so the ratio of social capital to private productive capital was higher in the early years after the war (1949–57) than in the late 1960s.

In fact, according to *The Report on War Damage* (Keizai Antei Honbu 1949), the average rate of damage of social overhead capital, including harbours, canals, bridges, railways, gas and electricity supply installations, telecommunications, and water supply, was only 9.3%, compared with the overall average of 25%. Furthermore, damage to social overhead capital accounted for only 5.3% of the total war damage. As the social overhead capital here excludes ships and public buildings, the real rate may be higher than the above figures suggest, though not significantly so as the share of these assets in the social capital was not significantly large.

In these circumstances, for several years after the war the government stimulated investment in directly production-related social overhead capital more than in projects geared primarily to improving the quality of life. This is confirmed by Fig. 14.3.

With the rapid growth in Japanese industry in the post-war period this trend changed, and the share of government investment in production-related social overhead capital exceeded that of the other, particularly in the period 1949–56. More specifically, the share of investment in production-related social overhead capital in total government investment was 40.2% in 1945–9, 67.2% in 1949–56, 57.8% in 1957–61, and 54.8% in 1962–5. Thus, the share was highest in the period 1949–56.

As private investment gained momentum after 1955, by which time the economy had required the pre-war peak of economic activities, the growth of government investment lagged behind that of private investment. Accordingly, the balance between social and private productive capital was broken, with social capital lagging far behind.

It is difficult to find an optimal ratio of social capital to private productive capital. From the producers' view point, however, it is certainly better to have more production-related social capital. According to Table 14.18, the ratios of all production-related social capital (owned by the government) to private capital and the ratio of public utility capital to manufacturing capital grew continuously to 1953 and declined thereafter.

The wide availability of production-related social capital benefited both X- and Z-sectors through pure external and pecuniary external effects as well. Table 14.19 indicates that the Z-sector received greater pecuniary external effects than the X-sector as its inputs of social capital services were greater.

The demand-creating effects of investment in social overhead capital also favoured the Z-sector more. As is well known, an increase in government investment creates a demand for the products of various industries because of the interdependence of industries. The net effect is the increase in final and

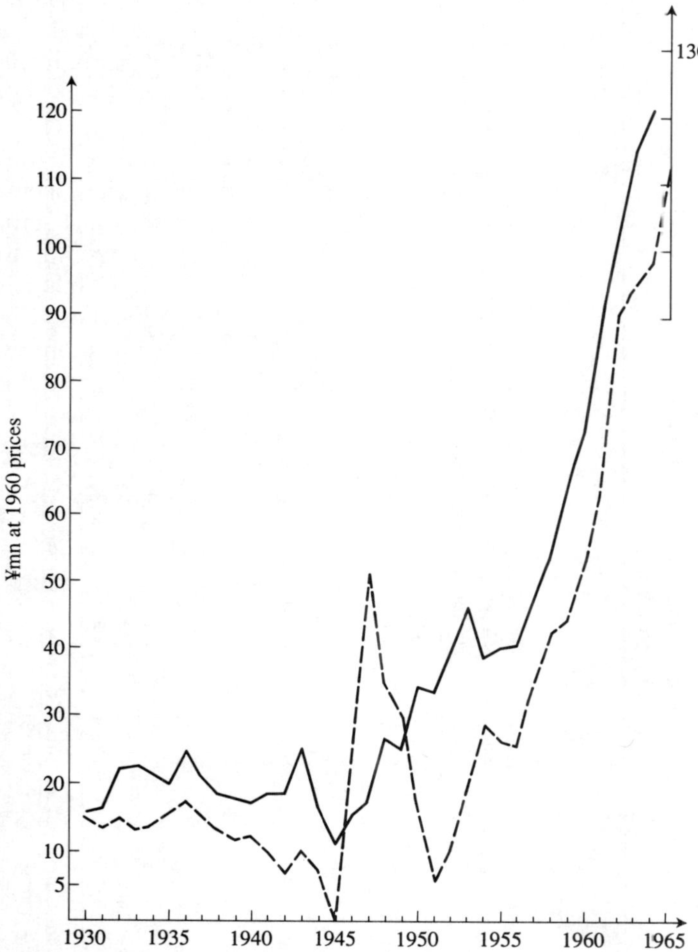

——— Production-related social capital investment;
--------- Other social capital investment.

Source: Takeuchi (1967) : 337.

FIG. 14.3 Evolution of government investment (¥m. at 1960 constant prices).

intermediate demand less the leakage through imports. Table 14.20, which shows how much demand for industrial products is generated by a one-unit increase in government investment, shows clearly that the heavy industries, particularly machinery, metals, and chemicals, benefited more from demand stimulation due to government investment than the light industries. Thus, given the input–output structure of the Japanese economy in the 1950s, government investment stimulated the *Z*-sector more, both by improving efficiency and by generating demand.

TABLE 14.18. Social capital and private capital (1960 prices, ¥bn.)

	By owner						By industry			Ratios	
	Social capital			Private capital							
	(1)	(2)	(3)	(4) Manufacturing	(5) All industries	(6) Non-agriculture	(7) All industries	(8) Manufacturing	(9) Public utilities	(1)/(5)	(9)/(8)
1952	5,573	3,312	8,885	3,806.2	12,966.7	10,756.3					
1953	5,821	3,312	9,211	4,046.8	13,618.5	11,244.0	18,810.6	4,046.8	4,610.6	0.4274	1.14
1954	6,087	3,562	9,468	4,348.1	14,545.0	11,998.4	19,892.3	4,348.1	4,917.1	0.4185	1.13
1955	6,334	3,713	10,047	4,592.8	15,402.4	12,658.7	20,864.2	4,592.8	5,233.6	0.4112	1.14
1956	6,617	3,867	10,485	4,962.5	16,474.0	13,559.6	22,056.6	4,962.5	5,621.8	0.4017	1.13
1957	6,945	4,108	11,053	5,583.9	17,947.1	14,851.6	23,620.8	5,583.9	6,021.9	0.3870	1.08
1958	7,314	4,428	11,741	6,260.8	19,512.4	16,230.8	25,257.3	6,260.8	6,498.2	0.3748	1.04
1959	7,725	4,769	12,494	6,958.6	21,253.0	17,737.7	27,145.8	6,958.6	7,048.9	0.3635	1.01
1960	8,207	5,185	13,392	8,210.1	23,955.1	20,126.1	29,960.9	8,210.1	7,675.8	0.3426	0.93
1961	8,805	5,717	14,523	9,617.0	27,128.6	22,978.4	33,732.7	9,671.0	8,818.1	0.3246	0.91
1962	9,519	6,506	16,025	11,338.3	30,717.5	26,174.9	38,097.9	11,338.3	10,041.1	0.3099	0.89
1963	10,372	7,354	17,727	13,062.2	34,470.3	29,505.2	42,858.0	13,062.2	11,436.0	0.3009	0.88
1964	11,275	8,250	19,525	14,923.2	38,850.8	33,443.9	47,886.2	14,923.2	13,071.8	0.2902	0.88

Notes: Column (1) is made up of road, harbour, and riparian works; agriculture, forestry, and fisheries; national railways, telegraphs, and telephones; column (2) represents government capital less item in (1); column (3) = (1) + (2).

Sources: Keizai Kikakucho (1967*a*, 1969); Takeuchi (1967).

TABLE 14.19. Input–output coefficient of public utilities by industries

	Food	Textiles	Wood and wood products	Pulp, paper, printing, publishing	Chemical products	Coal products	Petroleum products	Rubber products	Leather and products
Gas supply									
1951	007752	003446	—	007951	000795	502352	000364	—	—
1954	009193	003695	—	010973	000834	398279	000518	—	—
Electricity supply									
1951	043610	051831	018859	031429	081522	001366	001253	003282	000719
1954	040284	055747	015680	038818	072522	001124	001478	003241	000656
Transport and communications									
1951	022176	019933	013755	018180	023279	007348	000961	002875	001591
1954	023512	021849	010943	021868	023983	006128	001421	002800	001422
Total public utilities									
1951	025090	024467	014040	019929	031861	026021	000985	002827	001386
1954	025624	026557	011259	024148	030840	021280	001394	002757	001241

Table 14.19 (*contd.*)

	Ceramics and pottery	Iron and steel	Non-ferrous metals	Machinery	Other mfg.	Gas	Electricity	Transport and communications
Gas supply								
1951	020176	003976	001226	021535	—	—	—	000895
1954	015773	004304	000969	021766	—	—	—	001104
Electricity supply								
1951	021074	085782	019724	028486	002692	002296	011850	048477
1954	018693	071644	016387	025962	002939	002385	012011	056181
Transport and communications								
1951	010389	021692	005638	016434	003050	002092	008505	035304
1954	009233	019750	004980	014824	003245	002141	008314	032987
Total public utilities								
1951	012516	031414	007754	018597	002871	002042	008712	036081
1954	011019	027460	006650	016896	003064	002093	008570	035417

Note: Values are input–output coefficient multiplied by 1,000,000.

Source: MITI (1957*b*).

TABLE 14.20. Production induced by one unit of government investment

	Final demand	Intermediate demand	Import of goods and services (including tariff)	Total production
Agriculture, forestry and fisheries	—	0.052	0.034	0.018
Mining	—	0.088	0.066	0.022
Light industries	—	0.270	0.005	0.265
Heavy industries	0.133	0.862	0.036	1.960
Chemicals, petroleum, coal products	—	0.180	0.014	0.166
Metals	—	0.418	0.013	0.405
Machinery	0.133	0.264	0.009	1.389
Construction	0.857	0.010	—	0.867
Public utilities	0.001	0.121	—	0.122
Commerce and services	0.009	0.139	—	0.147
Not classified elsewhere	—	0.084	—	0.084
TOTAL	1.000	1.625	0.141	2.485

Source: Takeuchi (1967).

The fact that investment in social overhead capital favoured the *Z*-sector more was very important, given that the prospects for the Z-sector seemed very uncertain at the end of the war. Thus, the investment in social overhead capital actually pushed (K_x, K_z) north and thereby facilitated the take-off of the *Z*-sector.

14.3. Import Restrictions and Tariff Policy

It is totally irrelevant to evaluate Japanese trade policies only from the tariff structure because strict controls on quantity were in fact more effective. Sometimes the tariffs were even redundant because the restriction on import quantities were so severe. From 1945 to 1951, foreign trade was controlled entirely by the occupation forces, who also supervised major economic policies. The Japanese government had no power.

The revised tariff law (*Kanzei Teiritsuho*) which was enacted in May 1951 was the result of four-round negotiations between the government and the occupation forces. As tariffs were in general set low by this law, Japanese imports might have expanded drastically had there not been controls based on foreign exchange allocation and quantitative import restrictions. These were in fact the major policy instruments protecting domestic industries in the period before the take-off of the *Z*-sector, and they remained in force until 1960.

When Japan joined the GATT (General Agreement on Tariffs and Trade) in September 1955, the Japanese government offered concessionary tariff treatment on 248 items: for 75 items tariffs were reduced, and for 173 items tariff rates remained unchanged (see Nihon Kanzei Kyokai 1959: 90–1). At this time the tariffs on cars were lowered, but the margin was small. Through out the negotiations for tariff reduction thereafter, the number of items for concessionary treatment was expanded but the effects on real imports remained small because of the quantitative restraints that were effective until the government announced the trade liberalization programme in 1960.

After 1959, some tariff rates were cautiously raised in anticipation of the removal of quantitative restraints. The revision of tariff rates in June 1961, when the Brussels Tariff Nomenclature was adopted, was the result of such cautious manoeuvres: tariffs were unchanged on 70 items, lowered on 386 and raised on 251. Generally speaking, the tariffs increases were designed to protect newly growing industries. Thus, the emphasis shifted from quantitative restraints to tariff protection. After the mid-1960s the emphasis shifted to trade liberalization and tariff reduction, but this period lies beyond our coverage. From the end of the war, to the mid-1960s, therefore, the government controlled imports chiefly through quantitative restraints. When the imports were complementary to, rather than a substitute for, domestic products, the government gave preferential tax treatment for the import of machinery necessary for the growth of the *X*-sector. After the *Z*-sector had grown somewhat (largely thanks to government support), tariffs were imposed on *Z*-sector goods to protect the newly developing domestic *Z*-sector. Most industrial machinery was eventually affected by this policy evolution.

14.3.1. The tariff revision of 1951

In principle the tariff structure from 1945 to 1951 was one of high tariffs on luxuries and consumer goods. Agricultural products were also highly protected, except when they were industrial materials and intermediate goods for industries when low tariffs (around 15%) were imposed. In 1951 a revised tariff law was enacted after four rounds of negotiation with the allied government. A summary of the tariff rates is given in Table 14.21. Details of the structure of the tariffs can be found in Nihon Kanzei Kyokai (1959) and Oka (1964). In terms of our own analytical framework, their results can be paraphrased as shown below.

1. As the general price level rose disproportionately during and immediately after the war, many *ad quantum* tariffs were replaced by *ad valorem* tariffs to make the tariff policy more effective.
2. Tariff rates were kept low as a result of pressure from the occupation forces and in anticipation of joining the GATT. Thus, maximum rates were 50% (exceptionally higher on luxuries), 20% tariffs were applied to 170 items, and 15% rates to 217 items.

TABLE 14.21. Customs Tariff Law in 1951

Commodities	Tax rate(%)	Notes
Section 1		
Vegetables and animals	none	except natural yeasts
Section 2		
Cereals, flour, and starch		
Rice	15%	
Wheat	20%	
Section 3		
Beverages, tobacco		
Tea	35%	
Fruit with added sugar	35%	
Section 4		
Skin, hair, bones, and horn	0–50%	depending on domestic production or degree of processing
Section 5		
Oils, fats, and their products	0–20%	same as Section 4
Section 6		
Medicines, chemicals, cosmetics	0–25%	non-manufactured products
Section 7		
Dyeing materials, pigments, paint	50%	methyl alcohol, cosmetics
	0–25	depending on domestic production
Section 8		
Yarns and thread, pulp, fibre	0–7.5%	materials no tax; yarn 7.5%
	20%	silk yarn
	25%	synthetic products
Section 9		
Fabrics	10–30%	30% for highly processed, generally 15%
Section 10		
Clothing and clothing accessories	20–50%	
Section 11		
Paper, paper products	0–25%	no tax for educational goods, average is 10–15%
Section 12		
Minerals and products	0–20%	many items are no tax or 5%
Section 13		
Pottery and glass	10–50%	10% for materials, 50% for highly processed goods, many items are about 20%

Table 14.21 (*contd.*)

Commodities	Tax rate(%)	Notes
Section 14		
Base metals and metals	0–15%	20–30% for highly purified metals
Section 15		
Metal products	15–20%	50% for goldsmiths' and silversmiths' wares
Section 16		
Watches and machinery	15%	
	15–50%	40–50% for consumers' goods, 15% for productive goods
Section 17		
Miscellaneous	0–10%	generally
	50%	miscellaneous goods

Source: Ohkurasho (n.d.): 262–304.

3. Some foodstuffs, such as rice and wheat, and industrial machines were exempted from tariffs.
4. Tariffs on raw materials except agricultural products were set at zero or very low.
5. Tariffs on consumer goods were higher (30%–50%), with 50% tariffs being imposed on most consumer luxuries.
6. Tariffs on the intermediate products of heavy industries were generally low, mostly 15%.

It is clear that the tariff structure of 1951 was not designed to protect the *Z*-sector but rather to reduce the import of consumer goods and finished goods for which import substitutes were available. Protection for the *Z*-sector was provided through fiscal subsidies and preferential finance, as already examined.

Under the fixed exchange rate of $1.00 = ¥360, Japan exported *X*-sector goods and imported *Z*-sector goods as well as some *S*-sector goods, with the balance of payments staying mostly in the red. For this reason, quantitative restraints as well as high tariffs were imposed on imports of goods other than industrial materials and machines. (If we were to convert the quantity restriction on *Z*-sector imports into tariff, the real tariffs on *Z*-sector goods would be much higher, but this would need a separate empirical study.) In consequence, whereas in 1953 foods and industrial materials accounted for 67.5% of total imports, and 86.7% if the import of mineral fuels was included, the corresponding figures for 1960 were 46.4% and 77.9% respectively. This change can be attributed to the government's import policy.

To look at commodity distribution of foreign trade since 1953 as shown in Table 14.22, export of X-sector goods accounted for 59.5% of total export in 1953 and 51.8% in 1960. The importance of X-sector exports was as high as in contemporary LDCs.

14.3.2. Participation in the GATT

In June 1955, Japan joined the GATT. Table 14.23 shows selected tariff concessions to the United States among the reduction of tariffs which was applied to 75 items in total. In general, Japan's concessions were on Z-sector goods and agricultural products, but not on X-sector goods, reflecting the lower level of US interest in the latter. In practice, however, imports were further constrained by the severe quantitative restraints.

After joining the GATT, Japan secured quantitative restraints on imports, and tariffs became more important measures to control imports in the 1960s. While the period after 1960 lies beyond our coverage, the evolution of tariff rates on selected goods is shown in Table 14.A.1, which consists of the data for selected goods of the X- and Z-sectors.

From Table 14.24, which outlines the revised tariff rates enacted in June 1961, it appears that the government used tariffs to compensate for the reduction of quantitative restrictions, in that tariffs were raised on agricultural and Z-sector products. There was also a shift from non-tariff measures to tariffs for controlling the import of X-sector goods. In general, however, non-tariff trade barriers remained higher than in Western industrial nations until the early 1970s.

Trade policies thus played important roles in supporting industrial take-off pushing (K_x, K_z) first north-east and then more towards the north.

14.4. The Supply of Food and Labour

As already explained, an increase in the food supply has two implications for economic development: it tends to keep real wages low by stabilizing the prices of S-sector goods, and when it is due to increases in productivity it expands the supply of labour to industry. Immediately after World War II, Japan faced the following situations in these regards: first, a shortage of foods so extreme throughout the country that it had to depend on aid and food import until 1948; and an enormous population to feed because of the repatriation of Japanese from overseas. In one sense, then, Japan suffered the same population pressure of many LDCs, but there was an important difference in that Japan's 'overpopulation' was relatively well educated and had extensive industrial experience.

Accepting Minami's (1970) conclusion that Japan shifted to a limited supply of labour around 1960, we will examine in what follows labour supply in the post-war period from another point of view.

TABLE 14.22(*a*). Post-war foreign trade: exports

Exports	1953		1955		1960		1965	
	$m.	%	$m.	%	$m.	%	$m	%
X-sector	758	59.5	1178	58.6	2101	51.8	2946	34.9
Foods	120	9.4	126	6.3	256	6.3	344	4.1
Textiles	460	36.1	749	37.2	1223	30.2	1582	18.7
Others	178	14.0	303	15.1	622	15.3	1020	12.1
Z-sector	481	37.7	781	38.8	1852	45.7	5324	63.0
Chemicals	72	5.6	103	5.1	181	4.5	547	6.5
Non-ferrous metal products	28	2.2	42	2.1	68	1.7	84	1.0
Metal products	192	15.1	387	19.2	568	14.0	1718	20.3
Machinery	189	14.8	249	12.4	1035	25.5	2975	35.2
TOTAL	1,275	100.0	2,011	100.0	4,055	100.0	8,452	100.0

Source: Nippon Ginko (1970).

TABLE 14.22(*b*). Post-war foreign trade: imports

Imports	1953		1955		1960		1965	
	$m.	%	$m.	%	$m.	%	$m.	%
Foods	624	25.9	625	25.3	548	12.2	1470	18.0
Textile materials	666	27.6	586	23.7	762	17.0	847	10.4
Other materials	336	13.9	491	19.9	774	17.2	1354	16.6
Subtotal	1626	67.5	1702	68.9	2084	46.4	3671	44.9
Metal materials	173	7.2	186	7.5	673	15.0	1019	12.5
Mineral fuels	289	12.0	289	11.7	742	16.5	1626	19.9
Subtotal	462	19.2	475	19.2	1415	31.5	2645	32.4
Chemicals	93	3.9	112	4.5	265	5.9	408	5.0
Machinery	161	6.7	132	5.3	435	9.7	760	9.3
Other goods	67	2.8	50	2.0	293	6.5	684	8.4
Subtotal	321	13.3	294	11.9	993	22.1	1852	22.7
TOTAL	2,410	100.0	2,471	100.0	4,491	100.0	8,169	100.0

Source: Nippon Ginko (1970).

TABLE 14.23. Selected tariff concessions by Japan to the US, 1955

Commodity	General tax	Concession tax
Pineapple (canned)	35%	25%
Grapes (dried)	20%	10%
Asparagus (airtight cans)	25%	20%
Cellulose acetate	30%	27.5%
Artificial resins and artificial plastic materials	30%	22.7%
Textiles from yarn of man-made fibres	25%	22.5%
Lubricating oil	30%	22.5%
Motor vehicles for transport of persons (large and middle size)	40%	35%
Aeroplanes (with 4 engines)	15%	10%
Radio receivers	20%	18%
Television receivers (with cathode-ray tube of maximum length of image not less than 23")	30%	25%
Cameras	40%	30%
Cinematographic cameras (for films of a width not less than 16 mm)	30%	20%
Clocks	30%	27%
Statistical machines	15%	10%
Pneumatic tyres for motor vehicles	30%	25%

Source: Oka (1964): 194.

14.4.1. Population pressure immediately after the war

The number of Japanese returning from abroad, including from former colonies, at the end of World War II was some 6.5 million, at a time that the total population of Japan's four main islands was about 72 million. As many farms had been destroyed, this 9% increase in population was an enormous burden on the food supply. What was worse, the rice crop of 1945 was extremely poor, and there was a large-scale shortage of supply despite strict controls on compulsory sales to the government. The government, therefore, extended the compulsory sales order to include wheat and potatoes too, and generally controlled the food markets for several years after the war. It was only in black market prices that the market mechanism revealed itself.

Let us examine these circumstances through numerical data. As Table 14.25 shows, the average population of the four main islands of Japan in the period 1936–40 was about 70.5 million, but it increased to 75.8 million in 1946, 78.1

TABLE 14.24. Outline of revised tariff rates (June 1961)

Section	Items	No. of revised tariff items: Reduction	Raised	Revision continuing	Total	Main commodities taxed
1	Live animals; animal products	7	22	45	74	Butter, cheese, powered milk
2	Vegetable products	26	19	117	162	Small red beans, broad beans, walnuts, buckwheat, rape seed and soya beans, dried edible seaweeds
3	Animal and vegetable fats and oils and their products	5	2	37	44	Soyabean and groundnut oil
4	Prepared foodstuffs; beverages, spirits, and vinegar; tobacco	17	13	84	114	Sugars, curry preparations, prepared baking powders
5	Mineral and mineral products	4	7	85	96	Graphite
6	Products of the chemical and allied industries	35	93	329	457	Petroleum products, santonin, egg albumins, glass glues derived from bones
7	Artificial resins and plastic materials; rubber, synthetic rubber, and articles thereof	4	8	55	67	Natural rubber, rubber products, synthetic rubber
8	Raw hides and skins, leather, furs and articles thereof; handbags	0	18	20	38	
9	Wood and articles of wood; wood charcoal; cork; manufactures of straw, esparto, plastic materials	4	16	37	57	Russian wood and cork products
10	Paper and paper board; paperpulp	8	17	38	63	Processed paper, building board

Table 14.24 (*contd.*)

11	Textiles and textile articles	41	38	148	227	Woollen yarn, woven fabrics, cotton fabrics, handkerchiefs, lace
12	Footwear, headgear, umbrellas, sunshades, walking-sticks; artificial flowers	2	22	9	33	
13	Articles of stone, plaster, cement, asbestos, mica, and similar materials; ceramic products; glassware	18	18	32	68	Glass products, abrasive paper
14	Pearls, precious stones and metals	3	9	20	32	Silver
15	Base metals and their products	25	24	199	248	High-speed steel, nickel and products, aluminium, magnesium
16	Machinery, electrical machines	34	5	143	182	Work machines, electric generators, motors, converters
17	Vehicles, aircraft and parts; vessels and associated transport equipment	5	6	43	54	Vehicle parts
18	Clocks and watches	11	21	66	98	Wrist watches
19	Arms and ammunition; parts thereof	0	1	7	8	
20	Miscellaneous manufactured articles	4	21	45	70	
21	Works of art, collectors' pieces, and antiques	0	4	2	6	
TOTAL		253	380	1,550	2,183	

Source: Oka (1964).

TABLE 14.25. Supply of main foodstuffs

	Imports		Domestic production			Population ('000s)	Rice (per capita)		Farm price index of rice (1954–6 =100,000)	Agricultural prices index (total) (1954–6 =100,000)
	Wheat[a] ('000t)	Rice[a] ('000t)	Wheat ('000t)	Rice ('000t)	Barley[a] ('000t)		Domestic production (kg.)	Total supply (kg.)		
1936–40 (annual average[b])	157.7	175.9	1,454.9	9,881.0	758.2	70,482	140.2	142.7	339.8	354.1
1946	340.2	15.6	(1,095.2)	9,207.9	(761.6)	75,800	121.5	121.7	—	—
1947	694.3	2.8	(1,163.6)	8,797.8	(805.1)	78,101	112.6	112.7	—	—
1948	695.1	42.3	1,207.0	9,965.9	845.4	80,010	124.6	125.1	—	—
1949	1,955.0	135.5	1,304.4	8,482.9	954.8	81,780	103.7	105.4	42,677	56,749
1950	1,573.0	671.6	1,338.6	9,650.9	897.2	83,200	116.0	124.1	55,355	62,327
1951	1,653.6	798.8	1,490.2	9,041.7	1,058.2	84,500	107.0	116.5	73,355	80,159
1952	1,661.8	978.5	1,537.7	9,922.8	1,078.4	85,800	115.7	127.1	84,871	85,619
1953	1,686.5	1,079.1	1,374.3	8,238.6	1,099.3	87,000	94.7	107.1	103,161	99,099
1954	2,187.1	1,432.3	1,516.0	9,113.4	1,261.3	88,200	103.3	119.6	101,032	102,698
1955	2,287.5	1,246.4	1,468.0	12,384.8	1,148.2	89,276	138.7	152.7	101,032	99,696
1956	2,277.0	759.6	1,375.4	10,898.7	1,132.9	90,172	120.9	129.3	97,936	97,310
1957	2,239.5	347.2	1,330.3	11,464.2	1,129.3	90,928	126.1	129.9	101,710	99,918
1958	2,280.1	505.5	1,281.2	11,992.9	1,132.5	91,767	130.7	136.2	101,226	99,890
1959	2,412.2	277.4	1,416.2	12,501.0	1,241.5	92,641	134.9	137.9	101,419	99,134
1960	2,678.0	174.6	1,530.9	12,858.0	1,207.1	93,419	137.6	139.5	101,419	100,572

Note: Figures in parentheses are estimates.

[b] Pre-war imports of four main islands from foreign countries, and excluding imports from Korea and Taiwan.

[a] Conversion factor rice 1 "koku"=150 Kg; wheat 1 "koku"=136.9 Kg; barley 1 "koku"=108.8 Kg.

Source: Nihon Ginko (1966); Umemura *et al.* (1966).

million in 1947, and 80 million in 1948. Thus, the population expanded by 9.5 million in less than ten years. The majority of this growth took place after 1945, i.e. over four years, due to the return of soldiers and other Japanese from abroad.

Due to the loss of the colonies and deterioration of farm land, however, domestic production of foodstuff drastically declined at this time. Per capita production of rice, for instance, dropped sharply, from 140.2 kilogrammes in 1936–40 to 121.5 kilogrammes in 1946, and 116.5 kilogrammes in 1950, five years after the end of the war. Wheat aid relieved the shortage immediately after the war and was expanded thereafter. Rice was also imported, and by 1950 the per capita rice supply had risen to 124.1 kilogrammes. As the availability of wheat and rice increased the supply situation was eased somewhat, but food prices continued to rise rapidly until 1953, reflecting general tendency of tight supply.[11]

In response to the tight food supply, the government raised its purchasing prices annually. Table 14.26 shows the government prices for purchasing foods from farmers and selling them to consumers between 1946 and 1955. Because of serious shortage of supply, however, black-market prices increased faster than the official prices did. On the other hand land reform, the second round of which was implemented in October 1946, had created a number of independent farmers, and the increases in the prices at which the government purchased foods from farmers was an effective incentive for farmers to increase production. After hitting a peak in 1954, rice imports actually started to decline.

While it is conjectured that the terms of trade of both *X*- and *Z*-sectors deteriorated *vis-à-vis* the *S*-sector in the period 1946–54 due to the rapid

TABLE 14.26. Purchasing and selling price of rice by government (¥) (per 60kg of raw rice)

	Purchasing price	Selling price
1946	220	189
1947	702	793[a]
1948	1,458	1,882[b]
1949	1,759	2,140
1950	2,419	2,326[c]
1951	2,820	2,695[d]
1952	3,000	3,263[e]*
1953	3,384	3,588[f]*
1954	3,704	4,051[f]*
1955	3,902	4,046[f]*

[a]22 Dec. [b]15 Nov. [c]18 May

[d]1st Jan. [e]1st Aug. [f]1st Jan.

*Third-class rice

Source: Shokuryocho (1955, 1956).

increase of food prices, rigorous investigation is difficult because economic statistics were not compiled well before 1953. We therefore cannot draw strong conclusions about this period.

Increases in the prices of food and agricultural products do not favour capital accumulation in heavy industry. Many LDCs found themselves facing this difficulty before the Green Revolution of the 1960s. The Japanese response was to provide both fiscal and financial support to heavy industries until agricultural productivity increased enough to change the situation from one of food crisis to food surplus. (The evolution of agricultural productivity is shown in Table 14.A.2.)

14.4.2. The supply of industrial labour

As a result of land reform the unit of agricultural production became very small. In some senses this set a limit to improvements in efficiency (although labour productivity rose due to the increased use of chemical fertilizer and farm machinery) but on the other hand the land reform made price incentives more effective because producers could benefit from the increases in government purchasing prices.

In fact, although the area of cultivated land increased only slowly until 1960 and declined absolutely thereafter, per capita production increased significantly, with an annual average increase of 6.7% for 1950–55 and 2.1% for 1955–60 (see Table 14.A.2). In the same periods, the number of people engaged in agriculture decreased at an annual rate of 0.7% and 2.8% respectively, and subsequently rose to an annual rate of 3.6% (1960–3).

To return from the long-run trend to the immediate post-war situation, however, Table 14.27 shows the percentage of people employed in primary

TABLE 14.27. Composition of employees ('000s)

	Primary	Secondary	Tertiary	Agriculture	Manufacturing
1930	14,490	5,993	8,788	13,742	4,702
	(49.4)	(20.4)	(30.2)	(46.8)	(16.0)
1940	14,192	8,419	9,403	13,363	6,845
	(44.0)	(26.1)	(29.9)	(41.5)	(21.2)
1950	17,208	7,812	10,568	16,102	5,690
	(48.3)	(21.9)	(29.8)	(45.2)	(16.0)
1955	16,111	9,222	13,928	14,892	6,902
	(41.0)	(23.5)	(35.5)	(41.0)	(17.6)
1960	14,237	12,764	16,682	13,121	9,553
	(32.6)	(29.2)	(38.2)	(30.0)	(21.9)

Note: Figures in parentheses show the ratio of industrial employees to total.

Source: Nippon Ginko (1966).

employment before the war and in selected years in the post-war period. From 49.4% in 1930 and 44% in 1940, it rose to 48.3% in 1950 (when many of those who returned from abroad worked in the primary sector because factory jobs were still not available), and still remained as high as 41% in 1955, only a little lower than in 1940.

As the growth of industry accelerated, the picture began to change. By 1960 the rate of employment in the primary sector had declined to 32.6%, while the share of the secondary sector in total employment had increased to 29.2%. Thus, in 1960 the employment structure became far more oriented towards industry than in the last years of the pre-war period.

Let us now consider the allocation of labour between the *X*- and *Z*-sectors. Table 14.28 indicates the changes in the number of employees and in real wage rates in the *X*- and *Z*-sectors over time, although the industry coverage is not complete. Even so, it is some indication of the distribution of employment between our *X*- and *Z*-sectors. In 1953 employment in the *X*-sector was 2.75 million and in the *Z*-sector 2.25 million. As early as 1956, however, *Z*-sector employment surpassed that of the *X*-sector, and despite some cyclical fluctuations continued to grow rapidly whereas *X*-sector employment stagnated.

TABLE 14.28. Number of employees and wage rates

	Employees ('000s)		Annual average wage (per capita, ¥'000)		Real wage rate (in terms of 1954-6 agricultural prices)	
	X-sector	*Z*-sector	*X*-sector	*Z*-sector	*X*-sector	*Z*-sector
1953	2,755	2,253	117.7	196.7	119.9	200.3
1954	2,812	2,388	123.9	204.8	126.3	208.8
1955	2,778	2,506	129.5	210.2	137.5	223.1
1956	2,818	2,843	131.8	236.8	144.2	259.1
1957	2,843	3,180	135.5	243.8	141.1	253.9
1958	2,885	3,467	142.5	252.9	152.7	271.1
1959	2,749	3,627	158.9	272.4	168.5	288.9
1960	2,782	3,992	173.5	294.5	173.5	294.5
1961	2,934	4,314	196.9	327.6	176.4	293.5
1962	3,041	4,632	227.6	351.0	189.5	292.3
1963	3,110	4,879	254.2	389.4	199.1	304.9
1964	3,118	5,063	283.5	431.4	210.5	320.3
1965	3,140	5,210	316.6	467.0	217.1	320.3

Notes: *X*-sector=foods, textiles; *Z*-sector=chemicals, primary metals, metals, machinery, electrical and transport machinery, paper and pulp.

Source: Keizai Kikakucho (1969).

14.4.3. Real wage rates and the labour supply

The critical test of whether an economy has shifted from a state of unlimited labour supply to a more limited supply of labour is whether the labour supply from the *S*-sector to industry is still infinitely elastic with regard to the changes in real wage rates. If the supply has become elastic only in finite scale, then the supply has become 'limited'. While there are many ways to test this point (e.g. by estimating the supply function of the labour supply), it is sufficient to show that the real wage rate in the *S*-sector is higher than marginal labour productivity in the sector. Minami (1970) established that this turning point occurred in 1960.

The next task was to investigate when the supply of skilled labour to the *Z*-sector became limited, rather than the general supply of skilled labour. To do this, we first follow the movements of money wage of agriculture which is given in Table 14.A.2. It grew by an annual average of 19.4% in the period 1948–50 and then slowed to an annual rate of 7.5% in 1950–5 and 4.3% in 1955–60. After 1960, when the general shortage of skilled labour began, money wages increased rapidly again, with an annual average rate of 19.8% for 1960–3.

We cannot state definitively what happened with real wages in 1948–50 because of the lack of statistics, but they do not seem to have increased substantially in 1948–50, even though money wages rose significantly because of the drastic increases in food prices. We have no data on producers' prices of foods for 1948, but as they rose at an annual rate of 9% in 1950–5 and 0.8% in 1955–60, the real wage rates remained rather stable, though there was a sign of a rise in the latter five years. The increasing trend of real wage rates became more definite after 1960. While agricultural prices rose at an annual rate of 7.9%, due partly to the government price support policy, money wages increased by 19.3% annually, giving 10% an increase of more than in real wage rates. This is clear indirect evidence of the shift in the supply situation.

We now turn specifically to the *Z*-sector and investigate whether the *Z*-sector faced wage increases in trying to expand employment. Fig. 14.4 shows the real wages in the *X*- and *Z*-sectors from 1953 to 1965. A constant difference in the real wage rates in the two sectors could be attributed to the differences in skill requirement and would not imply a difference in the supply of labour to the two sectors. If the supply to the *Z*-sector was more limited than to the *X*-sector, there should be a steady increase in the wage differential. Fig. 14.4 suggests that this was not the case in the period 1953–65. In fact, although the number of employees in the *Z*-sector expanded rapidly after 1955, the real wage rate of the *Z*-sector relative to that of the *X*-sector never increased to any significant degree.

We therefore, conclude that there was no shortage of skilled labour specific to the *Z*-sector, before the mid-1960s.

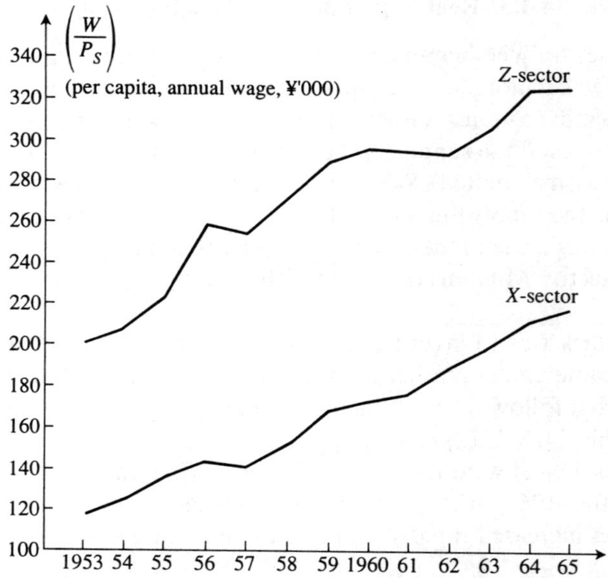

FIG. 14.4 Wage rate of manufacturing (per capital annual wage, ¥'000, in terms of agricultural prices).

TABLE 14.A.1. Evolution of tariff rates of selected goods

	Agricultural goods					
	Wheat			Wheat flour		
	General tax(%)	MFN tax	Temporary tax	General tax(%)	MFN tax(%)	Temporary tax(%)
1960	20%		tax exemption	25%		12.5%for exported sodium glutamate
	(204)		(31 Mar.1961)	(212–1)		
1961	20%		tax exemption	25%		cont.
	(1001)		(31 Mar.1962)	(1101)		
1962	20%		tax exemption (31 Mar.1963)	25% (1101–1)	12.5% for some items	cont.
1963	20%		tax exemption (31 Mar.1964)	25%	cont.	cont.
1964	20%		tax exemption (31 Mar.1965)	25%	cont.	
1965	20%		temporary			temporary
1966	20%		tax exemption (31 Mar.1967)	25%	cont.	
1967	20%		tax exemption (31 Mar.1968)	25%	cont.	
1968	20%		tax exemption (31 Mar.1969)	25%	cont.	
1969	20%		tax exemption (31 Mar.1970)	25%	cont.	
1970	20%		tax exemption (31 Mar.1971)	25%	12.5%	

Table 14.A.1 (*contd.*)

X-sector

	Silk yarn			Cotton fabrics		
	General tax	MFN tax	Temporary tax	General tax	MFN tax	Temporary tax
1960	20%			10%		
	(821)			(901)		
1961	15%			10%		
	(5004)			(5508–3)		
1962	15%			10%		
1963	15%			10%		
1964	cont.			cont.		
1965						
1966	cont.		temporary			temporary
1967	cont.			cont.		
1968	cont.	12%		cont.	8.8%	
1969	cont.	(1)12%(31 Dec.1969)	(1)12%	cont.	(1)8.8%(31 Dec.1969)	
		(2)10.5%(31 Dec.1970)	(2)10.5%		(2)8.2%(31 Dec.1970)	
1970	cont.	(1)10.5%(31 Dec.1970)	(1)10.5%	cont.	(1)8.2%(31 Dec.1970)	(1)8.2%
		(2)9%(31 Dec.1971)	(2)9%		(2)7.6%(31 Dec.1971)	(2)7.6%

Table 14.A.1 (*contd.*)

Statistical Appendix (cont.) *X*-sector				*Z*-sector		
	Tea			Steel board		
	General tax	MFN tax	Temporary tax	General tax	MFN tax	Temporary tax
1960	35% (303)			15% (1405–5)		tax exemption for some items
1961	35% (0902–2–(2))		commodity tax 10% Oolong tea	15% (7313)		
1962	35%		cont.	15%	15% for some items	
1963	35%		cont.	15%	cont.	
1964	cont.		cont.	cont.	cont.	
1965						
1966	cont.		5% for Oolong tea	cont.	cont.	temporary
1967	cont.			cont.	cont.	
1968	cont.	29%		cont.	(1)15%(30 June1968) (2)12%(31 Dec.1969)	
1969	cont.	(1)29%(31 Dec.1969) (2)26%(31 Dec.1970)		cont.	(1)12%(31 Dec.1969) (2)10.5%(31 Dec.1970)	
1970	cont.	(1)26%(31 Dec.1970) (2)23%(31 Dec.1971)	(1)26% (2)23%	cont. cont.	(1)10.5%(31 Dec.1970) (2)9%(31 Dec.1971)	

Table 14.A.1 (*contd.*)

Statistical Appendix (cont.) *Z*-sector

	Passenger cars			Synthetic resins		
	General tax	MFN tax	Temporary tax	General tax	MFN tax	Temporary tax
1960	40% (1642–1)			30% (666–1)	20%	
1961	40% (8702–1)	35% for some items	commodity tax 50–15%	20% (3902–1)		
1962	40%	cont.	40–20%	20%		
1963	40%	cont.	cont.	20%	20% (3902–2(1))	
1964	cont.	cont.	40–15%	cont.	cont.	
1965						
1966	cont.	cont.	cont. temporary	cont.	.	15% (3902–1–(2))
1967	cont.	cont.		cont.	cont.	
1968	cont. (8702–1–(1))	36%		cont. (3902–1–(1))	16%	
1969	cont.	(1)36%(31 Dec.1969) (2)34%(31 Dec.1970)		cont.	(1)16%(31 Dec.1969) (2)14%(31 Dec.1970)	
1970	cont.	(1)34%(31 Dec.1970) (2)32%(31 Dec.1971)	20% (31 Mar. 1971)	cont.	(1)14%(31 Dec.1970) (2)12%(31 Dec.1971)	

Note: Codes in general tax column indicate BTN number. MFN means most favoured nation.

TABLE 14.A.2. Agricultural employees and wage rates

	Number of employees in agriculture, forestry and fisheries ('000s)		Agricultural production in terms of agricultural prices (1934–6 prices, ¥ mil.)		Number of employees in agriculture ('000s)		Cultivated area ('00 chou)		Gross capital stock of agriculture (¥m. at 1934–6 prices)		Wage rate in agriculture (males, ¥day)		Farm price index of agricultural products (1954–6 = 100)	
1948	16,370	—	—		15,850	—	58,610	—	5,761	—	174	—		
1949	18,080	10.4	—		16,780	5.8	58,840	0.3	5,975	3.7	293	68.3	59.1	—
1950	17,410	−3.8	2,962	—	15,990	−4.8	59,070	0.3	5,985	0.1	248	−15.4	64.5	9.1
1951	16,170	−7.2	2,987	0.8	15,250	−4.7	59,270	0.3	6,234	4.1	252	1.6	82.3	27.5
1952	16,370	1.2	3,379	13.1	15,550	1.9	59,420	0.2	6,364	2.0	277	9.9	87.3	6.0
1953	17,190	5.0	3,001	−11.2	15,440	−0.8	59,530	0.1	6,501	2.1	305	10.1	103.5	18.5
1954	16,790	−2.4	3,347	11.5	15,280	−1.1	59,840	0.5	6,658	2.4	349	14.4	102.5	−1.0
1955	17,180	2.3	4,102	22.5	15,410	0.8	60,310	0.7	6,819	2.4	357	2.2	99.3	−3.2
1956	(16,040)	(−4.5)	3,891	−5.2	15,020	−2.6	60,630	0.5	7,218	5.8	363	1.6	98.1	−1.3
	16,750	−2.6												
1957	(15,610)	(−2.7)	4,111	5.6	14,710	−2.1	60,950	0.5	7,354	1.8	388	6.8	100.3	2.2
	15,210	−9.2												

Table 14.A.2 (*contd.*)

1958	14,710	-3.3	4,256	3.5	14,060	4.5	61,150	0.3	7,565	2.8	394	1.5	99.6	-0.7
1959	14,070	-4.4	4,413	3.6	13,800	-1.9	61,230	0.1	7,821	3.3	410	4.0	99.8	0.2
1960	13,910	-1.2	4,551	3.1	13,390	-3.0	61,220	-0.1	8,101	3.5	440	7.3	103.3	3.5
1961	13,530	-2.8	4,655	2.2	13,030	-2.7	61,360	0.2	8,712	7.5	530	20.4	111.7	8.1
1962	13,110	-3.5	4,846	4.1	12,640	-3.0	61,320	-0.1	9,341	7.2	646	21.8	125.1	11.9
1963	12,400	-5.5	4,735	-2.3	12,000	-5.1	61,110	-0.4	n.a.	n.a.	756	17.0	130.0	3.9
Annual average														
1948–50	3.1		—		0.4		0.4		1.9		19.4		—	
1950–55	-0.2		6.7		-0.7		0.4		2.6		7.5		9.0	
1955–60	-4.1		2.1		-2.8		0.3		3.5		4.3		0.8	
1960–63	-3.8		1.3		-3.6		-0.1		—		19.8		7.9	

Notes: Values on the right hand of each column show the annual change percent. As for the numbers of employees in 1956, 1957, those in first rows of the years are linked with 1959–63, those in second rows of the years are linked with 1948–55. Chou = 2.451 acres

Source: Naikaku Tokeikyoku (1957).

Notes

1. Details of economic restoration are to be found in Keizai Kikakucho (1959–63, 1962). What follows draws mostly on the volume on *Foreign Trade and Balance of Payments* (1962) in this series.
2. On gap analysis and absorptive capacity, see Chenery and Strout (1965). Gap analysis focuses on gaps such as trade balances and those between domestic investment and savings that become a barrier to high economic growth. Absorptive capacity is the capacity to effectively use capital, domestic or external; it reflects the management, production and industrial capability of a nation. If it is low, an economy even with high domestic savings cannot grow fast because industrial investment stagnates.
3. On Japanese policies towards capital importation, see e.g. Keidanren (1966), Komiya (1969).
4. This is defined as value of direct and indirect damage over asset value under no damage assumption.
5. As $\dot{K}_z < 0$ does not necessarily mean that the actual point was below the minimum critical effort curve we added the word 'probably'. If the upturn of (K_x, K_z) after 1951 was self-sustained, it implies that the actual point was above the minimum critical effort curve. But we conjecture that the upturn was due to the strong government support.
6. The new supply of fixed investment funds deriving from government loans (cols. 5 and 8 of Table 14.14) was deflated by the price indexes of investment goods (1955 = 100) to obtain government loans at constant price. These were then subtracted from the net capital stock used in Fig. 14.1 to obtain the points (K_x, K_z) of Fig. 14.2, which represent the state that would have been observed in the absence of government support.
7. This may appear to have favoured the *X*-sector, but most machinery that benefited from this was for the synthetic fibre industry, which is inherently a *Z*-sector industry. Encouragement of the synthetic fibre industry also encouraged the chemical industries.
8. The origin of this system is found in the revision of income taxes in 1913, which introduced a clause exempting from income taxes those manufacturers producing goods specified as important in the emperor's ordinance.
9. While the function of the capital invested may be public, the origin of the investment may be private. Because of data constraints, however, we confine ourselves to social overhead capital owned and managed by the government.
10. We have already made a similar analysis of the pre-war period in sec. 10.3.3.
11. The index of producers' price of foods in Umemura (1961) is a weighted average of government prices and black-market prices. Therefore, although there were many government controls, this index is considered a good reflection of market forces.

English References

Agarwala, A. N., and Singh, S. P. (eds.) (1963), *The Economics of Underdevelopment*, Oxford University Press, New York.

Baran, P. A. (1957) *Political Economy of Growth*, Monthly Review Press, New York.

Boserup, M. (1963) 'Agrarian Structure and Take-off', in W. W. Rostow (ed.), *The Economics of Take-off into Sustained Growth*, Macmillan, London.

Chenery, H. B. and Strout, A. M. (1965) 'Foreign Assistance and Economic Development', AID (Agency for International Development), Discussion Paper (revd.).

Clark, C. (1951) *The Conditions of Economic Progress*, 2nd edn., Macmillan, London.

Dobb, M. (1963) *Economic Growth and Underdeveloped Countries*, International Publishers, New York.

Fei, J. C. H., and Ranis, G. (1964) *Development of the Labor Surplus Economy: Theory and Policy*, Economic Growth Center, Yale University, New Haven, Conn.

Gerschenkron, A. (1952) 'Economic Backwardness in Historical Perspective', in B. Hoselitz (ed.), *The Progress of Underdeveloped Countries*, Chicago University Press, Chicago.

Inada, K. (1971) 'Economic Development in the Monocultural Economies', *International Economic Review*, 12/2.

International Bank (1955) 'The Economic Development of Malaya', *Report of a Mission Organized by the International Bank for Reconstruction and Development*, Johns Hopkins University Press, Baltimore, Md.

Lange, O. (1961) *Economic Development Planning and International Cooperation, Lectures to the Central Bank of Egypt*, Central Bank of Egypt, Cairo.

Leibenstein, H. (1957) *Economic Backwardness and Economic Growth: Studies in the Theory of Economic Development*, John Wiley, New York.

Lewis, W. A. (1955) *The Theory of Economic Growth*, Irwin, Homewood, Ill.

—— (1958) 'Unlimited Labor: Further Notes', *Manchester School of Economic and Social Studies*, 26 (Jan.).

—— (1963) 'Economic Development with Unlimited Supply of Labour', *Manchester School of Economic and Social Studies*, 1954; repr. in A. N. Agarwala and S. P. Singh (eds.), *The Economics of Underdevelopment*, Oxford University Press, New York.

Lockwood, W. W. (1954) *The Economic Development of Japan*, Princeton University Press, Princeton.

—— (1965) *The State and Economic Enterprise in Japan*, Princeton University Press, Princeton.

Maizels, A. (1963) *Industrial Growth and World Trade*, Cambridge University Press, Cambridge.

Meier, G. M. (1953) 'The Problem of Limited Economic Development', *Economia Internazionale*, 4; repr. in A. N. Agarwala and S. P. Singh (eds.), *The Economics of Underdevelopment*, Oxford University Press, New York, 1963.

Myrdal, G. (1968) *Asian Drama I. II. III.*, Pantheon, New York.

Nakamura, J. I. (1966) *Agricultural Production and the Economic Development of Japan*, Princeton University Press, Princeton.

Nurkse, R. (1952) 'Some International Aspects of the Problem of Economic Development', *American Economic Review,* Papers and Proceedings.

Ohkawa, Kazushi (1965*a*) 'Agriculture and Turning Points', *Developing Economies*, 3 (Dec.).

—— (1965*b*) 'A Century of Japanese Economic Growth' in W. W. Lockwood (ed.), *The State and Economic Enterprise in Japan*, Princeton University Press, Princeton.

Prebisch, R. (1964) *Towards a New Trade Policy for Development*, UNCTAD, New York.

Ranis, G. and Fei, J. C. H. (1961) 'A Theory of Economic Development', *American Economic Review* (Sept.).

Rostow, W. W. (1960) *The Stages of Economic Growth; A Non-Communist Manifesto*, Cambridge University Press, Cambridge.

Schultz, T. W. (1964) *Transforming Traditional Agriculture*, Yale University Press, New Haven, Conn.

Solow, R. M. (1956) 'Contribution to the Theory of Economic Growth', *Quarterly Journal of Economics*, 70 (Feb.).

UNCTAD (1968) *Trade Prospects and Capital Needs of Developing Countries*, United Nations, New York.

Vernon, R. (1966) 'International Investment and International Trade in Product Cycle', *Quarterly Journal of Economics*, 80 (May).

Japanese References

Akamatsu, Kaname (1944) *Sinkeizai Chitsuzyo no Keisei Genri*, repr. in 'Sekai Keizai Ron', Kunimoto Shobo, Tokyo.

—— (1956) *Wagakuni Sangyo Hatten no Ganko Keitai* (Flying-Geese Pattern in Japan's Industrial Development), Hitotsubashi University.

Akasaka, Keiko (1964) 'Seizogyo Jyugyoshasuu no Suikei: 1919–1942' (Estimates of Number of Employees in Manufacturing Industries, 1919–1942), Hitotsubashi University.

Arisawa, Hiromi (ed.) (1959) *Gendai Nihon Sangyo Koza*: 1–7 (Courses on Industries of Contemporary Japan), Iwanami Shoten, Tokyo.

Asahishimbun (1930) *Nihon Keizai Tokei Sokan* (Overview of Japanese Economic Statistics), Asahi Shimbumsha, Tokyo.

Baba, Masao, and Tatemoto, Masahiro (1967) 'Nihon niokeru Gaikoku Boeki to Keizai Seicho' (Foreign Trade and Economic Growth in Japan), in M. Shinohara and S. Fujino (eds.), *Nihon no Keizai Seicho* (Economic Growth in Japan), Nihon Keizai Shimbumsha, Tokyo.

Echigo, Kazunori (1969) *Kibo no Keizaisei* (Economy of Scale), Shinhyoron, Tokyo.

Emi, Ko-ichi (1971) *Choki Keizai Tokei 4: Shihonkeisei* (Long-Term Economic Statistics 4: Capital Formation), Toyo Keizai Shinposha, Tokyo.

—— and Shionoya, Yuichi (1965) *Choki Keizai Tokei 7: Zaiseisishutsu* (Long-Term Economic Statistics 7: Fiscal Expenditures), Toyo Keizai Shinposha, Tokyo.

Fujino, Shozaburo (1965) *Nihon no Keikijunkan* (Business Cycles of Japan), Keiso Shobo, Tokyo.

Gaishi Donyu Nenkan Henshu Iinkai (1962–7) *Gaishi Donyu Nenkan* (Annual Report of Foreign Capital Imports), Tokyo.

Hayashi, Takehisa (1965) *Nihon niokeru Sozei Kokka no Seiritsu* (Establishment of Tax-Based Government in Japan), Tokyo Daigaku Shuppan-kai, Tokyo.

Iida, Ken-ichi (ed.) (1964) *Gendai Nihon Sangyo Hattatsu-shi 4: Tekko* (History of Development of Industries of Contemporary Japan: Iron), Tokyo.

Ikuzawa, Kenzo (1970) *Kaitei SITC Bunrui niyoru Senzen no Boeki Kozo: Showa 10-nen no Boeki Tokei no Kumikaesagyo* (Pre-War Trade Structure based on Revised SITC: Recompilation of 1935 Trade Statistics), Economic Research Institute, Kyoto University, Kyoto.

Imai, Noriyoshi (1965) *Gendai Nihon no Dokusen Shihon 6: Shikin Chotatsu to Zaisei Toyushi* (Monopolistic Capital of Contemporary Japan 6: Fiscal Investment and Loan Programmes), Shiseido, Tokyo.

Iwanami Shoten (1968) *Kindai Nihon Sogo Nenpyo* (Chronological Table of Modern Japan), Iwanami Shoten, Tokyo.

Kajinishi, Mitsuhaya (ed.) (1964) *Gendai Nihon Sangyo Hattatsushi 11: Sen-i* (*Jo*) (History of Development of Industries in Contemporary Japan 11: Textiles), Gendai Sangyo Hattatsushi Kenkyukai, Tokyo.

—— (1965) *Nihon Keizaishi Taikei 5: Kindai (Jo)* (Japanese Economic History 5: Modern Period, vol. i), Tokyo Daigaku Shuppankai, Tokyo.

Kaneko, Eiichi (1964) *Gendai Nihon Sangyo Hattatsushi 9: Zosen* (History of Development of Industries in Contemporary Japan 9: Shipbuilding), Gendai Nihon Sangyo Hattatsushi Kenyukai, Tokyo.

Keidanren (1966) *Gaishi Donyu nikansuru Seido to Seisaku* (System and Policies regarding Capital Importation), Tokyo.

Keizai Antei Honbu (1949) *Senso Higai Chosa Shu* (Collected Research Documents on War Damage), Govt. of Japan, Tokyo.

Keizai Kenkyusho (1953) *Keizai Tokei* (Economic Statistics), Iwami Shoten, Tokyo.

Keizai Kikakucho (1959–63) *Sengo Keizaishi: Zaisei kin-yuu; Keizai Seisaku; Kokumin Shotoku; Boeki to Kokusaishushi; Sokan Hen* (Economic History of Post-War Japan: vols. for Finance and Public Finance; Economic Policy; National Income; Foreign Trade and Balance of Payments; Overview), Govt. of Japan.

—— (1962) *Shihon Sutokku to Keizai Seicho* (Capital Stock and Economic Growth), Govt. of Japan.

—— (1967*a*) *Keizai Bunseki 17: Soshihon Sutokku no Suikei; Minkan Kigyo* (Economic Analysis 17: Estimates of Gross Capital Stock: Private Capital, 1952–64), Govt. of Japan.

—— (1967*b*) *Keizai Bunseki 20: Shihanki Soshihon Sutokku no Suikei* (Economic Analysis 20: Quarterly Estimates of Gross Capital Stock), Govt. of Japan.

—— (1967*c*) *Keizaikenkyu Chosa Shiryo 2: Choki Keizai Tokei no Seibi Kaizen ni kansuru Kenkyu; Shihon Sutokku, Rodoryoku, Chinginritsu* (Documents for Economic Studies 2: Studies for Improvement of Long-Term Economic Statistics: Capital Stock, Labour Forces, and Wage Rates), Govt. of Japan.

—— (1969) *Keizai Bunseki 27: Saikin no Sangyo Betsu Seisansei no Doko* (Economic Analysis 27: Recent Trends in Productivity by Industrial Sector), Govt. of Japan, Tokyo.

—— (1970) *Kokumin Shotoku Tokei Nenpo: 1970* (Annual Report on National Income), Govt. of Japan.

Koizumi, Akira (1965*a*) *Nihon Keizai Taikei 2: Nihon no Sangyo* (Japanese Economy 2: Industries), Seirin Shoin Shinsa, Tokyo.

—— (1965*b*) *Nihon Keizai Taikei 5: Nihon no Zaisei Kin-yu* (Japanese Economy 5: Public Finance), Seirin Shoin Shinsa, Tokyo.

Kojima, Kiyoshi (1958) *Nihon Boeki to Keizai Hatten*, Kunimoto Shobo, Tokyo.

Komiya, Ryutaro (1969) 'Chokusetsu Toshi to Sangyo Seisaku' (Direct Foreign Investment and Industrial Policies), in H. Niida and A. Ono (eds.), *Nihon no Sangyo Seisaku* (Industrial Policies of Japan), Iwanami Shoten, Tokyo.

Koyama, Hirodate (1943) *Nihon Gunji Kogyo Hattatsu-shi* (History of the Development of the Arms Industry in Japan), Nihon Sangyo Kiko Kenkyu, Ito Shoten, Tokyo.

Matsui, Kiyoshi (ed.) (1963) *Kindai Nippon Boeki-shi 1–3* (History of Foreign Trade in Modern Japan 1–3), Yuhikaku, Tokyo.

Meiji Zaiseishi Hensankai (1926) *Meiji Zaisei-shi* (History of Public Finance of the Meiji Era), Meiji Zaiseishi Hakkosho, Tokyo.

Minami, Ryoshin (1965) *Choki Keizai Tokei 12: Tetsudo to Denryoku* (Long-Term Economic Statistics 12: Railways and Electricity), Toyo Keizai Shinposha, Tokyo.

—— (1970) *Nihon Keizai no Tenkanten* (Turning-Point of the Japanese Economy), Sobun-sha, Tokyo.

MITI (1952) *Nihon Boeki no Genjyo* (Present Situation of Japan's Foreign Trade), Kaigai Shijo Chosakai, Tokyo.

—— (1955) *Showa 30-nen Tsusho Hakusho: Nihon Boeki no Genjyo* (White Paper on Trade, 1955: Present Situation), Tsusho-Sangyo Chosakai, Tokyo.

—— (1957*a*) *Sangyo Gorika Hakusho* (White Paper on Modernization of Industries), Nikkan Kogyo Shimbumsha, Tokyo.

—— (1957*b*) *Nihon Keizai no Sangyo Renkan Bunseki* (Inter-Industry Analysis of Japanese Economy), Toyo Keizai Shinposha, Tokyo.

—— (1962) *Kogyo Tokei 50 Nenshi* (Fifty-Year History of Census of Manufactures), Govt. of Japan.

—— (1963) *Honpo Kogyo no Susei 50 Nenshi: 1905–1960* (Fifty-Year History of Japanese Mining Industry), Tusho Sangyo Chosakai, Tokyo.

—— (1964) *Honpo Kogyo no Susei 50 Nenshi* (*Zokuhen*) (Fifty-Year History of Japanese Mining Industry), Tusho Sangyo Chosaki, Tokyo.

Mitsubishi Keizai Kenkyusho (1930–) *Honpo Jigyo Seiseki Bunseki* (Analysis of Japanese Business), various edns. from 1930, Tokyo.

Monbu-sho (1969) *Waga Kuni no Kyoiku no ayumi to Kongo no Kadai* (Development of Japanese Education and Future Tasks), Ministry of Education, Govt. of Japan.

Mori, Kazuo (1972) '*Taisho 3 -nen, Showa-10 nen no Sangyo Renkanhyo; Shisan*' (Estimation of Input–Output Tables of Japan in 1914 and 1935), repr. in K. Takahashi (ed.), *Nippon Kindaika no Kenkyu: Gekan* (Studies in the Modernization of Japan), Tokyo Daigaku *Shuppan-kai*, Tokyo.

Nagoya-shi Sangyobu (1936) *Kogyo Chosa* (Census of Manufactures), Nagoya, 1936.

Naikaku Tokeikyoku (1881–1939) *Nippon Teikoku Tokei Nenkan 1–59 1881–1939* (Annual Statistics of the Japanese Empire 1–59, 1881–1939), repr. 1962–7, Govt. of Japan.

—— (1934) *Showa 5 nen no Kokufu oyobi Kokumin Shotoku* (Wealth and National Income in 1930), Govt. of Japan.

—— (1948) *Showa 10 nen no Kokufu oyobi Kokumin Shotoku* (Wealth and National Income in 1935), Govt. of Japan.

—— (1957) *Rodoryoku Chosa Kaizan Kekka Hokoku* (Report on Revised Census of Labour), Govt. of Japan.

Nakajima, Tsuneo (ed.) (1967) *Gendai Nihon Sangyo Hattatsu-shi 18: Shokuhin* (History of Development of Industries in Contemporary Japan 18: Food Processing), Govt. of Japan.

Nakayama, Ichiro (ed.) (1954) *Nihon Keizai no Kozo Bunseki 1–2* (Structural Analysis of the Japanese Economy 1–2), Toyo Keizai Shinposha, Tokyo.

—— (1959) *Nippon no Kokufu Kozo* (Structure of Japan's National Wealth), Toyo Kezai Shinposha, Tokyo.

—— (1961) *Shihon Chikuseki to Kinyu Kozo* (Capital Accumulation and Financial Structure), Toyo Kezai Shinposha, Tokyo.

—— (1969) *Nihon Keizai no Hatten: Kogyoka to Mirai* (Development of the Japanese Economy: Industrialization and the Future), Ushio Shuppansha, Tokyo.

Nihon Kanzei Kyokai (1959) *Nihon no Kanzei* (Import Tariffs of Japan), Nihon Kanzei Kyokai, Tokyo.

Nihon Kokuseijiten Kankokai (1953) *Nihon Kokusei Jiten 1* (Dictionary of Japanese Politics 1), Rengo Shuppansha, Tokyo.

Nihon Tokei Kenkyusho (1958) *Nihon Keizai Tokei Shu* (Japanese Economic Statistics), Nihon Hyoron Shinsha, Tokyo.

—— (1960) *Nihon Tokei Hattatsu-shi* (History of the Development of Japanese Statistics), University of Tokyo Press, Tokyo.

Nippon Gakujutsu Shinkokai (ed.) (1951) *Tsusho Joyaku to Tsusho Seisaku no Hensen* (*Joyaku Kaisei Kankei Nippon Gaikosho Bessatsu*) (Evolution of Commerce Treaties and Foreign Trade Policies: Documents on Revision of External Commerce Treaties), Sekai Keizai Chosakai, Tokyo.

Nippon Ginko (1966) *Honpo Keizai Tokei* (Japanese Economic Statistics), Bank of Japan, Tokyo.

Nippon Ginko Tokeikyoku (1964) *Oroshiuri Bukka Shisu: Meiji 20–37* (Wholesale Price Indices: 1887–1904), Bank of Japan, Tokyo.

—— (1966) *Meiji Iko Honpo Shuyo Keizai Tokei* (Major Economic Statistics of Japan since the Meiji Era), Bank of Japan, Tokyo.

—— (1967–9) *Keizai Tokei Nenpo* (Annual Report on Economic Statistics), Bank of Japan, Tokyo.

—— (various issues) *Keizai Tokei Geppo* (Economic Statistics Monthly), Bank of Japan, Tokyo.

Noshomusho Somukyoku Hokokuka (1959) *Noshomu Tokei Hyo* (Statistics of Agriculture and Commerce), Keio Shobo, Tokyo.

Ohe, Shinobu (1968) *Nippon no Sangyo Kakumei* (Industrial Revolution of Japan), Iwanami Shoten, Tokyo.

Ohkawa, Kazushi (1955) *Nogyo no Keizai Bunseki* (Economic Analysis of Agriculture), Daimeido, Tokyo.

—— (1956) *Nippon Keizai no Seichoritsu* (Growth Rate of the Japanese Economy), Iwanami Shoten, Tokyo.

—— (1960) *Kajyo Shugyo to Nihon Nogyo* (Excess Employment and Japanese Agriculture), Shunjyusha, Tokyo.

—— (1962) *Nihon Keizai Bunseki: Seicho to Kozo* (Analysis of Japan's Economy: Growth and Structure), Shunzyusha, Tokyo.

—— (1963) *Nihon Nogyo no Seicho Bunseki* (Analysis of Growth of Japanese Agriculture), Daimeido, Tokyo.

—— and Rosovsky, H. (1966) *Nihon niokeru Keizai Seicho no 100 nen* (A Century of Japanese Economic Growth) in W. W. Lockwood (ed.), 'Nihon Keizai Kindaika no Hyakunen', Nihon Keizai Shimbum, Tokyo.

—— *et al.* (1966) *Choki Keizai Tokei 3: Shihon Sutokku* (Long-Term Economic Statistics 3: Capital Stock), Toyo Keizai Shinposha, Tokyo.

—— (1967) *Choki Keizai Tokei 8: Bukka* (Long-Term Economic Statistics 8: Prices), Toyo Keizai Shinposha, Tokyo.

Ohkurasho (1938) *Meiji Taisho Zaiseishi* (History of Public Finance of the Meiji-Taisho Era), Govt. of Japan.

—— (1970) *Zaisei Tokei: Showa 45 Nendo* (Statistics of Public Finance: Fiscal Year 1970), Govt. of Japan.

Ohkurasho Showa Zaiseishi Henshushitsu (1955) *Showa Zaiseishi 3, 18* (Showa History of Public Finance), Toyo Keizai Shinposha, Tokyo.

Ohkurasho (n.d.) *Honpo Kanzei Seido Enkaku Shiryo* (Documents on Institutional Development of Japanese Import Tariffs), Govt. of Japan. Year of edition is unclear.

Oka, Shigeo (1964) *Sengo Nihon no Kanzei Seisaku* (Tariff Policy in Post-War Japan), Nippon Hyoronsha, Tokyo.

Sengo Keizaishi Hensan-shitsu (1962) *Sengo Keizaishi: Boeki Kokusaishushi Hen* (Economic History of Post-War Japan: Trade and Balance of Payments), Govt. of Japan.

Shinohara, Miyohei (1949) *Koyo to Chingin* (Employment and Wage Rates), Jitsugyo no Nippon-sha, Tokyo.

—— (1961) *Nihon Keizai no Seicho to Junkan* (Growth and Cyclical Movements of Japanese Economy), Sobunsha, Tokyo.

—— (1967) *Choki Keizai Tokei 6: Kojin Shohi Shishutsu* (Long-Term Economic Statistics 6: Personal Consumption), Toyo Keizai Shinposha, Tokyo.

—— and Fujino, Shozaburo (eds.) (1967) *Nihon no Keizai Seicho* (Economic Growth of Japan), Nihon Keizai Shimbumsha, Tokyo.

Shionoya, Yu-ichi (1966) *Nihon no Kogyo Seisan Shisu: 1874–1940* (Index of Industrial Production of Japan: 1874–1940), Keizaigaku Zenshyu, 13, Chikuma Shobo, Tokyo.

Shokosho (1944), *Kojo Tokei Hyo* (Census of Manufactures), Keio Shobo, orig. pub. 1909–40.

Shokuryocho (1955) *Shokuryo Kanri Tokei Nenpo* (Annual Statistics of Food Control), Govt. of Japan.

—— (1956) *Shokuryo Kanri Tokei Nenpo* (Annual Statistics of Food Control), Govt. of Japan.

Showa Dojinkai (1957) *Wagakuni Kanzenkoyo no Igi to Taisaku* (Importance of Full Employment and Policies for its Achievement in Japan), Shiseido, Tokyo.

—— (1960) *Wagakuni Chingin Kozo no Rekishiteki Kosatsu* (Historical Reflections on Japanese Wage Structure), Shiseido, Tokyo.

Takahashi, Kamekichi (1968) *Nihon Kindaikeizai Keiseishi 3* (History of the Establishment of the Modern Japanese Economy 3), Toyo Keizai Shinposha, Tokyo.

Takeuchi, Yoshio (ed.) (1967) *Nihon no Shakai Shihon: Genjnobunseki to Keikaku* (Social Capital of Japan: Present Situation and Plans), Kashima Kenkyusho *Shuppan-Kai*, Tokyo.

Tamaki, Hajime (1967) *Gendai Nihon Sangyo Hattatsu-shi 29: Soron* (History of the Development of Industry in Contemporary Japan 29: Summary), Gendai Nihon Sangyo Hattatsu-shi Kenkyu Kai, Tokyo.

Tokei Kenkyukai (1970) *Keiryo Moderu Kenkyu Shiryo 6: Keizaikeikaku notameno Keiryo Moderu nikansuru Kenkyu* (Documents for Studies of Econometric Models 6: Studies for Econometric Model Building in Economic Planning), Tokei Kenkyukai, Tokyo.

Toyo Keizai Shinposha (1926) *Meiji Taisho Zaisei Shoran* (Statistics of Budget in Meiji and Taisho Eras), Toyo Keizai Shinposha, Tokyo.

—— (1927) *Meiji Taisho Kokusei Soran* (Census of Population, Meiji and Taisho Eras), Toyo Keizai Shinposha, Tokyo.

—— (1935) *Nihon Boeki Seiran* (Detailed Statistics of Japanese Trade), Toyo Keizai Shinposha, Tokyo.

—— (1950) *Showa Sangyo-shi 1–3* (History of Industries in the Showa Era 1–3), Toyo Keizai Shinposha, Tokyo.

Tsugei, Yoshio (1969) *Nihon Sangyokigyo-shi Gaisetsu* (Introduction to Japanese Manufacturing Enterprises), Zeimu Keiri Kyokai, Tokyo.

Tsujimura, Kotaru (1963) 'Koyo Kozo to Rodo Bunpairitsu' (Structure of Employment and Labour Share of Income) in Ryutaro Komiya (ed.), *Sengo Nippon no Keizai Seicho* (Japan's Post-War Economic Growth), Iwanami Shoten, Tokyo.

Ueno, Hiroya (1965) *Nihon Keizai no Seicho Moderu* (Growth Model of the Japanese Economy), Toyo Keizai Shinposha, Tokyo.

Umemura, Mataji (1961) *Chingin, Koyo, Nogyo* (Wage Rates, Employment and Agriculture), Daimeido, Tokyo.

—— (1966) *Choki Keizai Tokei 9: Noringyo* (Long-Term Economic Statistics 9: Agriculture and Forestry), Toyo Keizai Shinposha, Tokyo.

—— (1968) 'Yugyoshasu no Sin Suikei, 1871–1920' (New Estimates of Employment, 1871–1920), in *Keizai Kenkyu*, 19/1, Hitotsubashi University, Iwanami Shoten, Tokyo.

Watanabe, Tsuinehiko and Egaitsu, Norio (1967) 'Gijutushinpo to Keizaiseicho' (Technical Progress and Economic Growth), in K. Motoo (ed.), *Keizai Seicho to Shigen Haibun* (Economic Growth and Resource Allocation), Iwanami Shoten, Tokyo.

Yamada, Isamu (1969) *Nihon Keizai no Kozohendo to Yosoku* (Changes in the Structure of the Japanese Economy and Forecast of its Future), Shunjusha, Tokyo.

Yamaguchi, Kazuo (1956) *Meiji Zenki Keizai no Bunseki* (Analysis of the Japanese Economy in the Early Meiji Era), University of Tokyo Press, Tokyo.

Index

CARDIFF UWCC LIBRARY